BASIC PRINCIPLES
OF ANALYTICAL
ULTRACENTRIFUGATION

BASIC PRINCIPLES
OF ANALYTICAL
ULTRACENTRIFUGATION

Peter Schuck
Dynamics of Macromolecular Assembly Section
Laboratory of Cellular Imaging and Macromolecular Biophysics
National Institute of Biomedical Imaging and Bioengineering
National Institutes of Health, Bethesda, Maryland

Huaying Zhao
Dynamics of Macromolecular Assembly Section
Laboratory of Cellular Imaging and Macromolecular Biophysics
National Institute of Biomedical Imaging and Bioengineering
National Institutes of Health, Bethesda, Maryland

Chad A. Brautigam
Department of Biophysics
The University of Texas Southwestern Medical Center
Dallas, Texas

Rodolfo Ghirlando
Laboratory of Molecular Biology
National Institute of Diabetes and Digestive and Kidney Diseases
National Institutes of Health, Bethesda, Maryland

CRC Press
Taylor & Francis Group
Boca Raton London New York

CRC Press is an imprint of the
Taylor & Francis Group, an **informa** business

CRC Press
Taylor & Francis Group
6000 Broken Sound Parkway NW, Suite 300
Boca Raton, FL 33487-2742

First issued in paperback 2020

© 2016 by Taylor & Francis Group, LLC
CRC Press is an imprint of Taylor & Francis Group, an Informa business

No claim to original U.S. Government works

Version Date: 20151001

ISBN 13: 978-0-367-57514-4 (pbk)
ISBN 13: 978-1-4987-5115-5 (hbk)

Visit the Taylor & Francis Web site at
http://www.taylorandfrancis.com

and the CRC Press Web site at
http://www.crcpress.com

To T. the Beautiful and L. the Great

Peter Schuck

To my PhD adviser, Dr. Susan Pedigo, who brought me into the biophysical world

Huaying Zhao

To my wife, Lisa, and to all the teachers who have inspired me

Chad A. Brautigam

To Heini Eisenberg who introduced me to the joys of analytical ultracentrifugation

Rodolfo Ghirlando

Contents

Foreword   xiii

Preface   xvii

Symbol Description   xix

CHAPTER   1 ▪ Analytical Ultracentrifugation Basics   1

     1.1   PHYSICAL PRINCIPLES   1

         1.1.1   Basic Experimental Setup   1

         1.1.2   Sedimentation Velocity   2

         1.1.3   Sedimentation Equilibrium   14

     1.2   DETECTION PRINCIPLES   16

         1.2.1   Absorbance   17

         1.2.2   Rayleigh Interferometry   19

         1.2.3   Fluorescence   22

CHAPTER   2 ▪ The Sedimenting Particle   23

     2.1   PARTICLE–SOLVENT INTERACTIONS   24

         2.1.1   Molar Mass, Buoyancy, and Macromolecular Partial-Specific Volumes   24

         2.1.2   Density Increments and Preferential Binding Parameters   27

         2.1.3   Effective Partial-Specific Volumes and the Sedimenting Particle   29

         2.1.4   Measuring Hydrodynamic Friction   31

         2.1.5   Charge Effects   34

     2.2   PARTICLE–PARTICLE INTERACTIONS   36

         2.2.1   Dynamic Interacting Systems and Reversible Associations   36

             2.2.1.1   Conformational States   36

             2.2.1.2   Self-Associating Systems   38

		2.2.1.3	Heterogeneous and Mixed Associating Systems	42
		2.2.1.4	Solvent-Mediated Protein Interactions	45
	2.2.2		Sedimentation of Concentrated Solutions	46
		2.2.2.1	Sedimentation Equilibrium at High Concentrations	46
		2.2.2.2	Sedimentation Velocity at High Concentrations	49
2.3			SEDIMENTATION EFFECTS ON MACROMOLECULAR MIGRATION	51
	2.3.1		Pressure	53
		2.3.1.1	Compressibility of Solvent and Particles	53
		2.3.1.2	Pressure Dependence of Interactions	55
	2.3.2		Hydrodynamic Drag Acting on the Particle — Do Molecules Bend or Align in Sedimentation?	55
	2.3.3		Dynamic Hydrostatic and Viscous Interactions from Co-Solute Sedimentation	57

CHAPTER   3 ▪ Sample Preparation and Ancillary Characterization   59

3.1			COMMON SAMPLE REQUIREMENTS	59
	3.1.1		Purity	59
	3.1.2		Concentrations	62
		3.1.2.1	Concentration Ranges for Different Detection Systems	62
		3.1.2.2	Concentration Considerations for Ideal Sedimentation	65
		3.1.2.3	System-Dependent Concentration Choices	66
		3.1.2.4	Considerations for Sample Preparation	67
	3.1.3		Loading Volumes	68
	3.1.4		Stability	69
3.2			CHOICE OF SOLVENT AND BUFFER	71
	3.2.1		Measuring Solvent Density	74
	3.2.2		Measuring Solvent Viscosity	74
3.3			DETERMINING THE PARTIAL-SPECIFIC VOLUME	75
	3.3.1		Ultracentrifugal Approaches	76
		3.3.1.1	Prior Knowledge of Molar Mass or Hydrodynamic Shape	76
		3.3.1.2	Density Contrast by Sedimentation Equilibrium	78
		3.3.1.3	Density Contrast by Sedimentation Velocity	83
		3.3.1.4	The Neutral Buoyancy Method	86

3.3.2   Densimetry   87

3.3.3   Theoretical Predictions   91

    3.3.3.1   Using Traube's Rules for the Prediction of Partial Molar Volumes from Chemical Structures   92

    3.3.3.2   Prediction of Protein Partial-Specific Volumes from Amino Acid Composition   92

    3.3.3.3   Partial-Specific Volumes of Glycoproteins and Other Protein Conjugations   94

    3.3.3.4   Prediction of Solvent Effects on Protein Partial-Specific Volumes   95

    3.3.3.5   Prediction of the Partial-Specific Volume of Nucleic Acids   96

CHAPTER 4 ■ Data Acquisition   99

4.1   RADIAL DIMENSION   100

   4.1.1   Radius Calibration   100

    4.1.1.1   Reference Positions in the Counterbalance of a Reference Cell   100

    4.1.1.2   Imaging of Periodic Precision Masks   102

    4.1.1.3   Linearity of Radial Position   103

   4.1.2   Meniscus and Bottom   105

    4.1.2.1   Absorbance   107

    4.1.2.2   Interference   108

    4.1.2.3   Fluorescence   109

    4.1.2.4   Computational Determination   112

   4.1.3   Radial Resolution   114

   4.1.4   Time-Invariant (TI) Noise   117

    4.1.4.1   Physical Origin   117

    4.1.4.2   Accounting for TI Noise   122

   4.1.5   Radial Gradients of Signal Magnification   126

4.2   TEMPORAL DIMENSION   126

   4.2.1   Rotor Acceleration and Effective Sedimentation Time   127

   4.2.2   Scan Repetition Rate and Number of Scans Required in the Analysis   131

   4.2.3   Speed of Data Collection   133

   4.2.4   Ensuring Temporal Accuracy   136

   4.2.5   Radial-Invariant (RI) Noise   137

		4.2.5.1	Physical Origin	138
		4.2.5.2	Accounting for RI Noise	139
	4.2.6	Time-Dependent Signal Intensity		141
4.3	SIGNAL DIMENSION			144
	4.3.1	What Are We Detecting? The Choice of Optical System		144
	4.3.2	Signal Increments		147
		4.3.2.1	Absorbance/Pseudo-Absorbance	147
		4.3.2.2	Interference	149
		4.3.2.3	Fluorescence	151
	4.3.3	Signal Linearity		152
		4.3.3.1	Experimental Limits of Signal Linearity	152
		4.3.3.2	Effect of Nonlinear Signals on Observed Sedimentation Parameters	154
		4.3.3.3	Linearizing Data Transformation	156
	4.3.4	Buffer Mismatch Signals		156
	4.3.5	Noise		159
		4.3.5.1	Statistical Noise	162
		4.3.5.2	Residuals Bitmaps	164
		4.3.5.3	Other Measures for Randomness of the Residuals	166
4.4	SPECTRAL DIMENSION			169
	4.4.1	Spectral Resolution		171
	4.4.2	Wavelength Reproducibility of Radial Scans		175
	4.4.3	Wavelength Scans		175
	4.4.4	Multi-Wavelength and Multi-Signal Detection		176

| CHAPTER | 5 ▪ The Sedimentation Experiment | | | 181 |

5.1	THE IMPORTANCE OF AVOIDING CONVECTION			181
5.2	ROTORS AND SAMPLE CELL ASSEMBLIES			182
	5.2.1	Centerpieces		182
		5.2.1.1	Sector-Shaped Solution Columns for Unimpeded Sedimentation	183
		5.2.1.2	Synthetic Boundary Centerpieces	185
		5.2.1.3	Band Centrifugation and Active Enzyme Centrifugation	188
		5.2.1.4	Centerpiece Material and Design	190
	5.2.2	Windows		191

	5.2.3	Rotors		192
		5.2.3.1	The Choice of Rotors	192
		5.2.3.2	Rotor Stretching and Thermal Behavior	193
	5.2.4	Cell Assembly		196
		5.2.4.1	Leaks	200
		5.2.4.2	Ageing and Baseline Blanks	202
		5.2.4.3	Resuspension, Disassembly, Cleaning, and Storage	205
5.3	TEMPERATURE			206
	5.3.1	Temperature Measurement and Control		207
	5.3.2	Temperature Equilibration		208
	5.3.3	Temperature-Dependent Experiments		211
5.4	VACUUM			212
5.5	ROTOR SPEEDS			213
	5.5.1	Rotor Speeds in Sedimentation Velocity		213
	5.5.2	Rotor Speeds and Solution Column Lengths in Sedimentation Equilibrium		215
5.6	STARTING THE RUN			219
5.7	COMMENCING DATA ACQUISITION AND RUN DURATION			221
	5.7.1	Scan Parameters for Sedimentation Equilibrium		221
	5.7.2	Overspeeding and Expected Time to Equilibrium		222
	5.7.3	Sedimentation Velocity		225
5.8	STOPPING THE RUN			226

CHAPTER 6 ■ Control And Calibration Experiments   227

6.1	DATA DIMENSIONS			229
	6.1.1	Temporal Accuracy		229
		6.1.1.1	Scan Velocity	230
	6.1.2	Temperature Calibration		230
	6.1.3	Radial Calibration		232
	6.1.4	Spectral Calibration		234
	6.1.5	Photometric Calibration		234
	6.1.6	Rotor Speed		235
	6.1.7	Mass and Density		235
6.2	STANDARD CONTROL EXPERIMENTS			237

APPENDIX   A ■ Macromolecular Partial-Specific Volumes   239

    A.1   PROTEINS   239
    A.2   NUCLEIC ACIDS   241
    A.3   CARBOHYDRATES   243
    A.4   DETERGENTS AND LIPIDS   243

APPENDIX   B ■ Solvent Properties   245

    B.1   WATER   245
    B.2   HEAVY WATER: D_2O 247
    B.3   HEAVY OXYGEN WATER: $H_2{}^{18}O$ AND $D_2{}^{18}O$ 248
    B.4   ORGANIC AND OTHER NON-AQUEOUS SOLVENTS   248
    B.5   CO-SOLUTES   250

APPENDIX   C ■ Refractive Index Increments   251

APPENDIX   D ■ Solution Column   253

Bibliography   255

Index   295

Foreword

This eminently readable book tells the outcome – to present date – of a journey of scientific discovery: A saga in which new territory is explored, peaks conquered, and guideposts left for those who choose to follow. Sedimentation analysis of macromolecular systems is the field in which Dr. Peter Schuck and his fellow investigators have been involved, and to which they have made contributions not only through their own laboratories' investigations, but through the widespread use made of their algorithms and programs by scientists throughout the world. In this present volume we have for our information, guidance (and just plain scientific enjoyment) an up-to-date statement of what an investigator can hope to achieve in the present state of knowledge. Such guidance is most welcome: along with the widespread and worldwide use of the SEDFIT bulletin board and the frequent dedicated training sessions which provide a platform for dissemination of optimal procedures, ideas, findings and opinions.

Scientific knowledge, though, sits in a historical context. The authors here have given credit to 'what has gone before', and to the major volumes which have been published in earlier periods (Svedberg and Pedersen's volume[1] and Schachman's book/monograph[2]). It may be a little unusual today — even 'unfashionable' — to find value in science published more than a decade ago, but for myself I would wish that every would-be young biophysical scientist would read through Cheng and Schachman's remarkable paper (1955 — referenced in this book, and freely available on-line[3]). In this work the authors followed a carefully planned line of enquiry and experimentation to confirm that macromolecular solutes such as proteins could be treated as dispersions of 'hydrodynamic particles' whose properties can be regarded as scale-independent. It speaks volumes for the quality of their experimental work that when they report a study on the concentration-dependence of the sedimentation of polystyrene spheres, they obtain a value for the c-dependence (at high dilution) of the sedimentation rate that comes to within 1% of the best theoretical estimate of this parameter for spheres based upon fluid dynamics, computed decades later (Brady and Durlovsky[4]).

[1] T. Svedberg and K.O. Pedersen, *The Ultracentrifuge*, Oxford University Press, London, 1940.

[2] H.K. Schachman, *Ultracentrifugation in Biochemistry*, Academic Press, New York, 1959.

[3] P.Y. Cheng and H.K. Schachman, Studies on the validity of the Einstein viscosity law and Stokes law of sedimentation, *J. Polym. Sci.*, vol. 16(18), pp. 19–30, 1955, doi:10.1002/pol.1955.120168102.

[4] J.F, Brady and L.J. Durlofsky, The sedimentation rate of disordered suspensions, *Phys. Fluids*, vol. 31(4), pp. 717–727, 1988, doi:10.1063/1.866808.

A particular beauty of sedimentation analysis as a discipline lies in the fact that just a single (differential) equation describes everything. But this 'Lamm' equation cannot be solved directly in the general case. The work of Dr. Schuck and his team is founded upon their demonstration that the fitting by non-linear least squares methods of sets of solutions of the Lamm equation is a stable procedure. Distributions of hydrodynamic parameters, in particular of sedimentation coefficients (s-values), are output. These 'c(s) vs s' distributions have become normative. With many cell scans logged over time the final dataset is highly information-rich, and the time-invariance of most of the noise structure facilitates its effective removal within the software (SEDFIT) environment. By floating other parameters, starting from a solute frictional ratio, within the fit, further information can be yielded as regards solute molecular weight and interaction parameters. Optimal routines are clearly and critically described in this volume.

I will take a moment to recall how great the change is that the use of on-line computation with software sets such as SEDFIT/SEDPHAT has brought to 'prior art'. I began my personal research life in sedimentation analysis working with a Phywe Air-Turbine Analytical Ultracentrifuge, which lacked even rotor temperature control. Data acquisition was achieved using manual scanning of photographic records, and this persisted even into the Beckman Model E era. Computational analysis was with rotary-mechanical calculating machines. Fitting datasets using an 'over-determined set of non-linear simultaneous equations' (*sic!*) was not even a dream of the future. Yet within a decade or two the whole field of sedimentation analysis and indeed of biophysical analysis in general has been transformed by the advent of powerful computational hardware and of equally powerful analytical algorithms and associated software. This growth in activity has provided workers with tools which while simple in concept call for guidance in their use. We in our field have been fortunate indeed to have had over the years scientists of distinction who have been prepared not only to make advances, but to disseminate 'good practice' in application to their colleagues.

Clearly I am talking here with the present authors and their volume in mind; 'c(s) plots computed via SEDFIT' have become the mainstay of a mass of published work, and a foundation for a broader understanding of the nature and properties of macromolecular systems, both in biological and in materials sciences. We have here an authoritative guidance to the wide range of procedures and modes of analysis which are possible. Perhaps I may speculate a little as to how this whole mode of analysis has become so popular?

Excellence is a necessary but not a wholly sufficient explanation for a scientific finding or a new mode of application becoming widely accepted. I suggest that something which I call 'immediate impact' is a necessary part of the package. What do I mean by 'immediate impact'? I will describe what I mean by describing something from my own scientific experience. 'Ostensive definition' is, I believe, the technical term.

Long ago, as part of my extensive involvement with muscle/motility research, I addressed the problem of the basic structure of the myosin thick filament of vertebrate skeletal muscle. A self-assembling structure, interfacing with an ordered array of thin filaments, it must possess rotational symmetry, to satisfy x-ray and EM structural data. Yet was this a 2-, 3- or 4-fold symmetry? Despite having an ability to prepare purified filaments in low quantity, the apparently simple problem of finding out their mass/unit length gave us an unexpected headache. For reasons which still remain obscure today, the frictional behavior of even 'synthetic filaments' made from pure myosin was found to be seriously anomalous, and depended steeply on the presence of divalent cation, especially of Mg^{++} (Persechini and Rowe, 1984)[5]. We are talking of s-values varying by up to 30%. With Charlie Emes, a graduate student in my lab, an approach was developed which basically accepted the presence of this anomaly, and coped with it, using a neat bit of logic. The answer was simple: 3-fold rotational symmetry it must be. No problems in publishing this finding, in a paper[6] I am very proud of, but it all went over like the proverbial lead balloon. Partly this may have been that the formidable figure of the eminent enzymologist Bill Harrington of Johns Hopkins loomed over our work. Bill was convinced that the symmetry had to be either 2- or 4-fold: All tied up with his view that myosin existed in solution as dimers — the existence of which was denied by both myself[7] and Sara Szuchet[8] of the Yphantis Laboratory. The low-angle x-ray people actually favored 3-fold, but in a heavily qualified manner, so that was not much help.

However, just a couple of years later, another graduate student (Maria Maw) starting work in my lab on the electron microscopy of thick filaments, one day brought to me an image which seemed to show a native thick filament clearly splitting into three sub-filaments. She admitted that she had gotten the preparation procedure for negative staining wrong — she had used a water rinse in place of a solvent rinse. I asked her to do it again just like that, and when she had found and recorded 200 such images I would get seriously excited. She did of course exactly that, and we soon had a paper written and happily grabbed by *Nature*[9] (the only journal of that name in those distant days). That settled the argument. Oh — and for that water rinse, if you take away the charge shielding of a structure held together by charge–charge interactions, it is no great surprise if it starts to fall apart. I did start out my research career as a colloid scientist!

And this is my ostensive definition of immediate impact. If you can show a simple

---

[5]A.J. Persechini and A.J.Rowe, Modulation of myosin filament conformation by physiological levels of divalent cation, *J. Mol. Biol.*, vol.172(1), pp. 23–39, 1984, doi:10.1016/0022-2836(84)90412-1.

[6]C.H. Emes and A.J. Rowe, Frictional properties and molecular weights of native and synthetic myosin filaments from vertebrate skeletal muscle, *Biochim. Biophys. Acta*, vol. 537(1), pp.125–144,1978, doi:10.1016/0005-2795(78)90608-6.

[7]C.H. Emes and A.J. Rowe. Hydrodynamic studies on the self-association of vertebrate skeletal muscle myosin, *Biochim. Biophys. Acta*, vol 537(1), pp. 110-124, 1978, doi:10.1016/0005-2795(78)90607-4.

[8]S. Szuchet, Effect of purification procedures on the self-association of myosin at high ionic strength, *Arch. Biochem. Biophys.*, vol. 180(2), pp. 493–503, 1977, doi:10.1016/0003-9861(77)90064-9.

[9]M.C. Maw and A.J. Rowe, Fraying of A-filaments into three subfilaments, *Nature*, vol. 286(5771), pp. 412–414, 1980, doi:10.1038/286412a0.

image or convincing graph of what you are talking about then a few equations and sound interpretations are not the whole thing in the game. Quality of analysis and interpretation has to be a starting point. But output which makes an immediate impact is the vital, last stage in any fruitful investigation. The work of the four scientists who have authored this volume excel in all of this, and this book will be as vital a tool for novices as much as for senior workers seeking guidance on the more recondite areas of analysis.

And for the future, where may we be heading? There are trends already established for the increasing level of study of non-biological systems (polymers), and for the use of the AUC within the bio/pharma industry. The acceptance of c(s) profiles as a matrix-free 'gold standard' method orthogonal to column-based technology has undoubtedly encouraged interest and commitment within the bio/pharma sector, and we can expect that area of activity to increase. I am also certain that the present achievement levels of the AUC hardware/software can be surpassed. The instrument provides data at a remarkably high level of precision, particularly when Rayleigh interference optics are being employed. The ultimate level of random 'shot noise', shown to be ± 0.002 fringe, can probably not be surpassed, even in a newly designed optical system: But when the total signal is usually in the range 1–300 fringe, the theoretically available precision (signal/noise ratio) leaves many biophysical analytical instruments well behind. Most ways of probing systems in solution involve the pertubation of basic physical parameters of the system: exposure to magnetic fields, temperature jumps, particle–photon interactions, for example. A pertubation of a centrifugal field ('g-force') is so simple and basic an approach that there should always be a place for its employment. The authors of this book have given us, not a compendium, but a monograph based upon their extensive experience in the field of sedimentation analysis — which incidentally they have been at pains to relate to results from other biophysical techniques, such as dynamic light scattering. Readers and potential readers: If you already know about sedimentation analysis you will find interest and information alike in its pages. If you are a newcomer to the area, you should take advantage of the opportunity to read a well-written and authoritative account, and become excited by its possibilities.

Arthur J. Rowe, Nottingham University, U.K.

Preface

Analytical ultracentrifugation (AUC) consists of the application of a high gravitational field to a solution of particles and the real-time detection of the evolving spatial concentration gradients. When applied to macromolecules or nanoscopic particles, ultracentrifugation data can provide rich information on their shape, solvation, composition and size-distribution, as well as allow for a detailed view of their reversible single- or multi-component interactions over a wide range of affinities. After almost a century of methodological development, stimulated by ever-changing emphasis in applications, as well as substantial instrumental and computational advancements, a wealth of theoretical and experimental knowledge of sedimentation has been accumulated. Unfortunately, no systematic textbook or comprehensive monograph on AUC has been published since the initial work by Svedberg and Pedersen 1940, and the seminal detailed methodological summary by Schachman, 1959. This has rendered AUC a discipline that is hard to master without direct access to experienced laboratories where it is routinely practiced, and hampers this powerful technique from once again becoming a mainstream tool in the molecular sciences.

The goal of the present book is to provide a description of the basic principles in theory and practice, sufficiently comprehensive for the reader to confidently practice AUC, and to be aware of its full potential and possible pitfalls. The book aims to help the reader gain a solid understanding of the basic concepts, and to facilitate further reading of the referenced detailed topics, with appreciation for their historic and current relevance. The emphasis is experimental, and more detailed descriptions of the theoretical frameworks and data analysis strategies are planned in forthcoming volumes. Although we always strived to provide the most important and historically accurate references, we recognize that ambiguities exist, and apologize for any perceived omissions or limitations in our knowledge.

The first chapter introduces the basic principles and technical setup of an analytical ultracentrifugation experiment, together with a brief description of the optical systems used for detection. The ultracentrifugation experiment is subsequently explored in Chapter 2 from a macromolecular standpoint to arrive at a detailed physical picture of the sedimentation process, from which to derive the relevant macromolecular parameters. Next, we recapitulate important practical aspects for conducting an experiment, including sample preparation (Chapter 3), details on data acquisition and data structure (Chapter 4), and the practical execution of the centrifugal experiment (Chapter 5). Instrument calibration and quality control experiments are outlined in Chapter 6. Tables of often useful data for AUC, including the properties of common macromolecules and solvents, are assembled in the appendices.

Throughout, to enrich the utility of the book and illustrate the facility of practical application of ideas ranging from simple to advanced, specially marked textboxes highlight how the topic at hand corresponds to AUC-related functions in the widely used data analysis programs SEDFIT and SEDPHAT, which can be freely obtained from the website of the Dynamics of Macromolecular Assembly Section of the National Institute of Biomedical Imaging and Bioengineering at sedfitsedphat.nibib.nih.gov. Although the book is conceived as a standalone reference, it also provides a broader background to our workshops on AUC and related biophysical techniques at the National Institutes of Health.

We hope this book fills a gap in the literature of biophysical methodology, and will offer the reader interesting and useful material.

Peter Schuck
Huaying Zhao
Chad A. Brautigam
Rodolfo Ghirlando

This work was supported by the Intramural Research Programs of the National Institute of Biomedical Imaging and Bioengineering, and the National Institute of Diabetes and Digestive and Kidney Diseases at the National Institutes of Health.

SYMBOL DESCRIPTION

1 index 1 to denote solvent component in multi-component mixtures

a chemical activity

a subscript 'a' to indicate the macromolecular component

$A(r)$ radial dependence of the absorbance

$a(r,t)$ radial- and time-dependent signal

$a^*(r,t)$ radial- and time-dependent signal after consideration of various optical detection effects

β in the context of nonlinear detection, the nonlinearity exponent

$\beta(t,\omega)$ time- and/or rotor-speed dependent baseline signal offset that is radially constant ('RI noise')

b bottom radius (distance from center of rotation to the distal end of the solution column)

$b(r)$ radial-dependent baseline signal offset that is temporally constant ('TI noise')

B_1 preferential binding parameter for water

B_3 preferential binding parameter for co-solute

c molar concentration

$c(s)$ diffusion-deconvoluted differential sedimentation coefficient distribution

$c_B^*(c_A)$ phase transition line (in concentration of the larger component B) of the vanishing undisturbed boundary in the effective particle model

χ^2 measure for fit quality

$\delta_{i,j}$ Kronecker symbol, $\delta_{i,j} = 1$ if $i = j$, else $\delta_{i,j} = 0$

δ_{beam} effective beam diameter in the fluorescence detector

$\delta(r,t)$ the statistical noise of each data point at radius r and time t

Δl optical pathlength difference

ΔJ fringe displacement

D diffusion coefficient

D_{norm} normalized volume of spectra basis in MSSV

dn/dw refractive index increment based on weight-concentration (in the literature often referred to as dn/dc)

ε molar extinction coefficient

$\varepsilon_{\text{molar}}$ molar extinction coefficient

$\varepsilon^{(IF)}$ molar effective fringe increment

$d\varepsilon/dr$ spatial gradient of specific signal increment in fluorescence detection

η solvent viscosity

η_0 standard viscosity (of water at $20°C$ in 1 atm)

f hydrodynamic translational friction coefficient

f_0 translational friction coefficient of the equivalent compact, smooth sphere with same mass and density as the particle

F_b buoyancy force

F_f frictional force

F_{sed} sedimentation force

$\phi(r,s)$ radial- and s-value dependence of the incident photon flux in FDS detection

γ chemical activity coefficient

$I(r)$ radial dependence of the transmitted light intensity in the sample sector

$I_0(r)$ radial dependence of the transmitted light intensity in the reference sector

j_{sed} sedimentation flux

j_{diff} diffusion flux

k in density contrast experiments with heavy water, the relative

	increase in mass due to H-D exchange	PD	subscript to denote a quantity in the 'play dough' formalism of a solid, unhydrated, inert object of uniform density
κ	solvent compressibility coefficient		
k_B	Boltzmann constant	ρ	solvent density
K	association equilibrium constant	ρ_0	standard density (of water at 20 °C in 1 atm)
K_β	nonlinearity constant		
K_D	dissociation equilibrium constant	r	radius (distance from the center of rotation)
k_{off}	chemical off-rate constant	r_0	reference radius (arbitrarily chosen)
k_s	non-ideality coefficient of sedimentation		
k_D	non-ideality coefficient of diffusion	r_{DH}	in density contrast experiments with heavy water, the molar ratio of D to (H + D)
λ	wavelength	R	gas constant
l	optical pathlength	R^*	in density contrast experiments, the ratio of buoyant molar masses or viscosity corrected sedimentation coefficients
μ	chemical potential		
m	meniscus radius (distance from the center of rotation to the proximal end of the solution column)		
		R_0	radius of the equivalent compact, smooth sphere with same mass and density as the particle
m_a	mass of particle		
M	molar mass	R_S	Stokes radius
M_a	molar mass of the macromolecular component	r.h.s.	right-hand side (of an equation)
		rms	root mean square
M_b	buoyant molar mass	$rmsd$	root mean square deviation
$M_{app,b}$	apparent buoyant molar mass	s	sedimentation coefficient
$M_{b,PD}$	buoyant molar mass in the 'play dough' formalism	s_0	ideal sedimentation coefficient in the limit of infinite dilution
M_{PD}	molar mass in the 'play dough' formalism of an unhydrated object of certain shape and density	$s_{A \cdots B}$	sedimentation coefficient of the reaction boundary in a rapidly interacting system, as denoted in the framework of the effective particle model
M_{PZ}	molar mass of a polymeric macroion P jointly with z monovalent counterions		
		$s(r,t)$	spatio-temporal evolution of signal
M_S	molar mass of a mono-valent salt		
M_1	mass of bound water per mol of protein	s_w	signal weighted average sedimentation coefficient
ω	rotor angular velocity	$s_{20,w}$	sedimentation coefficient corrected to standard conditions
ϕ'	effective partial specific volume		
Π	osmotic pressure	SP	subscript to denote a quantity in the 'sedimenting particle' formalism
$p(r)$	radial distribution of the pressure in the solution column		
(p)	superscript to denote a non-diffusing particle	t	time (but in appendix B referring to the temperature in °C)

$t^{(sed)}$	effective sedimentation time	v	particle absolute velocity
T	absolute temperature	v_{scan}	velocity of the scanner
$\bar{v}$	partial-specific volume	w	weight concentration
$\bar{v}_a$	partial-specific volume of the macromolecular component	w_a	weight concentration of a macromolecule 'a'
$\bar{v}_{PD}$	partial-specific volume in the 'play dough' formalism	ξ	preferential binding parameter
		xp	subscript to denote 'experimental'
$\bar{v}_{SP}$	partial-specific volume in the 'sedimenting particle' formalism	z	number of charges (in different context, also denoting end-to-end distance of worm-like chain)
$\bar{v}_1$	partial-specific volume of bound water		

Analytical Ultracentrifugation Basics

T HE GOAL of this first chapter is to provide an initial overview of the basic experimental setup in analytical ultracentrifugation (AUC), a consideration of the forces involved and a first introduction of the fundamental equations. This will be linked with a brief description of the optical detection systems to set the stage for more detailed considerations in the subsequent chapters.

1.1 PHYSICAL PRINCIPLES

An analytical ultracentrifuge consists of an optical detection system integrated into an ultracentrifuge, allowing for the real-time detection of the evolution of the concentration distribution of particles subjected to centrifugation. Two major experimental methods are employed in analytical ultracentrifugation, which differ in the applied centrifugal force: sedimentation velocity (SV) and sedimentation equilibrium (SE). Many excellent reviews and monographs have been written during the long history of this technique (among them, [1–8]), and in the following we only recapitulate the basic principles.

1.1.1 Basic Experimental Setup

Analytical ultracentrifugation was pioneered in the early 20th century by Theodor (The) Svedberg[1] [9, 14, 15]. Prior to that, principles of sedimentation equilibrium in solution had been discovered by Jean Perrin [16, 17], but experiments had been

---

[1]Theodor (The) Svedberg developed the oil-turbine ultracentrifuge for the study of concentration gradients of dissolved particles, and received the Nobel Prize in Chemistry 1926 "for his work on disperse systems" [9]. For a scientific biography see [10]. Other significant contributors in the technical development of analytical ultracentrifuges were Jesse Wakefield Beams and Edward Greydon Pickels, who developed air-turbine ultracentrifuges [11], initially for different purposes, and later ultracentrifuges with electrical drives. The electrical drives were widely adopted, and ultimately led to the widespread use of analytical and preparative ultracentrifuges. For a detailed historical account of the development of the analytical ultracentrifuge, see the work of Boelie Elzen [12, 13].

confined to large particles for which analytical sedimentation in the earth's gravitational field (~ 1 g) was sufficient. In the analytical ultracentrifuge, the choice of rotor speed provides a convenient opportunity to scale the magnitude of the gravitational field from 100 g to 300,000 g, corresponding to rotor speeds of 1,000 and 60,000 rpm with current analytical rotors. This flexibility permits the study of particles over a large size-range, spanning molar masses from 100 Da up to 1 GDa.[2]

The implementation of AUC is conceptually very simple. The rotor spins in an evacuated, temperature-controlled chamber[3] that isolates the sample solution and suppresses temperature-driven convective flows. The sample is loaded in a cell assembly placed in the rotor such that it is sandwiched between quartz or sapphire windows that are transparent to the optical detection system. The detection system probes the concentration distribution of the sample solution in a radial direction, with light traveling through the sample in a direction parallel to the axis of rotation.

All but the earliest analytical ultracentrifuges share the basic design depicted in Fig. 1.1. Usually, the centrifugal cell contains a centerpiece with two sector-shaped solution columns, one for the sample and the other for the matching solvent buffer used as an optical reference. Most of the current detection systems are mounted within the evacuated rotor chamber, although sometimes light is guided out of the chamber for detection such as in the case of the Spinco Model E and Svedberg's original instrument.

1.1.2 Sedimentation Velocity

An SV experiment is basically the observation of the free fall of particles in solution under the influence of a strong gravitational field: in the reference frame of the spinning solution column, the centrifugal force is equivalent to a gravitational force.[4] This force is proportional to the square of the rotor speed $F_{\text{sed}} = m_a \omega^2 r$, where m_a is the particle mass, ω the rotor angular velocity, and r the distance from the center of rotation. During sedimentation, the macromolecules are subject to buoyancy forces that oppose the gravitational force. Based on Archimedes' principle, the magnitude of the buoyancy force is equal to the gravitational force on the displaced solvent. Sedimentation, neutral buoyancy, or flotation may be observed, depending on the relative densities of the immersed particle and solvent. Even though this adds a level of complexity to the sedimentation experiment, this property can be exploited to study the particle density or composition via contrast variation, in a manner analogous to scattering techniques [20]. Thus, the particle partial-specific

---

[2]We will be expressing molar masses in Daltons (Da), as equivalent to 1 g/mol. Likewise, when expressing the absorbance of solutions we will use OD units interchangeably with AU.

[3]The vacuum system of the ultracentrifuge was developed by Beams and Pickels in the 1930s and 1940s [12, 18]. It allows for a reduction in friction heat generated at high speeds and thereby enables the maintenance of constant temperature.

[4]Under usual experimental conditions, both the earth's gravitational field and the Coriolis force are negligible [10].

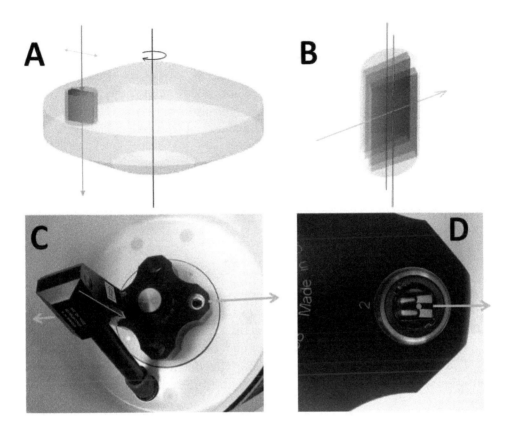

Figure 1.1 Geometry of analytical ultracentrifugation. *Panel A*: Schematics of an analytical ultracentrifugation rotor, with a sector-shaped sample volume (dark gray) in a cylindrical sample cell, and the light path in the optical system (red arrow), which is triggered with the revolution of the rotor (black curved arrow), scanning the concentration distribution in radial direction (blue double sided arrow). *Panel B*: Schematic side view of a double sector cell assembly, with two liquid solution columns (often one used as a sample and one as optical reference) and light paths of optical detection (red lines), with the direction of the gravitational field indicated by the blue arrow. *Panel C*: Picture of a 4-hole An-60 Ti rotor inside the chamber of an analytical Optima XL-A ultracentrifuge (Beckman Coulter), with the attachment for the absorbance optics installed. *Panel D*: Top view of the sample cell inside the rotor (with the red dot indicating a viewpoint along the direction of the red arrow in Panel A). It contains ~200 μL of an aqueous sample in a 12 mm centerpiece, with the meniscus in both sample and reference compartment visible approximately at half height.

volume, $\bar{v}$, and the solvent density, ρ, become relevant quantities in the buoyancy force $F_b = -m_a\bar{v}\rho\omega^2 r$, with $m_a\bar{v}\rho$ representing the mass of solvent displaced. As noted later (Section 2.1), a determination of what should be considered the volume of the sedimenting particle is not trivial, as this can include contributions from hydration, weakly bound co-solutes, and locally altered solvent, among others.

The sum of the centrifugal and buoyancy forces is matched by a frictional force, which arises from the hydrodynamic translation of the particle migrating in solution (Fig. 1.2). It is due to the work required to move solvent molecules to create space for the sedimenting particle and to move the solvent molecules in the zone of hydrodynamic drag. The frictional force is opposed to migration, and its magnitude is proportional to the absolute velocity of migration v, taking the form $F_f = -vf$, where f represents the hydrodynamic translational friction coefficient.[5]

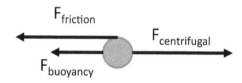

Figure 1.2   Frictional, centrifugal, and buoyancy forces acting on a particle during SV.

It is more convenient to express the sedimentation velocity in terms that reflect the particle's molecular properties, independent of the applied gravitational field, $\omega^2 r$. This is accomplished by normalizing the particle's velocity relative to the gravitational field, thus defining the sedimentation coefficient:

$$s = \frac{v}{\omega^2 r} \tag{1.1}$$

It is measured in units of Svedberg, abbreviated S, with $1\,S = 10^{-13}$ sec. Following custom, the sedimentation coefficient will also be alternately referred to as the 's-value'. The sign of the s-value can be positive or negative, dependent on the relative densities of the particle and solvent resulting in either sedimentation (positive sign) or flotation (negative sign). The balance of forces leads to a relationship of s in terms of the molecular mass and friction:

$$s = \frac{m_a\,(1 - \bar{v}\rho)}{f} \tag{1.2}$$

---

[5] As an example for the magnitude of the frictional force, a BSA monomer with a s-value of ~4.3 S will sediment with a velocity of ~0.8 μm/sec at 50,000 rpm and experience a frictional force of ~0.05 fN.

SEDFIT – In the Options ▷ Calculator menu, the frictional force can be calculated for a sedimenting particle under given experimental conditions.

We can express the frictional coefficient using Stokes' law and the Stokes radius R_S as

$$f = 6\pi\eta R_S \tag{1.3}$$

where η is the solvent viscosity and R_S is the radius of an equivalent sphere that has the same frictional coefficient as the particle under consideration (but not necessarily the same mass).[6,7] R_S can be expressed relative to the radius R_0 of a hypothetical solid and smooth spherical particle that has the same mass and density as our sedimenting particle. We may imagine a solvent free, compact particle composed of play dough modeling material;[8] which when compacted into a solid sphere would have a radius R_0, and corresponding translational friction coefficient f_0. The frictional ratio $f/f_0 = R_S/R_0$ describes the excess friction our original particle exhibits relative to that sphere arising from rearranging its mass in the most compact form. The value of f/f_0 is frequently utilized as a measure of shape asymmetry, but it should be noted that values > 1.0 do not necessarily imply asymmetric particle contours.[9] Together with Eq. (1.2), this leads to the following relationship for the sedimentation coefficient:

$$s = \frac{m_a\left(1 - \bar{v}\rho\right)}{6\pi\eta R_0\left(f/f_0\right)} \tag{1.4}$$

Accordingly, aside from the dependence on solvent density and viscosity, s is a molecular constant, reporting on the molecular mass, partial-specific volume, and shape. To remove the dependence on the solvent properties, it is customary to convert the experimental s-values to values that would be observed if the experiment were carried out in water at $20\,°C$ having a density ρ_0 of 0.9982 g/mL and a viscosity η_0 of 1.002×10^{-3} Pa·sec (or 0.01002 Poise). This now allows a comparison of s-values

---

[6]Cheng and Schachman [21] have experimentally confirmed that the Stokes' law of sedimentation and Einstein law of viscosity indeed hold for microscopic particles, although derived on the assumption that particles are large relative to the molecules of the solvent medium.

[7]This holds true for stick boundary conditions. In organic solvents slip boundary conditions apply, and the frictional coefficient becomes $f = 4\pi\eta R_S$.

[8]We envisage a malleable material of constant density and well defined surface that can be of different shape, but at constant mass and uniform density.

[9]For example, we can imagine a ball of play dough of radius r_0 was formed into a particle with a hollow spherical shell of outer radius $r_1 = 2r_0$; it would be round but have a frictional ratio 2.0. Also, the value of the frictional ratio is dependent on the definition of the sedimenting particle (see Chapter 2) and whether, for example, solvation is included. Customarily, hydration is not included, such that tightly bound water will appear to create additional macromolecular 'shape asymmetry' when determined using f/f_0. Typical values for proteins range from 1.2 to 1.5 for shapes that are globular to moderately elongated [22, 23], but can be as small as 1.1 for some γ-crystallins with compact shape and low hydration [24, 25].

determined in different buffers and temperatures. Using these standard conditions, $s_{20,w}$ is defined:

$$s_{20,w} = s_{xp} \frac{\eta_{xp}}{\eta_0} \frac{(1 - \bar{v}_0 \rho_0)}{(1 - \bar{v}_{xp} \rho_{xp})} \qquad (1.5)$$

(with the subscript 'xp' indicating the experimental values).[10]

SEDFIT – A function in the `Options ▷ Calculator` menu transforms experimental s-values to $s_{20,w}$-values. For a complete sedimentation coefficient distribution, provided that all species visible have the same $\bar{v}$, this transformation can be done in the `Options ▷ Size Distribution Options` menu.

An example for the s-values as a function of particle mass and shape for proteins with a range of frictional ratios in aqueous solutions is given in Fig. 1.3.

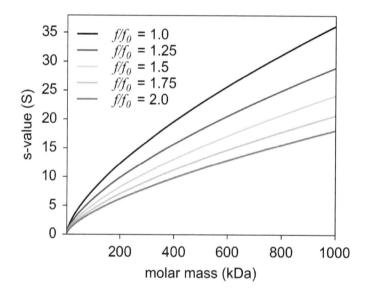

Figure 1.3 Dependence of the approximate s-value of proteins in aqueous solvents on the protein molar mass and frictional ratio, approximated as $s_{20,w} \approx 0.012 M^{2/3} (1 - \bar{v}\rho) \bar{v}^{-1/3} (f/f_0)^{-1}$ with s in units of S, M in Dalton, $\bar{v}$ in mL/g (assumed here to be 0.73 mL/g) and ρ in g/mL (assumed here to be the standard density of 0.9982 g/mL). Shown is a family of curves with different f/f_0 values. The black curve with $f/f_0 = 1.0$ represents the fastest possible sedimentation velocity for a particle of the given mass and density. Frictional ratios in excess of 2.0 are rare (but not impossible) for folded proteins, but higher values will occur for worm-like chains or rod-like particles, such as nucleic acids and chromatin arrays.

[10]Note this relationship is incorrect in ref. [26].

In the simplest depiction of sedimentation, we can calculate the trajectory of a single, non-diffusing particle, $r^{(p)}(t)$, by using the definition of the s-value Eq. (1.1) to determine the position-dependent velocity, which leads to a differential equation:

$$\frac{dr^{(p)}}{dt} = s\omega^2 r^{(p)} \qquad (1.6)$$

It has the solution:

$$r^{(p)}(t) = r_0 e^{s\omega^2 t}$$

$$r^{(p)}(t) = m e^{s\omega^2 t} \qquad (1.7)$$

where r_0 is the particle position at $t = 0$. With the starting position taken as the meniscus position m, Eq. (1.7) may be used to describe the propagation of a sedimentation boundary of non-diffusing species. Since the centrifugal field increases with radius, it produces an acceleration that increases with time, such that particles are expelled exponentially from the center of rotation (Fig. 1.4).[11]

This simple, single-particle model omits environmental forces that act on the particle, such as those responsible for diffusion, as well as forces arising from the presence of other macromolecules. The latter include long-range electrostatic forces, steric repulsion under high macromolecular concentrations, or short-range attractive forces leading to transient complex formation. Their influence and analysis are frequently of great importance and their treatment will be the topic of Section 2.2.

However, it is possible at this stage to relate the process of sedimentation to diffusion via the common assumption that both share the same translational friction coefficient (for limitations of this assumption, see e.g. Section 2.3). In such a case, the Stokes–Einstein relationship $D = k_B T/(6\pi\eta R_S)$ (with k_B the Boltzmann constant and T the absolute temperature) can be used, jointly with Eq. (1.2) to arrive at the Svedberg equation [27]:

$$\frac{s}{D} = \frac{M(1 - \bar{v}\rho)}{RT} \qquad (1.8)$$

where M denotes the particle molar mass, and R is the gas constant. The Svedberg equation is very important in that it relates the three most fundamental quantities that can be measured directly in sedimentation experiments: the sedimentation coefficient (obtained from the migration of the sedimentation boundary with time

---

[11]In the case of flotation, characterized by a negative sedimentation coefficient s due to particle densities lower than the solvent density, Eq. (1.7) describes the particle migrating toward the center of rotation with a decreasing absolute velocity.

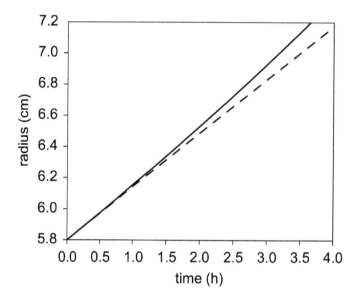

Figure 1.4 Radial position as a function of time for a 6 S species sedimenting at 50,000 rpm, starting at a position at 5.8 cm and following Eq. (1.7) (solid line). For comparison, the trajectory that would correspond to a constant velocity is shown as a dashed line. Over the radial range covered within the geometric constraints of the current analytical ultracentrifuge, the exponential acceleration is qualitatively not large, but nonetheless significant for all quantitative considerations.

in SV), the diffusion coefficient (obtained from the spread of the sedimentation boundary with time), and the molar mass (obtained from the exponential gradient in SE, see below).

> SEDFIT – Even though the default description of macromolecular sedimentation and diffusion in the discrete non-interacting species model is phrased in terms of molar mass and sedimentation coefficient, from which the diffusion coefficients are calculated via the Svedberg equation (1.8) the Options ▷ Fitting Options ▷ Fit M and s function can toggle the program into a mode where the diffusion coefficient can be directly entered, and molar masses are then calculated implicitly in conjunction with the given sedimentation coefficient. Similarly, a switch of coordinates from diffusion coefficients into Stokes radii is possible when analyzing dynamic light scattering data, using the function Model ▷ Dynamic Light Scattering ▷ Discrete Stokes Radii.

For large macromolecules, macromolecular assemblies and many nanoparticles with R_0 greater than 10–15 nm, SV experiments are usually conducted under conditions such that the root-mean-square (r.m.s.) distance particles travel by diffusion is small or even negligible compared to the migration by sedimentation. However, for most macromolecules, the opposite is true under typical experimental conditions. Therefore, in revising the analyses presented in Fig. 1.2 and Fig. 1.4, it is more accurate to imagine an ensemble of molecules where the individual molecules mostly undergo a random walk from diffusion, which is biased by the centrifugal field. This is schematically depicted in Panel A of Fig. 1.5 but can be better visualized in a movie simulating biased random walk [28].[12]

We can now anticipate the main features of a real SV experiment, in which we study a large ensemble of particles,[13] and monitor their concentration as a function of time and radial distance [28]. In such an experiment, a sector-shaped solution column usually starts out with a uniform distribution of particles. In the frame of reference of the spinning solution column, we will refer to the direction of the centrifugal force as 'down' and its distal end as the 'bottom' (at radius b) of the centrifugal cell. At the upper end of the solution column is the meniscus (at radius m), representing the air-solution interface. The solution column in SV typically includes radial distances of ~6.0 cm to ~7.2 cm from the center of rotation.

When centrifugation starts, all molecules sediment at a velocity $v = s\omega^2 r$ that is dependent on their radial position, where they experience the exponential acceleration described above. Will the ensuing radial differences in velocity create a concentration gradient? To answer this, let us observe a volume element ΔV between radius r_1 and $r_2 = r_1 + \Delta r$, of a height h, and with an average width in the plane of rotation of $\bar{y} = \varphi \bar{r}$ (with φ the angle of the sector-shaped solution column, usually 2.4°, and the average radius $\bar{r} = r_1 + \Delta r/2$). This is depicted in Panel B of Fig. 1.5. Following a small increment of time dt, the radial boundaries of this volume element will have migrated to $r_1' = r_1 + s\omega^2 r_1 dt$ and $r_2' = r_2 + s\omega^2 r_2 dt$, i.e., to radii larger by a factor $(1 + s\omega^2 dt)$. The height stays unchanged, but the width has grown to $\bar{y}' = \varphi (\bar{r} + s\omega^2 \bar{r} dt)$, namely by the same factor $(1 + s\omega^2 dt)$. The volume element $\Delta V = (r_2 - r_1) h \bar{y}$ will still contain the same number of molecules, but based on the change in width and radii of the boundaries, the volume element will increase by a factor $\Delta V'/\Delta V = (1 + s\omega^2 dt)^2$, which is the square of the radial displacement factor. Importantly, the magnitude of this dilution during the sedimentation process is independent of radius, and will therefore not result in a radial concentration gradient. Furthermore, due to the sector-shaped solution column and the radial acceleration experienced by the migrating particles, the particle concentration will decrease over time,[14] such that:

$$\frac{c(t')}{c(t)} = \left(\frac{r(t)}{r(t')} \right)^2 \tag{1.9}$$

---

[12]From a set of experimental scans obtained during the sedimentation of a monodisperse single species, a comparison of the r.m.s. displacement by diffusion and sedimentation can be made from the width of the sedimentation boundary relative to its distance from the meniscus, see below.

[13]A typical sample for SV, 0.4 mL of a 1 μM solution, will contain ~2.4×10[14] macromolecules.

[14]More quantitatively, the relative change in concentration will be proportional to the increase in the size of the volume element $(dc/dt)/c = -(d\Delta V'/dt)/\Delta V$. After dropping quadratic terms, we find $(dc/dt)/c = -2s\omega^2$, from which $c(t) = c(t_0)e^{-2s\omega^2(t-t_0)}$ follows. This is a special case solution of the Lamm equation (Eq. 1.10 below) for the plateau region, i.e., in the absence of diffusion.

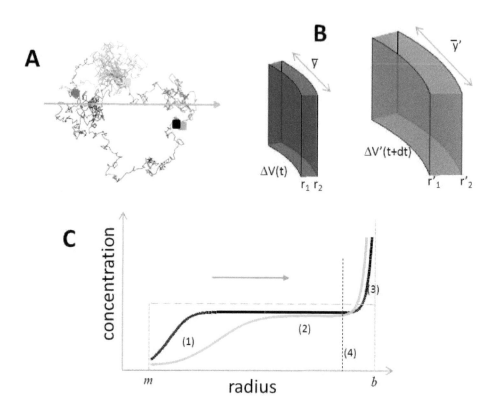

Figure 1.5 Schematic representation of the sources of concentration gradients in SV for an ensemble of diffusing molecules. *Panel A:* Except for very large particles (typically > 1000 kDa) the sedimentation of individual particles can be depicted as a bias in their diffusional random walk through solution, producing a net migration in the direction of the gravitational field (arrow). To illustrate this point, depicted here are paths of three different molecules subjected to random walk with bias, all starting at the red dot and ending at the different colored squares, based on the trajectories shown. This results in a diffusional spread of the sedimentation boundary. *Panel B:* As a result of migration, and because at larger distances from the center of rotation the average inter-particle distance increases both laterally and radially, the concentration drops with time. *Panel C:* Sketch of concentration profiles at different points in time (cyan later than black). A uniform concentration distribution from the meniscus m to the bottom b exists at the start of sedimentation (dashed line). At later times, four different features of the sedimentation progress can be discerned: (1) The sedimentation boundary separates the solvent plateau (the part of the solution column that has been cleared of particles, visible at later times in the vicinity of the meniscus) from the solution plateau (where the particle concentration exhibits no gradient). The boundary migrates with time, and with increasing time a greater part of the solution column will be cleared of particles. Note that the shape of the sedimentation boundary will show broadening with time due to diffusion. (2) The solution plateau concentration drops with time due to the effects of radial dilution. (3) The impermeable bottom of the solution column will generate a strong increase in concentration from the local accumulation of particles. Driven by this concentration gradient is a diffusional transport directed opposite to sedimentation. This is called the back-diffusion region. (4) The total number of particles soluble 'above' the region of back diffusion (separated by the dotted vertical line) will decrease with time due to the transport to the bottom.

This is referred to as the 'square dilution law' [29].

> SEDFIT – In the Options ▷ Calculator menu the magnitude of radial dilution can be calculated for given experimental conditions.

This region where no radial concentration gradient exists but the concentration drops with time is called the 'solution plateau' (see Panel C of Fig. 1.5). At some time during sedimentation, the region near the meniscus will have cleared of particles, exposing the so-called 'solvent plateau' that increases radially with time. The region between the solvent and solution plateau is the sedimentation boundary, which migrates with the velocity of the sedimenting particles. The midpoint tracks those molecules that were initially closest to the center of rotation, i.e., at the meniscus, and the boundary shape is rich in information.[15] This boundary usually undergoes broadening with time due to diffusion and polydispersity in particle size, and its shape is influenced by attractive and repulsive particle interactions. At the distal end of the solution column, referred to as the 'bottom,' there is an impermeable wall where particles either pellet or stay in solution, the latter resulting in steep concentration gradients that increase with time and are opposed by diffusional fluxes. The extent of this 'back-diffusion' zone depends strongly on the molar mass of the particles, increasing with smaller particle size (higher diffusion coefficients). The accumulation of molecules in this region observed during sedimentation is accompanied by the continual depletion of material at smaller radii. The converse holds true for flotation.

The manner in which the measured concentration gradients are related to the sedimentation coefficient is not trivial. For instance, in the case of a mixture of species having different sizes, the measured evolution of the concentration can exhibit multiple boundaries. Generally, larger particles sediment faster and undergo less diffusion while they migrate through the solution column, whereas smaller particles sediment slower but can migrate significantly due to diffusion. This is depicted in Fig. 1.6, which shows the concentration distributions at different points in time for a mixture of particles of 1 kDa, 10 kDa, 100 kDa, and 1 MDa sedimenting at a high rotor speed of 50,000 rpm, a typical rotor speed for SV studies of biological macromolecules. For the smallest species, the back-diffusion region extends over almost the entire solution column, and no clearly formed solution or solvent plateaus exist.

The entire process is captured concisely with the description of the temporal evolution of the balance between local sedimentation fluxes and diffusion fluxes in the Lamm equation [32]:

$$\frac{\partial c}{\partial t} = -\frac{1}{r}\frac{\partial}{\partial r}\left(cs\omega^2 r^2 - D\frac{\partial c}{\partial r}r \right) \tag{1.10}$$

---

[15]Technically, a better measure would be the second moment [30,31].

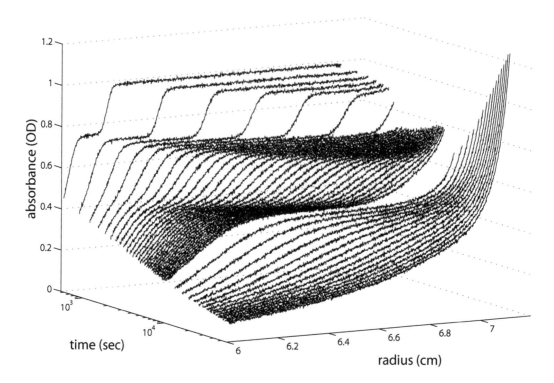

Figure 1.6 Calculated radial sedimentation profiles as a function of time for a hypothetical mixture of globular species of 1 kDa (0.35 S), 10 kDa (1 S), 100 kDa (5.9 S), and 1 MDa (28 S) sedimenting at 50,000 rpm, at equal loading signals of 0.25 OD, superimposed with random noise reflecting a realistic signal-to-noise ratio. The time-scale is logarithmic, in order to distinguish the species of different size. In addition to the different time-scales of sedimentation, dissimilar degrees of boundary spreading due to diffusion can be discerned, with the 1 MDa species exhibiting relatively little diffusion due to a low diffusion coefficient and the short time-scale of sedimentation, while the 10 kDa species exhibits larger root-mean-square displacement from diffusion than sedimentation, hence the broad boundary features at the longest times. An inspection of the plateau signals (i.e., the signals at radii higher than the leading edge of the boundary) demonstrates the slight dilution that takes place over time, even in regions not yet reached by the sedimentation boundary. This is due to the macromolecular sedimentation taking place in the radial direction throughout the whole sector-shaped solution column, lowering the intermolecular distances and hence the drop in concentration (see Fig. 1.5). The end (or 'bottom') of the solution column at 7.2 cm is not shown; the molecules sedimenting to the bottom of the solution column form very steep concentration gradients there. Only for the medium-sized and small protein do we see the steep concentration increase at the bottom. This is due to the higher diffusion coefficient of the smaller species opposing the accumulation of material at the bottom. This portion of the traces is referred to as 'back-diffusion.' For the smallest species, the back-diffusion region extends over the entire solution column, and no distinct sedimentation boundary can be discerned.

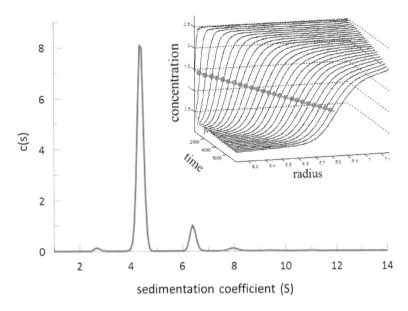

Figure 1.7 SV analysis of bovine serum albumin sedimenting at 50,000 rpm, recorded using Rayleigh interference optics (inset). Highlighted in red are the midpoints of the sedimentation boundaries, which, in conjunction with an estimate of the meniscus position, may be used to calculate a sedimentation coefficient of the main species. However, inclusion of the entire data set allows for a computational analysis that deconvolutes the sedimentation and diffusion processes, resulting in a differential sedimentation coefficient distribution $c(s)$ (blue solid line) [40]. This resolves the multiple oligomeric states of BSA, and presents their s-values and populations.

It was derived by Ole Lamm, a student of Svedberg, back in 1929 [32], but the lack of an explicit analytical solution hampered SV analysis throughout most of the 20th century; however, the Lamm equation can be solved efficiently and accurately using modern numerical algorithms [33].[16]

To establish a qualitative estimate for the power of sedimentation velocity to discriminate between particles of different size, we assume a spherical particle with $R_0 = R_S = R$, from which the mass is proportional to the third power of the radius, R^3, the friction is linear in R, and therefore the separation will increase with R^2. This provides a much better hydrodynamic resolution than many diffusion-based techniques, in which the resolution scales as R^{-1}. Furthermore, this also demonstrates that, for a given rotor speed, larger particles can be better resolved than smaller particles.

Modern analyses now consider the whole set of scans with signal profiles in their entire detectable range, and fit the entire evolution with numerical solutions to the Lamm equation (1.10). This has its roots in the early computerized simulation and

---

[16]Interconverting systems of molecules undergoing rapid chemical reactions on the time-scale of SV are described by systems of Lamm equations coupled by reaction fluxes [5]. For these, simple algebraic equations describing the salient features of the sedimentation boundary pattern exist, which relate to the underlying physics of the sedimentation/interconversion process [34].

analysis of SV in the 1960s and 1970s [35–39], and has significantly transformed SV in the last two decades with unprecedented precision and sensitivity [8]. In particular, it has become possible to utilize the observed time-course of the diffusional spread to deconvolute sedimentation from the effects of diffusion, much like the deconvolution of point-spread functions in optical imaging. In this manner it is possible to calculate sedimentation coefficient distributions that resolve species whose sedimentation boundaries are not visually separated [40]. This is illustrated in Fig. 1.7, which shows the differential sedimentation coefficient distribution $c(s)$ of a sample of bovine serum albumin (BSA) based on SV data collected at 50,000 rpm. Rather than determining the boundary mid-points and estimating an s-value via Eq. (1.7), it is now possible to deduce from the shape and evolution of the entire concentration profiles the presence of monomeric and oligomeric species of BSA. These can be visually discerned in the raw data only from the stretched appearance of the leading edge of the main boundary, because they are of low amplitude and diffusionally broadened. Computationally they are very well defined due to the large number of data points and high signal-to-noise ratio. The ability to carry out such computational analyses has significant implications for the design and setup of SV experiments, including the design and use of optical systems for data collection.

1.1.3 Sedimentation Equilibrium

SE experiments are conducted in the same instrument used for SV experiments. These two methods are complementary and share many common practical considerations. From a naïve point of view one could consider SE as a limiting state of SV attained after long times, and, in fact, sometimes for studies of very small molecules SV and SE can be just different parts of the same experiment.[17] However, for many theoretical and practical reasons, this perspective would be too narrow and lead one to overlook the fundamentally different nature of studying thermodynamic equilibrium by sedimentation velocity and sedimentation equilibrium.

Obvious key differences are: (1) equilibrium conditions can be theoretically derived from equilibrium thermodynamics, without reference to the dynamics through which equilibrium is attained and therefore allow modeling the sedimentation behavior of solutions that may be kinetically intractable; (2) since there is no net transport at equilibrium, kinetic considerations and hydrodynamic friction are irrelevant for an equilibrium analysis; (3) in SE, lower rotor speeds are used such that the back-diffusion region imposed by the impermeable bottom reaches throughout the solution column; and (4) the equilibrium experiment usually requires much longer times.

---

[17] A movie depicting the sedimentation process in the picture of biased diffusion for the case of large diffusion where particles approach steady-state of sedimentation and diffusion can be found at [28].

SEDFIT– Functions in the Options ▷ Equilibrium Tools menu allow for the determination of useful rotor speed(s) for SE experiments, the minimum time to attain equilibrium, and temporal rotor speed profiles that can accelerate the approach of equilibrium for a given system.

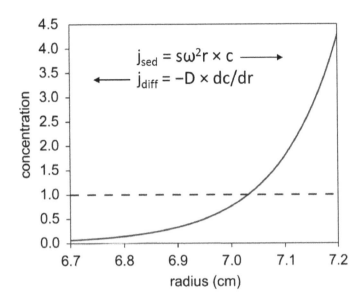

Figure 1.8 Principle of ideal sedimentation equilibrium, depicting the sedimentation and diffusion fluxes that are matched when the concentration distribution $c(r)$ assumes a Boltzmann exponential. The curve corresponds to a 100 kDa protein initially at a unit concentration (dashed line) in a solution column reaching from the meniscus at 6.7 cm to the bottom at 7.2 cm, at a rotor speed of 10,000 rpm. A partial-specific volume of 0.73 mL/g and solvent density of 1.00 g/mL are assumed.

Using the same idealized picture presented above, we can derive the essence of SE concentration distributions $c(r)$ solely by considering the balance of diffusion and sedimentation (Fig. 1.8): The flux from sedimentation is $j_{sed} = cv = cs\omega^2 r$; this is opposed by the diffusion flux, which according to Fick's first law is $j_{diff} = -Ddc/dr$. Equilibrium requires their values to match such that $j_{sed} + j_{diff} = 0$. We can already see that, because j_{sed} is stronger at higher radii, an equilibrium profile with increasing slope must arise. Quantitatively, the matching condition of the fluxes leads to the differential equation $dc/dr = c(r)(s/D)\omega^2 r$, which is satisfied by the barometric Boltzmann exponential:

$$c(r) = c(r_0) \exp\left[M(1 - \bar{v}\rho)\frac{\omega^2}{2RT}(r^2 - r_0^2)\right] \tag{1.11}$$

where we have made use of the Svedberg equation (1.8) to replace the ratio s/D, and with r_0 denoting an arbitrary reference radius. The concentration at the reference radius $c(r_0)$ is determined by the initial conditions.[18] The reader will note that there is a close analogy in the exponential structure and derivation of this equation to the well-known barometric formula describing the density of the earth's atmosphere as a function of height above ground [16]. The key quantity directly governing the exponential is the buoyant molar mass M_b:

$$M_b = M \left(1 - \bar{v}\rho\right) \tag{1.12}$$

Thus, with knowledge of the partial-specific volume, the sedimentation equilibrium profile allows for a direct determination of the particle molar mass [42]. A more thorough thermodynamic derivation can be found in [43]. As will be described later, the SE distributions of interacting components linked by mass action law are sums of Boltzmann exponentials for each species.

There are many practical differences between SV and SE experiments. Frequently, SE data are collected at multiple rotor speeds, referred to as a 'multi-speed' SE experiment. As will be described in Section 5.5.2, observing the redistribution of material at different gravitational fields provides information on the total mass of soluble macromolecules. This is often crucial for an unambiguous interpretation of the exponential concentration profiles in terms of a molar mass distribution and/or binding constants of interacting species. SE requires diffusion fluxes to be equilibrated across the whole solution column; this is a slow process and therefore shorter solution columns are usually used.[19] This poses a requirement for sample stability of the time of SE data acquisition, which could be challenging for some macromolecules. This is an important difference from SV, where we primarily aim to observe the typically faster sedimentation fluxes, which is best done with longer solution columns.

1.2 DETECTION PRINCIPLES

The choice of an optical detection system offers a spectral dimension to data acquired in AUC. Many different detection systems for the real-time measurement of the concentration gradients in the solution column have been developed during the long history of AUC, adding greatly to the versatility of the method for the study of macromolecular samples. For a practical data analysis, it is important to recognize the selectivity, precision, noise structure, and limitations of the data acquisition for each optical system.

The detection systems developed include the early method of recording the optical density by photography [1,2,15], and later schlieren optics [1,2,44–46], updated

---

[18]This equilibrium expression resembles the barometric formula of the pressure of an ideal molecular gas [41]. That concentration distributions of particles in sedimentation equilibrium obey the barometric formula was discovered by Jean Perrin [16], who received the Nobel prize in physics in 1926 "for his work on the discontinuous structure of matter, and especially for his discovery of sedimentation equilibrium" [17].

[19]The length of the solution column influences the time required to reach equilibrium. Typical loading volumes are 130–160 μL, producing 4–5 mm columns (Appendix D), and equilibration times are usually on the order of days.

in the recent decades [47–53], several different types of interferometers in different implementations [2,54–62], the absorbance scanner [63–66], turbidity [67,68], and fluorescence detectors [69–71]. Traditional optical systems are reviewed in [72]. Also, for SE studies of proteins at high concentration, post-centrifugal fractionation has been described [73–77].

Currently, the most widely used are the commercial UV-visible absorbance [65] and the Rayleigh interferometric systems [60, 78]. A commercial fluorescence detection system has been recently developed [71,79]. The following descriptions will be confined to an initial overview of these three types of real-time sedimentation detectors, with more detailed descriptions to follow in Chapter 4.

1.2.1 Absorbance

The absorbance optical system functions like a double-beam spectrophotometer (Fig. 1.9) in which the Xenon flashlamp light source is pulsed synchronously with a sample or reference sector when in alignment with the optical system. Monochromatic light with a nominal bandwidth of 2 nm is produced with the aid of a diffraction grating, at a selected wavelength between 200–800 nm [65]. It passes through the reference and sample solutions, respectively, and to the detector, which consists of a photomultiplier covered by a slit that scans in the radial direction.

In this manner, radial scans of the transmitted intensities across the sample $I(r)$ and reference cell $I_0(r)$ are obtained. Although these raw intensity data can be made available for analysis, they are typically transformed into absorbance values $A(r) = \log_{10}(I_0(r)/I(r))$. Within a linear range up to ~1.5 optical density units (OD), but dependent on the detection wavelength and macromolecular extinction profile, the measured absorbance follows the Beer–Lambert law:

$$A = \varepsilon\, l\, c \qquad (1.13)$$

where ε is the molar extinction coefficient per cm, and c the molar concentration. The optical pathlength l is determined by the centerpiece type, usually 0.3 cm or 1.2 cm. Typical data acquisition noise is ~0.005 OD with a signal-to-noise ratio up to 300:1. The useful loading concentrations for SE experiments are typically above 0.05 OD. In contrast, due to the large number of data points collected in an SV experiment, lower detection limits apply when using modern data analysis tools. In fact, signal magnitudes may be lower than the statistical noise of data acquisition (allowing, as an extreme example, the analysis of ~10 nM of a 50 kDa protein by far-UV detection with a loading signal of ~0.005 OD_{230} [80]). Examples of absorbance SE and SV data are shown in Fig. 1.10.

Molar extinction coefficients for proteins based on the amino acid composition can be predicted in a Options ▷ Calculator function of SEDFIT following the method of Pace and co-workers [82].

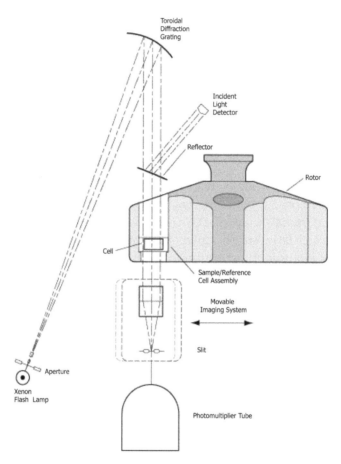

Figure 1.9 Schematics of the commercial absorbance optical system of the XL-A analytical ultra-centrifuge from Beckman Coulter (courtesy of M. Rhyner, Beckman Coulter, Inc.).

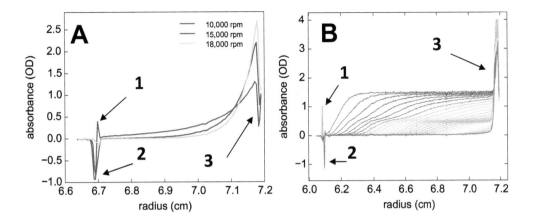

Figure 1.10 Typical absorbance data for SE at three different rotor speeds (*Panel A*) and SV at different time points (with the color temperature from purple to red indicating early to late scans) (*Panel B*). Arrows indicate the optical artifacts arising from the meniscus of the sample column (1), the meniscus of the reference column (2), and the artifacts from the bottom of the solution column (3). The graphs and many subsequent figures are plotted in GUSSI [81].

A key feature of this detection system is that it is selective to those solutes that have a non-vanishing extinction coefficient at the wavelength chosen. For example, most buffer components are transparent at the 280 nm absorption maximum of proteins. Furthermore, the selectivity and large range of possible wavelengths offers great opportunities for studies with multiple components that exhibit different absorbance profiles, as absorbance scans from different wavelengths can be acquired quasi simultaneously from the same solution during the same experiment, greatly enhancing their information content [66, 83–92]. As we will see in Chapter 4, an additional virtue of absorbance data is their relatively simple noise structure.

1.2.2 Rayleigh Interferometry

The Rayleigh interferometric imaging system has very different properties due to its detection of solution refractive index gradients. A coherent and collimated laser beam, wide enough to illuminate the entire solution column, is split and simultaneously passes through the sample and reference sectors (Fig. 1.11). A cylinder lens combines both beams to produce an interference pattern that is imaged on a charge-coupled device camera. The pattern of interference fringes reflects, at each radial position, differences in the optical pathlength, corresponding to differences in the solution refractive index that stem from the dissolved particles. Provided all other solution components are matched, then the recorded fringe shift ΔJ follows [93]:

$$\Delta J = \frac{\Delta l}{\lambda} = w \left(\frac{dn}{dw} \right) \frac{l}{\lambda} \tag{1.14}$$

where Δl is the optical pathlength difference, λ the wavelength, dn/dw the refractive index increment of the macromolecule under study, and w its weight con-

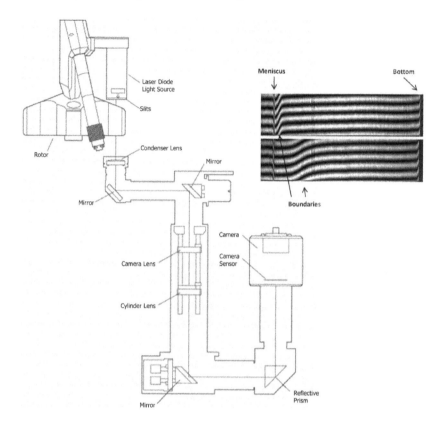

Figure 1.11 Schematics of the commercial Rayleigh interference optical system of the XL-I analytical ultracentrifuge from Beckman Coulter (courtesy of M. Rhyner, Beckman Coulter, Inc.). Included at top right are snapshots of the fringe picture at two different points in time.

centration.[20] The unit of ΔJ is commonly termed "fringes", and this term will be employed throughout this volume.

> **SEDFIT** In the Options ▷ Calculator ▷ Calculate vbar, dn/dc and extinction280 from protein sequence function refractive index increments (and molar fringe shift increments) of proteins can be predicted on the basis of their amino acid composition following the method of McMeekin and colleagues [94] as implemented in [95].

Typical interference data are shown in Fig. 1.12, and when compared to absorbance data, it is apparent that the interference data can have lower noise, with a signal-to-noise ratio as high as 1000:1. Data acquisition noise is on the order of a few

---

[20]In the literature the refractive index increment is usually referred to as dn/dc in weight units. To avoid confusion of concentration units, the weight concentration is referred to as w and the molar concentration as c throughout the book, and consequently the refractive index increment in weight units is denoted as dn/dw.

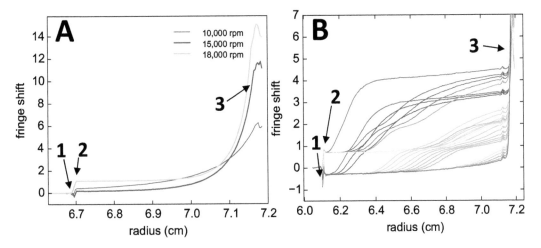

Figure 1.12 Typical interference data showing SE at three different rotor speeds (*Panel A*) and SV at different time points (with the color temperature from purple to red indicating early to late scans; *Panel B*). These data were acquired from the same experiment presented in Fig. 1.10, i.e., at the same rotor speed and with SV data collected at similar time-points. It is apparent that each scan is subject to an arbitrary vertical fringe shift. Again, arrows indicate the optical artifacts arising from the meniscus of the sample column (1), the meniscus of the reference column (2), and the artifacts from the bottom of the solution column (3).

thousandths of a fringe, permitting the detection of proteins at low μg/mL levels in aqueous solutions. In addition to the high sensitivity and practically unlimited signal linearity, an advantage of this detection system is the high speed of data acquisition, which can produce data at a rate of up to ~5 scans/min, approximately 5–10 times faster than the absorbance optical system.

Undoubtedly the biggest drawback of Rayleigh interferometry (and any refractometric technique) is that everything, including the solvent and buffer components, can contribute to the signal. Only with rigorous matching of the geometry and composition of the buffer between sample and reference sectors is it possible to eliminate co-solute signal contributions. Another complication is that the fringe displacement is only a relative measure and does not provide an absolute baseline for zero concentration (as is evident from the different offsets from scan to scan in Fig. 1.12). Furthermore, any surface unevenness on the nm scale (a few thousandths of a wavelength) in the optical elements will contribute a significant radial-dependent baseline profile, resulting in a complicated noise structure. Even though interpretation of the raw data does not appear to be as straightforward, these issues can be dealt with very effectively using computational methods, as will be described in Section 4.1.4. Thus, for many studies, this is the detection method of choice.

It is important to note that absorbance and interference optical data are usually collected concurrently to generate multi-signal data sets with greatly enhanced information content [85, 91, 96, 97].

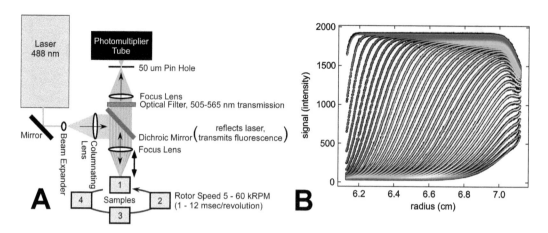

Figure 1.13 *Panel A*: Schematics of the commercial fluorescence detection system from Aviv Biomedical Inc. (Courtesy of G. Ramsey and J. Aviv, Aviv Biomedical Inc.). *Panel B*: An example for experimental SV data of EGFP.

1.2.3 Fluorescence

The commercial fluorescence detection system (FDS) consists of a scanning confocal fluorimeter, with a fixed excitation at 488 nm, and emission recorded in a bandpass of 505–565 nm [79,98](Fig. 1.13). The key features of fluorescence detection are high sensitivity, specificity and a large dynamic range. The lowest useful concentrations depend on the extinction coefficient and quantum yield of the fluorophore and its local environment. For enhanced green fluorescent protein (EGFP), commonly fused as a fluorescent tag, useful concentrations for sedimentation analysis can be as low as 1 pM [99] up to several μM [98, 100], or higher if used as a tracer. This superb sensitivity allows for binding constants in the low pM range to be measured [99]. After accounting for the characteristic data structure imposed by optical constraints, in the mid to high concentration range, excellent signal-to-noise ratios of 100:1–1000:1 can be achieved [100].

A drawback of FDS is that it may require extrinsic labeling, introducing the possibility of unintended modifications to the macromolecular properties, as well as quantum yields that may depend on protein oligomeric states and/or associated photophysical processes, thus requiring control experiments including those routinely carried out in bench-top fluorimeters [101, 102]. Fluorescence detection allows for selectivity (relative to unlabeled and non-fluorescent molecules) just like the absorbance system, but with significantly higher sensitivity. This selectivity has been particularly useful for the characterization of the oligomeric state of EGFP-tagged proteins directly in cell lysates [103], and of fluorophore-labeled antibodies in serum [104]. Another important consideration for the application of FDS at low concentrations of the proteins of interest is the need for carrier proteins to suppress surface adsorption.

Chapter 4 on data acquisition provides details on the practical use and sedimentation data properties for each of the optical systems discussed.

The Sedimenting Particle

CLOSER EXAMINATION reveals that the naïve concept of the sedimenting particle considered above needs to be refined in many ways. This second chapter provides an overview of what we can learn about a particle through AUC, besides its molar mass and friction. We will only present basic concepts, referencing the literature and deferring some in-depth treatments to later chapters.

When describing a particle, it is important to note that there is no such thing as an 'absolute contour,' even on a molecular scale, and we can only define its surface with reference to the method used for probing it. For instance, slightly different contours emerge when the particle is imaged by electron microscopy or atomic force microscopy.[1] In the former, the contour is defined by electron scattering and stain used for imaging, whereas in the latter this is defined by the imaging tip used and force generated. What is the relevant contour for AUC and how does the molar mass measured by this method compare to that from mass spectrometry or the chemical formula? The relevance of this question arises primarily because experiments are carried out in solution – the dissolved macromolecule may interact strongly with the solvent and co-solutes altering both its properties, as well as the structure of the solvent in its immediate vicinity. A better definition of the sedimenting particle in the context of the macromolecule being studied is thus required.

As a macroscopic method, AUC alone cannot provide any direct insights into the internal structure or composition of the sedimenting particle; the method is solely confined to the observation of the particle's hydrodynamic contour, its total mass and average density. Even though more detailed deductions from AUC data are possible, these require the introduction of prior, independent knowledge. We present two useful and complementary viewpoints for the study of macromolecules:

I. The sedimenting particle consists solely of the covalent chain and any non-dissociating macromolecular ligands. Applying the play dough formalism (as in Section 1.1.2), we only consider its non-hydrated mass as though the par-

---

[1]This does not refer to experimental uncertainties in either of the methods [105], but conceptually to the different properties that are interrogated by the techniques in ideal experiments.

ticle were a solid, inert object of certain shape with uniform density (this formalism will be denoted by the subscript 'PD').

II. The complete sedimenting particle (or 'equivalent particle' [106]) includes all material inside a region where the solvent surrounding the particle starts to differ in composition or density from that of the bulk solvent (this will be denoted with the subscript 'SP').

In the latter case, the total mass and volume of the sedimenting particle incorporates bound solvent and co-solutes, including water from hydration. Although the first situation is commonly assumed, it is the second situation that is closer to physical reality and naturally accounts for many effects observed in analytical ultracentrifugation. This is especially true for most biological macromolecules, which are polyelectrolytes that interact strongly with water and carry charges with dissociated counterions.

Another aspect of sedimenting particles discussed in the present chapter arises from the observation that macromolecules are not static rigid particles, but in a dynamic ensemble of states, with regard to chain conformation and oftentimes association state. Transient binding reactions between particles affects sedimentation velocity and the dynamics of this rapid equilibrium can be described through the concept of an 'effective sedimenting particle' reflecting the time-averaged states of this dynamic system [34]. Unlike sedimentation velocity, in sedimentation equilibrium chemical reactions and sedimentation will come to a joint equilibrium. Using both methods we can describe the reaction affinities, and to some extent, from sedimentation velocity, the kinetic rate constants of the reversible binding reactions among particles. Binding may also be mediated by solvent, linking protein-protein interactions with protein-solvent interactions. In addition to attractive forces, which are related to binding in a general sense, repulsive inter-particle interactions may also be observed. These arise from the mere fact that particles occupy a finite, non-overlapping volume, together with interactions from their mutual hydrodynamic flow field and electrostatics. At high concentrations these effects can dramatically alter the sedimentation behavior of the system.

Finally, in addition to attractive or repulsive interactions, the actual conditions applied during the sedimentation experiment may themselves change the properties or sedimentation behavior of the observed particles. Foremost are pressure effects arising in the solution column due to centrifugation. Another is hydrodynamic drag arising from the sedimentation process which may deform the particles under investigation. Finally, we will discuss altered sedimentation behavior of our particle of interest from secondary sedimentation effects on the solvent composition.

2.1 PARTICLE–SOLVENT INTERACTIONS

2.1.1 Molar Mass, Buoyancy, and Macromolecular Partial-Specific Volumes

Utilizing the play dough formalism, let us first consider a particle of molar mass M_{PD} and partial-specific volume $\bar{v}_{PD}$ (i.e., a particle density $\rho_{PD} = v_{PD}^{-1}$) in an aqueous solution with density ρ. In the case of an unmodified protein, the particle would consist of all amino acids, or more generally, the covalent chain of a macromolecule. Motivated by Archimedes' principle, the buoyant molar mass M_b introduced in Eq. (1.12) as $M_b = M(1 - \bar{v}\rho)$, would in this case be $M_{b,PD} = M_{PD}(1 - \bar{v}_{PD}\rho)$.

In the case of proteins, surface amino acids are hydrated with an average of 0.2 – 0.5 grams of water per gram of protein [22, 23]. Water molecules appear more or less tightly bound, with a local density that depends on the solvent exposed atoms [24, 107, 108]. Hydration water exhibits a structure different from bulk water due to hydrophobic interactions and electrostriction at the protein surface, generally leading to a slightly higher density than bulk water [109–114].

Let us reflect on the consequences of this mass of bound water M_1 by considering it to be part of the sedimenting particle, resulting in a total mass $M = M_{PD} + M_1$ (in this context denoting the solvent with index 1). The corresponding partial-specific volume is obtained from the quotient of the total volume and the total mass $\bar{v} = (M_{PD}\bar{v}_{PD} + M_1\bar{v}_1)/(M_{PD} + M_1)$. For example, a 'dry' protein with a molar mass of 100 kDa and a partial-specific volume of 0.73 mL/g having 0.3 g/g of hydration will sediment as a particle of 130 kDa with an effective partial-specific volume of 0.7923 mL/g (see below). As noted above, the sedimentation experiment measures the total buoyant molar mass, which has individual contributions from both protein and bound water $M_b = M_{PD}(1 - \bar{v}_{PD}\rho) + M_1(1 - \bar{v}_1\rho)$. However, if we consider the partial-specific volume of the hydration water $\bar{v}_1 = \rho^{-1}$, the contribution of the hydration term to the buoyant molar mass vanishes.[2] This is not surprising, as hydration is for all intents and purposes neutrally buoyant in water, and the formalisms for interpreting the experimental buoyant molar mass presented with and without hydration are equally correct when describing the sedimentation experiment in water.

More realistically, however, one should consider that protein sedimentation experiments are carried out in a buffered salt solution containing a sample that is brought to chemical equilibrium. Ideally, this chemical equilibrium can be achieved by equilibrium dialysis, though in practice, size-exclusion chromatography is often equally effective. This ensures that the buffer composition is well-defined.[3] Consider the case of phosphate buffered saline (PBS), a commonly used buffer with a density of ~ 1.005 g/mL at 20 °C. If the buffer salts are not part of the hydration shell, bound water would exhibit a slight density contrast with the solvent; neglecting hydration contributions, the buoyant molar mass will be reduced by 0.15 kDa (0.56%), amplified to an error of −0.56 kDa in the protein molar

---

[2]For simplicity this ignores the slight density difference between bulk and hydration water, which is valid to a first approximation. To some extent this may be accounted for already in the partial-specific volume of the macromolecule [114].

[3]This is also very useful when working with the interference optical detection system, as it allows for an optical matching of signals from the buffer salts that redistribute in the centrifugal field.

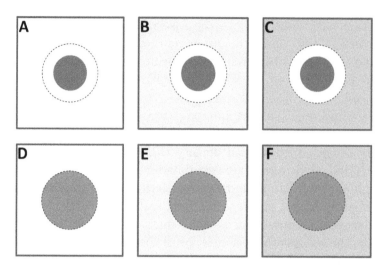

Figure 2.1 Cartoon of a hydrated particle (circles) in solutions (square) of different density (gray background of the squares). *Panels A–C*: The anhydrous particle in the play dough formalism (inner dark circle) and its hydration shell (indicated by dotted line). In water (*A*), the hydration shell is gravitationally transparent and under this condition it would not matter where the dotted line is drawn. In solutions of higher density, the hydration shell will contribute to the buoyancy if its solvent composition is different from that of the bulk solvent (*Panels B* and *C*). *Panels D–F*: A consistent picture emerges when the anhydrous particle and its hydration shell together constitute the 'sedimenting particle.' This particle has a larger radius and lower density than the anhydrous particle, but will appear to have the same size in all solvents.

mass.[4] This is ordinarily below the precision of the molar mass determinations from AUC and smaller than the uncertainty associated with possible errors in $\bar{v}_{PD}$ [3, 115], estimated to be ~1% when predicted from the amino acid composition [111] (see Section 3.3.3). However, in the presence of high concentrations of co-solutes that raise the buffer density, but do not enter the hydration shell, this change can become more significant. At a solution density of 1.1 g/mL, for example, this hydration effect is magnified to an error in the protein molar mass of ~ −15%, and ~ −48% at a solution density of 1.2 g/mL. Here, bound water contributes significantly to the density contrast of the solvated protein (compare Fig. 2.1). It is therefore important to carefully assign what constitutes the sedimenting particle — here the hydrated protein.[5] This example suggests that, unless we adopt the view of the complete sedimenting particle, accounting for the mass and buoyancy would require a partial-specific volume that depends on the solution density. This

---

[4]Note that absolute errors in the buoyant molar mass of proteins will be amplified in the molar mass by almost a factor 4 due to the transformation $M = M_b/(1 - \bar{v}\rho)$ with a partial-specific volume of 0.73 mL/g. Larger error amplification occurs for particles with densities closer to neutral buoyancy.

[5]Consideration of the hydration shell was introduced in the sedimentation analysis of tobacco mosaic virus by Schachman and Lauffer [116].

is captured only in an apparent partial-specific volume, denoted as ϕ', as described below.

The effects of preferential solvation also manifest themselves through the preferential binding or exclusion of other co-solute components from the immediate vicinity of the macromolecule. This becomes particularly important when studying highly charged macromolecules such as nucleic acids or halophilic proteins [43,117]. In these cases, preferential co-solute binding can cause significant changes in the buoyancy, and be highly relevant even at solvent densities close to those of water. Furthermore, changes in solvation occur upon protein folding, denaturation (e.g., when binding denaturants), and sometimes with protein interactions. This is of intense interest in the development of formulations for protein therapeutics, since many co-solutes that show preferential solvation act to either stabilize or destabilize the protein structure [118]. Even outside these research areas, the frequent use of glycerol to stabilize protein preparations poses problems for sedimentation experiments, as glycerol will raise the solution density and magnify solvation contributions to the buoyancy [119].[6]

As a result, a more thorough and thermodynamic description is necessary to understand the sedimenting particle, its solvation, and effects on buoyancy. Even though thermodynamic theory exclusively deals with macroscopic observables and recipes, it provides a rigorous framework for microscopic molecular models and the prediction of the macroscopic quantities.

2.1.2 Density Increments and Preferential Binding Parameters

The analysis of macromolecular preferential solvation is a highly developed field in theory and practice (e.g., [121–126]), and it is useful to recapitulate salient results for the further discussion of sedimentation. We will follow the approach pioneered by Casassa and Eisenberg [122] and separate the macromolecular component (here referred to with index 'a',[7] largely synonymous with the play dough formalism) from the co-solutes. When considering sedimentation equilibrium in such a thermodynamic multi-component framework, the buoyant molar mass arises from quotient of the density increment at constant chemical potential and the osmotic pressure concentration derivative [122]:

$$M_b = \left(\frac{d\Pi}{dw_a} \right)^{-1} \left(\frac{d\rho}{dw_a} \right)_{T,\mu_1,\mu_3} \tag{2.1}$$

---

[6]It should be noted that glycerol is more complex than other co-solutes, and the invariant particle model may not always apply [120]. The invariant particle model [106] makes the assumption that the solvation and other properties of the particle do not change with co-solute concentrations in the range studied.

[7]It is customary in the thermodynamic literature to denote the macromolecular component with index '2,' rather than the index 'a' adopted in the current work. The index 'a' has been retained both for consistency and to avoid confusion with symbols for other quantities discussed.

with osmotic pressure Π and macromolecular weight concentration w_a in g/L. The density increment $(d\rho/dw_a)_{T,\mu_1,\mu_3}$ is established experimentally at constant temperature and chemical potential of solvent μ_1 and co-solute μ_3, by equilibrium dialysis. In the limit of low concentrations, the inverse osmotic pressure concentration derivative is:

$$(d\Pi/dw_a)(RT)^{-1} = M_a^{-1} \tag{2.2}$$

from which we find the relationship between actual molar mass and measured buoyant molar mass:

$$M_b = M_a \left(\frac{d\rho}{dw_a} \right)_{T,\mu_1,\mu_3} \tag{2.3}$$

where the density increment now accounts for the buoyancy term. Since the density increment is an experimental recipe for densimetry at different macromolecular concentrations — in weight units — after dialysis, it follows from Eq. (2.1) – Eq. (2.3) that the ratio of the density increment and the measured change in osmotic pressure with concentration (as in SE) will provide the macromolecular molar mass, without hydration or solvation contributions. The mass so obtained is always identical to the species considered in the weight concentration determination [3]. This elegant approach, developed in the 1960s by Casassa and Eisenberg [43, 122, 127] has been an important milestone for determining the molar masses of polyelectrolytes, such as proteins and nucleic acids, in buffered salt solutions. Conversely if the macromolecular mass is known, perhaps from the sequence or mass spectrometry, the AUC experiment will allow for a determination of the density increment, which can be further interpreted in terms of particle-solvent interactions.[8] At a minimum, particle-solvent interactions need to be recognized as an important aspect of sedimentation in AUC, especially for the study of polyelectrolytes.

Let us examine the density increment more closely to understand what macromolecular determinants govern this quantity. In two-component solvents, the density increment is given by:

$$(d\rho/dw_a)_{T,\mu_1,\mu_3} = (1 - \bar{v}_a\rho) + (dw_3/dw_a)_{T,\mu_1,\mu_3}(1 - \bar{v}_3\rho) \tag{2.4}$$

where $(dw_3/dw_a)_{T,\mu_1,\mu_3}$ is the change in weight concentration of co-solute w_3 (in g/g) introduced in the dialysis bag due to the increase in the macromolecular weight concentration w_a, and $\bar{v}_3$ is the partial-specific volume of the co-solute [43, 122]. $(dw_3/dw_a)_{T,\mu_1,\mu_3}$ is also denoted as the preferential binding parameter ξ_3. While this parameter describes the properties of the solution in the dialysis bag, it seems to reflect only indirectly on the macromolecule itself. The change in co-solute concentration that occurs when bringing the solvated macromolecule into the solution depends highly on what the bulk co-solute concentration in that solution was prior to the introduction of the macromolecule. To focus better on the macromolecule,

---

[8]However, this only works well in the absence of significant thermodynamic nonideality and macromolecular self association or aggregation. Both or these factors would require additional terms in $(d\Pi/dw_a)$ or potentially involve mixtures of states exhibiting different $\bar{v}_a$.

one can consider the spatial distribution of solvent and co-solute molecules surrounding the macromolecule, and express preferential binding as a molar ratio of solvent and co-solute per macromolecule in a hypothetical domain close to the macromolecule [125, 126]. As reviewed by Ebel [124], this two-domain view is connected to a statistical thermodynamics approach, where preferential solvation parameters can be understood as deviations from bulk composition of the entire normalized radial distribution functions $g_{aJ}(r)$ of solvent or co-solute molecules relative to the macromolecule, extending to infinity via the Kirkwood–Buff integral $G_{aJ} = \int_0^\infty (g_{aJ}(r) - 1) \, 4\pi r^2 dr$ [124, 128]. The preferential binding parameter can be expressed as binding parameters B_3 and B_1 in terms of grams of co-solute and water, respectively, preferentially 'bound' to the macromolecule, with [120]:

$$(dw_3/dw_a)_{T,\mu_1,\mu_3} = B_3 - B_1 w_3^{bulk} \tag{2.5}$$

As described [120, 124] the contribution of hydration may be further refined by subdivision into exchangeable and non-exchangeable parts [125, 129, 130]. If B_3 and B_1 are constant, and independent of solvent composition, then the 'invariant particle model' applies [106]. This implies that $(dw_3/dw_a)_{T,\mu_1,\mu_3}$ as a function of solution density will be constant. In this case, the density increment may be expressed as:

$$(d\rho/dw_a)_{T,\mu_1,\mu_3} = (1 - \bar{v}_a\rho) + B_3 (1 - \bar{v}_3\rho) + B_1 (1 - \bar{v}_1\rho) \tag{2.6}$$

with B in g/g units of grams co-solute and water bound to the macromolecule [106, 120, 124].

2.1.3 Effective Partial-Specific Volumes and the Sedimenting Particle

Despite the failure of the naïve play dough formalism to describe the sedimentation of particles with preferential solvent interactions, it is often convenient to retain the familiar form of the buoyant molar mass in $M_a (1 - \phi'\rho)$ by substituting the density increment with an 'effective' or apparent partial-specific volume ϕ' that is operationally defined such that it leads from the macromolecular molar mass to the correct buoyant molar mass:

$$\phi' = \bar{v}_a - \rho^{-1} [B_3 (1 - \bar{v}_3\rho) + B_1 (1 - \bar{v}_1\rho)] \tag{2.7}$$

Consider a protein with a $\bar{v}_a$ of 0.73 mL/g and hydration term B_1 of 0.3 g/g in a solution of density 1.1 g/mL, the protein would appear to have an overall partial-specific volume of $\phi' = 0.7573$ mL/g. This is quite different from the partial-specific volume of the unhydrated macromolecule, as well as the partial-specific volume of the hydrated macromolecule $\bar{v}_{SP}$ (the latter being 0.7923 mL/g, see below). If the preferential interactions are either absent or the buoyancy term for the co-solute vanishes (like that assumed for hydration in water), then and only then does $\phi' = \bar{v}_a$. The effective partial-specific volume ϕ' is thus an operationally defined quantity rather than a particle property, as it is dependent on the solution density (and not only on the solvent interactions) even in the 'invariant particle' model. The reason for this is that it accounts for solvent interactions, but presumes that the total molar mass of the sedimenting particle to be that of the macromolecule only, which is conceptually inconsistent.

This can be remedied when considering the complete 'sedimenting particle,' identified by the subscript 'SP.' Using the preferential binding parameters introduced above, we can rewrite Eq. (2.3) and Eq. (2.6) as [124]:

$$M_a \left(\frac{d\rho}{dw_a} \right)_{T,\mu_1,\mu_3} = M_a \left(1 + B_1 + B_3 \right) \times \left(1 - \frac{\bar{v}_a + B_1 \bar{v}_1 + B_3 \bar{v}_3}{1 + B_1 + B_3} \rho \right) \quad (2.8)$$

and define $M_a(d\rho/dw_a)_{T,\mu_1,\mu_3} = M_{SP}(1 - \bar{v}_{SP}\rho)$ such that:

$$M_{SP} = M_a \left(1 + B_1 + B_3 \right) \quad (2.9)$$

and:

$$\bar{v}_{SP} = \frac{\bar{v}_a + B_1 \bar{v}_1 + B_3 \bar{v}_3}{1 + B_1 + B_3} \quad (2.10)$$

The inclusion of solvation effects in both the molar mass and the partial-specific volume finally provides a simple and physical description of the sedimenting particle. In this form, the partial-specific volume is independent of density and constant as long as the preferential solvation parameters are constant ('invariant particle model') (Fig. 2.1).[9] Under this formalism, the mass of the sedimenting particle is additive and based on its constituent components. At the same time, the partial-specific volume is the weight-average, consistent with expectations for any particle composed of multiple subdomains or composed of different chemical compounds, and consistent with our framework for describing binding events in AUC. Furthermore, this 'sedimenting particle' has the hydrodynamic properties expected for a single particle.[10]

Returning to our hypothetical hydrated protein of 100 kDa with 0.3 g/g hydration, we will have a total particle mass which is 1.3-fold the protein mass and a

---

[9] The sedimenting particle sheds new light on the nature of the apparent partial-specific volume ϕ', which is related to

$$\bar{v}_{SP} = \left(\phi' + \rho^{-1}(B_1 + B_3)\right)/(1 + B_1 + B_3)$$

or

$$\phi' = \bar{v}_{SP}(1 + B_1 + B_3) - \rho^{-1}(B_1 + B_3)$$

It can be shown easily that $\bar{v}_{SP} = \phi'$ at the point where the solution density provides neutral buoyancy $\rho = \bar{v}_{SP}^{-1}$. Therefore, the apparent partial-specific volume ϕ' changes continuously with solvent density from $\bar{v}_a$ at conditions where the hydration or solvation shell is neutrally buoyant to the value of the true sedimenting particle $\bar{v}_{SP}$ under conditions where the whole sedimenting particle is neutrally buoyant. (This assumes the density changes occur without changing the sedimenting particle; for example, density changes by using H_2O/D_2O mixtures would change both the mass and the hydration term and lead to different relationships.)

[10] The formalism of the sedimenting particle will also be extended later to the effective sedimenting particle of rapidly reacting systems. All 'sedimenting particles' have the property that they sediment and diffuse with single s- and D-values.

partial-specific volume $\bar{v}_{SP}$ of 0.7923 mL/g which may appear unfamiliarly high for a protein. However, it is easy to see that for any component i of the sedimenting particle for which $(1 - \bar{v}_i\rho)$ is zero, we may treat this component consistently as though B_i were zero and obtain the same thermodynamic results. For example, under usual dilute aqueous buffer conditions the solution density is close enough to that of the hydration shell that the latter is neutrally buoyant, such that we can neglect hydration, which brings us back to the standard values for both the molar mass and partial-specific volume.[11] However, it must be kept in mind that this simplification is valid only as long as neutral buoyancy is fulfilled for the hydration shell, and in the absence of preferential solvation (for example, from buffer salts). Equivalently, in the picture of the sedimenting particle, even though both the $\bar{v}_{SP}$ value and the M_{SP} value change with different estimates of hydration, the value assumed for B_1 is irrelevant (exactly compensating) as long as we are at conditions of neutral buoyancy for hydration. A slight deviation in density from this condition can be corrected for with even a moderately precise estimate for B_1, as this may provide a reasonable first approximation of hydration contributions. However, as will be discussed below, hydration remains important for hydrodynamic interpretations even under conditions of neutral buoyancy.

2.1.4 Measuring Hydrodynamic Friction

The effects of solvation on hydrodynamic friction are fundamentally different from the effects on buoyancy: whereas in the latter we can define an effective partial-specific volume, or adjust the solvent density to make certain solvent contributions to the sedimenting particle 'gravitationally transparent,' the same is not possible for hydrodynamic friction.[12] The interpretation of macromolecular shape based on the measured Stokes radius requires the definition of a well-defined contour and it would not make sense to define the particle such that it encompasses bulk solvent that does not co-migrate.

Nevertheless, for the interpretation of the observed sedimentation coefficients in terms of equivalent hydrodynamic shapes, the estimate of the degree of hydration is crucial. It is also important to recognize that thermodynamically bound water should be expected to be different from more tightly bound water governing hydrodynamic behavior [112, 131]. In fact, the detailed relationship between surface topology, chemical properties of different protein surface residues, molecular hydration dynamics and continuum hydrodynamics is a formidable problem [131, 132].

---

[11]More generally, density contrast variation experiments with the appropriate contrast agent are used to distinguish between the hydration and solvation effects.

[12]For example, in the case of protein/detergent complexes under detergent density matching conditions, the buoyant molar mass obtained will be that of the protein alone. However the hydrodynamic friction is that of the protein/detergent complex. Similarly, when working with buffers having a density close to 1.0 g/mL, where the hydration term is essentially matched, for purposes of the molar mass the particle contour can be extended indefinitely as all of the included solvent is neutrally buoyant. The frictional contour, however, is still very well defined.

TABLE 2.1 **Effect of Protein Hydration on the Thermodynamic and Hydrodynamic Sedimentation Parameters in Buffers of Different Density in the 'Complete Sedimenting Particle,' and the 'Play Dough' Formalisms.** Data assume a protein of $M_{PD} = 100$ kDa and $\bar{v}_{PD} = 0.73$ mL/g with $B_1 = 0.3$ g/g. The last row highlights the errors involved when the play dough formalism is used without accounting for the density dependence of ϕ', and instead assuming the partial-specific volume $\bar{v}_{PD}$ to be constant at 0.73 mL/g. Note that larger errors can be involved for particles having preferential interactions with co-solutes, including membrane proteins binding detergent, or highly charged macromolecules such as nucleic acids, and halophilic enzymes, among others.

	standard density 0.99823 g/mL	PBS 1.005 g/mL	high co-solute 1.1 g/mL	very high co-solute 1.2 g/mL
M_{SP} (kDa)	130	130	130	130
$\bar{v}_{SP}$ (mL/g)	0.7923	0.7923	0.7923	0.7923
$(f/f_0)_{shape,SP}$	1.0	1.0	1.0	1.0
ϕ' (mL/g)	0.73	0.7314	0.7573	0.7800
$\Delta M/M_{PD}$	0%	$\sim -0.56\%$	$\sim -15.2\%$	$\sim -48.4\%$
$\Delta s_{20,w}/s_{20,w}$	0%	$\sim -0.76\%$	$\sim -15.4\%$	$\sim -48.5\%$
$(f/f_0)_{shape,PD}$	1.0	1.01	1.18	1.94

Indeed, the prediction of a translational friction coefficient from protein structure requires the inclusion of a hydration term, usually accounted for by an empirical inflation of the van der Waals contour [133–135]. Similarly, an *ad hoc* correction is necessary for the interpretation of measured friction coefficients in terms of macromolecular shape. For example, a perfectly spherical protein with a $\bar{v}_{PD}$ of 0.73 mL/g and a compact hydration of 0.3 g/g would have an apparent frictional ratio of 1.12, considering the customary non-hydrated macromolecule as a reference. To reconcile the conceptual conflict that arises from a frictional ratio greater than unity for this compact spherical protein and regain a geometric interpretation, the frictional ratio has been subdivided into contributions from hydration and 'shape asymmetry,' $(f/f_0)_{shape}$ [1, 136]:

$$(f/f_0)_{PD} = (f/f_0)_{shape} \times \left(\frac{\bar{v}_{PD} + B_1\bar{v}_1}{\bar{v}_{PD}}\right)^{1/3} \tag{2.11}$$

On the other hand, if we consider the total mass and partial-specific volume of the complete sedimenting particle for the determination of the frictional ratio from a calculated f_0 and experimental sedimentation coefficient, we naturally arrive at the shape factor:

$$(f/f_0)_{SP} = (f/f_0)_{shape} \tag{2.12}$$

because the volume of the sedimenting particle $V_{SP} = M_{SP}\bar{v}_{SP}$ now accounts for the volume of both the protein and hydration, as does the radius $R_{0,SP}$.

This then raises the question as to which $\bar{v}$ should be used for the correction of experimental s-values to the standard conditions of water at 20°C? As described in Eq. (1.5) above, this transformation requires multiplication with the ratio $(1 - \bar{v}_0\rho_0)/(1 - \bar{v}_{xp}\rho_{xp})$ to correct for the different magnitude of buoyancy experienced by the particle. Here, again, the view of the complete sedimenting particle, if warranted by the 'invariant particle' model [106], is simple and most consistent. Consequently, the partial-specific volume of the complete sedimenting particle should be used under both conditions, e.g., if the experiment was carried out at 20 °C, it will be $\bar{v}_0 = \bar{v}_{xp} = \bar{v}_{SP}$.[13] For example, for our standard example protein with $\bar{v}_{SP}$ of 0.7923 mL/g (arising from $\bar{v}_{PD}$ of 0.73 mL/g and hydration of 0.3 g/g), s-values measured at a solution density of 1.1 g/ml would incur the correction factor of 1.628 to the standard conditions of water at 20 °C (with a density of 0.9982 mL/g). It can be shown that we arrive at the same factor in the play dough formalism where we would only consider the unhydrated amino acid under standard condition, $\bar{v}_{PD} = \bar{v}_0$ (0.73 mL/g), and account for the change in buoyancy with the apparent partial-specific volume ϕ' introduced above (Eq. 2.7), $\phi' = \bar{v}_{xp}$ (here 0.7573 mL/g).[14] However, a big error would be introduced when the altered buoyancy from the density contrast of hydration is not accounted for: incorrectly setting $\bar{v}_{PD} = \bar{v}_0 = \bar{v}_{xp}$ would lead to a correction factor of only 1.377, underestimating $s_{20,w}$ by ~15%.[15] Consequently, the determination of the effective partial-specific volume ϕ' using SE or densimetry for the experimental buffer conditions is very important (in the absence of knowledge of the B_1 and B_3) to arrive at the proper interpretation of the frictional ratio, which can be related to theoretical

---

[13]This also assumes, of course, that the experiment was conducted in the same primary solvent — H_2O. Caution is required, for example, if the solvent was changed to D_2O or H_2O/D_2O mixtures, as the sedimenting particle has now changed, since the hydration shell will contain D_2O, and H-D-exchange occurs changing the mass M_{PD} and partial-specific volume $\bar{v}_{PD}$ of the macromolecule to M_{PD*} and $\bar{v}_{PD*}$, respectively (see Section 3.3 on the determination of $\bar{v}$). In this case, the latter partial-specific volume is the correct one for the experimental conditions, if used in conjunction with the former for the standard conditions. Hydration will not contribute in either case.

[14]This identity of correction terms can be seen (for clarity only using hydration terms):

$$\frac{1 - \bar{v}_{SP}\rho_0}{1 - \bar{v}_{SP}\rho_{xp}} = \frac{1 - ((\bar{v}_{PD} + B_1\bar{v}_1)/(1 + B_1))\rho_0}{1 - ((\bar{v}_{PD} + B_1\bar{v}_1)/(1 + B_1))\rho_{xp}}$$

$$= \frac{1 + B_1(1 - \bar{v}_1\rho_0) - \bar{v}_{PD}\rho_0}{1 - (\bar{v}_{PD} - B_1(1 - \bar{v}_1\rho_{xp})/\rho_{xp})\rho_{xp}} = \frac{1 - \bar{v}_{PD}\rho_0}{1 - \phi'\rho_{xp}}$$

This derivation made use of the assumption that hydration is neutrally buoyant under standard conditions, but can be generalized easily if a second apparent partial-specific volume ϕ_0 under standard conditions is used to account for this hydration.

[15]In our example, the correction factor for obtaining $s_{20,w}$ from experimental data in PBS with a density of 1.005 g/mL would be 1.026. If, as is often the case, the contribution of the hydration shell to the buoyancy (i.e., the difference between $\bar{v}_{PD}$ and ϕ') is neglected in the transformation to $s_{20,w}$, a factor of 1.019 is obtained, underestimating $s_{20,w}$ by only 0.76%. However, this is larger than the experimental error and may still be significant in the precise interpretation of hydrodynamic shapes. For this reason it seems prudent to carry out SV experiments intended for detailed hydrodynamic interpretation under conditions close to standard conditions, i.e., with dilute co-solute and at a temperature of 20 °C.

shape models. A comparison of numerical values obtained in the different pictures is shown in Table 2.1.

2.1.5 Charge Effects

The sedimentation behavior of particles is profoundly influenced by their electrical charge, in several ways (beyond the consequences of preferential binding discussed above). This question has been addressed extensively in theory and experiment for polyelectrolytes such as DNA, or highly charged proteins [1,43,137–142]. Obviously, the condition of electroneutrality requires that macroions be balanced by dissociated ions of opposite charge, which will sediment in a manner that is coupled to the sedimentation of the macroion via Donnan equilibrium.[16] A quantitative description will depend on the definition of components [143]; for example the system could be described in terms of the electroneutral macroion with counterions, plus additional ions from the supporting electrolyte at some molar ratio (as would correspond to the experimental addition of macromolecular stock solution containing both macromolecule and supporting salt), or simply in terms of the mass of the macroion only [122]. The most natural description in turn will depend on the experimental procedures, as well as the *a priori* knowledge we have on the system. Either way, it is critical that the density increment or partial-specific volume be determined in a manner consistent with the adopted definition. It should be noted that the Casassa-Eisenberg framework of measuring density increments based on the macromolecular concentrations, measured in certain weight concentrations and brought to dialysis equilibrium, as outlined above, can be applied equally well to polyelectrolytes [43,141].

As outlined by Fujita [5], if we define an electroneutral macroion P carrying z charges jointly with the same number of mono-valent counterions (of total mass M_{PZ}), in the presence of a co-solute consisting of a monovalent salt as an electroneutral third component (of mass M_S), the apparent buoyant molar mass $M_{b,PZ,app}$ can be described as:

$$M_{b,PZ,app} = \frac{M_{b,PZ}\left(1 + \frac{z}{2}\frac{M_{b,S}}{M_{b,PZ}}\frac{w_{PZ}}{w_S}\right) - \frac{z}{2}M_{b,S}}{1 + (1+z)\frac{z}{2}\frac{M_{b,S}}{M_{b,PZ}}\frac{w_{PZ}}{w_S}} \tag{2.13}$$

The apparent buoyant molar mass of the macromolecular component is measured in the presence of g/mL concentrations w_{PZ} and w_S of the macromolecular component and the supporting salt, respectively, with buoyant molar masses of $M_{b,PZ}$ and $M_{b,S}$ [1,5]. There is a so-called primary charge effect arising from the term $0.5z\left(M_{b,S}/M_{b,PZ}\right)(w_{PZ}/w_S)$, resulting in a reduction of the macromolecular buoyant molar mass. For this reason, it is customary to add sufficient supporting electrolytes, so as to shield long-range electrostatic interactions and render this

---

[16] As shown by Braswell [142], it is possible to simultaneously record the distribution of ions (buffer salts) with interference optics, along with the sedimenting particle. The amplitude of the observed sedimentation boundary can then be interpreted in the context of the Donnan effect [142].

factor insignificant. This becomes particularly important in the study of polyelectrolytes and biological macromolecules. If no supporting electrolyte is added, the buoyant molar mass measured in SE from a refractive index gradient is reduced by a factor $1/(z+1)$ from the true value [140]. However, even in the limit of large molar excess of salt, the apparent buoyant molar mass will still be reduced by $(z/2)M_{b,S}$, referred to as a "secondary charge effect" [5].

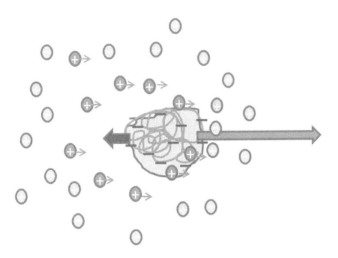

Figure 2.2 A macroion migrating in a cloud of solvated counterions (with ions from bulk in light blue, and ions contributing to the solvation shell in dark blue). Since the associated ions sediment more slowly than the macroions, a local electrophoretic force is thought to be generated that slows down the migration of the macroion. On the other hand, counterions associated with the macroion can exchange with those from the bulk salt solution ahead (e.g., the green one) [137].

While the previous considerations for the buoyant molar mass were established with SE experiments in mind, SV is similarly affected in a significant way by the co-migration of counterions. This was examined by Svedberg and Pederson [1] and subsequently in more detail by Pedersen [139]; the analysis was further refined by Alexandrowicz and Daniel [138, 144] and Eisenberg [137]. More recent treatments of this problem are based on solutions of the Navier–Stokes equation by Ohshima and colleagues [145] and numerical solutions of the Stokes–Poisson–Nernst–Planck equation by Keller et al. [146]. The basic phenomenology is this: In SV of charged particles, the unequal sedimentation velocity of the macroion and its slower counterions produces an electrical field and an electrophoretic migration opposite to sedimentation (Fig. 2.2). In the absence of supporting electrolyte, the sedimentation coefficient of the macroion is [137]:

$$s_P = \frac{M_{b,PZ}}{f_P + Zf_X} = \frac{D_P M_{b,PZ}}{RT\left(1 + Z\frac{D_P}{D_X}\right)} \tag{2.14}$$

where the subscript P and X refer to the macromolecule and counterion, respectively. The retardation may be significant, depending on the diffusion coefficient of the salt D_X and the macroion D_P. Based on the right-hand side of Eq. (2.14), it appears that for large particles this effect is not as pronounced as for smaller particles. This dynamic, primary charge effect can be overcome by use of sufficiently high electrolyte concentrations. However, it should be noted that a secondary charge effect may occur if the supporting salt cations and anions have different sedimentation coefficients, due to the remaining electrical field. This can be avoided by using a supporting salt with ions of nearly equal sedimentation coefficients, such as NaCl [137].

Finally, an entirely different category of charge effects arises from particles that are not invariant in shape, where changes in the macromolecular conformation occur due to intra-molecular electrostatic forces. The extent of such effects would be expected to depend on the charge density and ionic strength. For example, the persistence length of DNA (and therefore its hydrodynamic friction) is dependent on ionic strength.[17] Similarly, a charge effect that should be kept in mind as well is the possibility of charge-mediated modulation of protein interactions (i.e., ion binding linkage), as discussed below.

2.2 PARTICLE–PARTICLE INTERACTIONS

So far we have considered the consequences of solvation, but otherwise treated the particle of interest to be rigid and inert. This is unlikely to be the case for most particles, at any non-vanishing concentration. There are a variety of internal motions and particle-particle interactions, some obligate and others particular for certain systems, which can profoundly influence the sedimentation process. For example, a chemically distinct macromolecule, such as a protein, may assume different states corresponding to particular species (such as a monomeric or dimeric species) following the laws of reversible thermodynamics. Even rigid particles will always interact with each other through their mutual flow-field and impenetrably occupied volume.

The study of particle interactions is currently a major area of application of analytical ultracentrifugation. The present treatment is simply an overview of the expected phenomenology. In most AUC methods it is assumed that the macromolecules loaded in the SV experiment are in chemical equilibrium prior to the start of the ultracentrifugation run and this is usually essential to establish a well-defined experiment.

2.2.1 Dynamic Interacting Systems and Reversible Associations

2.2.1.1 Conformational States

Macromolecules can be thought of as an ensemble of interconverting states with different conformations, often associated with different Stokes radii. As a general

---

[17]The persistence length of DNA increases with decreasing ionic strength. The extent to which the hydrodynamic friction will change as a result will also depend on the size of the DNA.

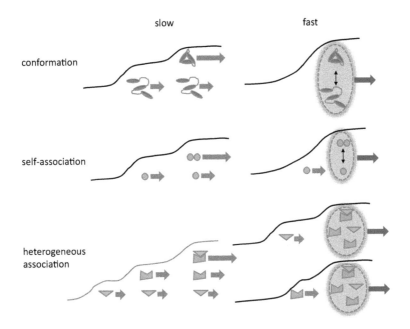

Figure 2.3 Schematic representation of sedimentation boundary shapes for macromolecular systems with slowly (left) or rapidly (right) interconverting states. The top row depicts a macromolecular component with conformational states exhibiting different translational friction, the middle row a self-associating system, and the bottom row a heterogeneous association between two macromolecular components forming a 1:1 complex.

rule, the observations made in SV experiments depend on the interconversion time-scale relative to the macroscopic migration. The characteristic time-scale of a SV experiment is typically 1,000 sec – 10,000 sec. If the rate of exchange between states is fast on this scale ($k_{off} > 10^{-2}/$sec), a time-averaged state will be observed, corresponding to a population average reporting on the energetics of the different states. On the other hand, if exchange is slow on this scale ($k_{off} < 10^{-4}/$sec), then the sedimentation properties of the different states may be hydrodynamically resolved in the form of separate sedimentation boundaries (Fig. 2.3, first row).[18] Only for a fairly narrow range of matching time-scales can we observe an intermediate behavior that allows for a computational analysis of the interconversion kinetic rates.

For example, for an unfolded protein assuming different 'random' conformations, an ensemble averaged state may be observed in SV, reflecting the thermodynamic equilibrium of states. However, folded states would be expected to be hydrodynamically resolved from unfolded chains in SV based on their stability and

---

[18]Even if the sedimentation coefficients of the different species are too similar to be visually resolved into distinct boundaries, the slowly equilibrating case will lead to an excess boundary spread corresponding to an apparent diffusion coefficient larger than either state, whereas the rapidly equilibrating case will exhibit an apparent diffusion coefficient that is an average diffusion coefficient of the states. An example of such behavior was studied by Werner, Cann, and Schachman for the ligand-induced allosteric transitions of aspartate transcarbamoylase, comparing the effect of different conformational change-inducing ligands at half-saturation [147, 148].

smaller hydrodynamic radius. SE experiments will be unaffected by macromolecular conformations, as long as the solvation and buoyancy factors are not significantly altered.[19]

2.2.1.2 Self-Associating Systems

In the case of self-associating macromolecules, different states arise from the reversible association of, say, monomeric macromolecules encountering and specifically forming oligomers of certain binding energy and life-times, which spontaneously dissociate back into their building blocks (e.g., monomers). Based on their molar masses, these oligomers will sediment faster, while assembled, than their building blocks. As described above, when the time-scale of chemical interconversion between oligomeric states is slow relative to the characteristic time-scale of sedimentation, hydrodynamic separation can be observed; whereas when the chemical interconversion is rapid, a time-averaged state is observed (Fig. 2.3, middle row) [152]. Again, there is a potential for intermediate behavior and an analysis of kinetic rate constants when the time-scales of sedimentation and the chemical reaction approximately match.

However, in comparison to conformational changes, the key difference in the case of self-association is that the likelihood of complex formation is dependent on macromolecular concentration. For rapidly interconverting systems, due to the different sedimentation coefficients of the oligomeric states, the evolution of the sedimentation profiles will be coupled with the instantaneous local population of states and therefore local concentration [153, 154]. The leading edge of the sedimentation boundary and the plateau region will be at concentrations only slightly lower than the loading concentration, and the macroscopically observed movement of this part of the boundary resembles the equilibrium population of states at the loading concentration and their time-averaged sedimentation coefficients. At the same time, if we consider the trailing part of the diffusion-broadened sedimentation boundary, the concentration will continuously become more dilute further away from the boundary midpoint. With greater dilution comes a lower probability of complex re-formation, and therefore slower sedimentation of the interconverting system, with the limiting velocity being that of the most disassembled state. These two extremes, taken together, make it clear that the sedimentation boundaries of rapidly self-associating systems will have a stretched appearance, at loading concentrations far above the equilibrium dissociation constant exhibiting a visually distinct bimodal shape.

In the interpretation of the sedimentation process of rapidly self-associating macromolecules, it is important to recognize that the sedimentation profiles are

---

[19]Related to this, a useful application of SV is the study of conformational changes induced upon small ligand binding. A classical example is the 3% decrease in the sedimentation coefficient of aspartate transcarbamoylase in the presence of ligand at saturating concentrations [149]; more recent examples include the study of riboswitches [150] and nucleotide dependent conformational changes of viral capsid proteins [151].

not a reflection of the sedimentation boundaries of stable, discrete species, but the locally time-averaged states. Only at sufficiently high loading concentrations that predominantly populate the most assembled state will the system of interconverting molecules migrate at a velocity close to that of the most assembled species. Likewise, only very low loading concentrations that make any transient complex formation highly unlikely allow the sedimentation boundary to reflect the physical properties of the monomeric species. At intermediate concentrations stretched sedimentation boundaries will be observed with intermediate migration velocity (Fig. 2.3). A key requirement when studying interacting systems is to explore a range of macromolecular concentrations to examine the change in their hydrodynamic ensemble properties. Thus, a concentration series carried out at different loading concentrations will allow for a quantitative analysis of the self-association. In the case of self-associating macromolecules with a rapid interconversion of oligomeric states, the shift in the equilibrium population of species is manifested by a shift in the time-average sedimentation velocity of each molecule with local concentration. For slowly interconverting states, separate boundaries reflecting the constant sedimentation properties of the individual species are observed at all concentrations, but with relative amplitudes that change according to loading concentration, if the samples were in equilibrium prior to the experiment.[20]

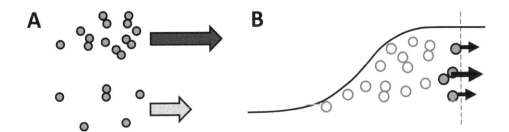

Figure 2.4 (*Panel A*) A self-associating system at high concentrations compared to the equilibrium dissociation constant will cause each molecule to be in the assembled state most of the time, sedimenting fast (top). By contrast, at low concentration the dissociated state predominates, and therefore all molecules sediment slower (bottom). (*Panel B*) One approach to quantify the sedimentation process is the signal-weighted average sedimentation coefficient [154, 155]. It is derived from the change in the total number of molecules 'above' (i.e., at lower radii than) an imaginary line in the plateau region (dashed line). In a small time-interval, only molecules in the shaded region adjacent to the dotted line will be able to cross it. The distribution of oligomeric states within this region is close to the equilibrium distribution at the loading concentration, since here only radial dilution of the plateau has taken place and there are no concentration gradients. Therefore, this analysis is independent of chemical conversion that may take place due to dilution in the sedimentation boundary.

---

[20]The presence of separate boundaries may be visually confounded by the presence of diffusion, dependent on the size of the particles.

While it has become possible to fit experimental sedimentation profiles directly with detailed mathematical models for coupled sedimentation and chemical reaction processes [154, 156, 157], a more traditional and very powerful approach is based on the concept of a single, signal-weighted average sedimentation coefficient, s_w, which describes the whole system [30, 154, 155, 158, 159]. With a definition based on the mass balance from total transport across a plane in the plateau region ahead of the sedimentation boundaries, this signal-weighted average sedimentation coefficient does not require analysis of the actual shapes of the sedimentation boundaries or their substructures (Fig. 2.4). Furthermore, because the interacting system is present in the plateau region at concentrations still close to the equilibrium loading concentrations, with negligible chemical fluxes and vanishing diffusion fluxes, the measured rate of transport in the plateau region reflects in a good approximation the equilibrium population of species. This turns the signal-weighted average sedimentation coefficient into a very robust measure of sedimentation independent of chemical conversion kinetics. Even though the plateau region does not provide a clue to macromolecular migration other than radial dilution, s_w can be determined easily from the decrease with time of the total amount of macromolecular material remaining 'above' (i.e., at smaller radii) a plateau radius, as measured by integration of the recorded sedimentation profiles (Fig. 2.4). Modern direct boundary analysis methods provide for rigorous and efficient ways to determine s_w computationally [154].

In this manner, the information of the loading concentration-dependent shift in oligomeric states can be extracted into an isotherm of signal-weighted average sedimentation coefficients as a function of loading concentrations. Let us consider a system where a monomer at concentration c_1 and a dimer at concentration c_2 are in a chemical equilibrium with association constant K following mass action law $c_2 = Kc_1^2$. If the monomer and dimer species have sedimentation coefficients s_1 and s_2, respectively, then a signal-weighted average sedimentation coefficient will be:

$$s_w = \frac{c_1 s_1 + 2Kc_1^2 s_2}{c_{tot}} \tag{2.15}$$

essentially independent of conversion kinetics, with c_{tot} and c_1 the total protomer and free monomer concentrations, respectively, in the loading mixture. (In Eq. (2.15) the absence of association-dependent signal changes is assumed, such that the extinction coefficient cancels out.)

How will macromolecular association alter sedimentation equilibrium? The basis for the study of reversibly self-associating macromolecules by SE is the consideration that both chemical and sedimentation equilibrium are simultaneously attained. In the example above, if the monomer sedimentation profile is described by the Boltzmann exponential in Eq. (1.11), then assuming that the dimer has twice the buoyant molar mass of the monomer, mass action law predicts that:

$$
\begin{aligned}
c_2(r) &= K c_1(r)^2 \\
&= K \left\{ c_1(r_0) \exp \left[M_1(1 - \bar{v}\rho) \frac{\omega^2}{2RT} \left(r^2 - r_0^2 \right) \right] \right\}^2 \\
&= K c_1(r_0)^2 \exp \left[2M_1(1 - \bar{v}\rho) \frac{\omega^2}{2RT} \left(r^2 - r_0^2 \right) \right] \\
&= c_2(r_0) \exp \left[M_2(1 - \bar{v}\rho) \frac{\omega^2}{2RT} \left(r^2 - r_0^2 \right) \right]
\end{aligned}
\tag{2.16}
$$

barring additional non-ideal interactions and significant changes in solvation upon oligomerization. Thus, the dimer also follows a Boltzmann exponential with twice the buoyant molar mass of the monomer, and what we measure will be the sum of monomer and dimer contributions:

$$
\begin{aligned}
c(r) =&\, c_1(r_0) \exp \left[M_1(1 - \bar{v}\rho) \frac{\omega^2}{2RT} \left(r^2 - r_0^2 \right) \right] \\
&+ K c_1(r_0)^2 \exp \left[2M_1(1 - \bar{v}\rho) \frac{\omega^2}{2RT} \left(r^2 - r_0^2 \right) \right]
\end{aligned}
\tag{2.17}
$$

where the concentration $c_1(r_0)$ at the reference radius will adjust such that total mass in the solution column is conserved [160]. Therefore, from a single sedimentation equilibrium profile it is not possible to distinguish whether the monomer and dimer are in a reversible equilibrium or in a mixture of non-interacting species. Again, it is essential to study a concentration series in SE experiments, to probe in a global analysis whether the populations of different species shift with loading concentration. Similarly, a shift in equilibrium populations can be effected by a change of rotor speed; as a consequence it is standard practice to study interacting systems at a range of rotor speeds (Section 5.5.2).

In the study of reversibly interacting systems, SE and SV are highly complementary and can be analyzed globally to determine the size of the oligomers formed and their interaction affinity [161, 162]. Generally, even though SE can define the average sizes very well, the definition of the oligomeric state by SE alone can remain ambiguous due to relatively poorer resolution and the correlation of fitting parameters when modeling Boltzmann exponentials. The added information from hydrodynamic separation in SV and the value of the observed sedimentation coefficients can often help greatly to resolve the oligomeric state, in particular with slow or intermediate exchange rates, and/or when Lamm equation modeling can be applied. Furthermore, whereas SE profiles can be very sensitive to impurities over a very wide size range [80], hydrodynamic separation in SV can highlight sample purity and ensure a robust and reliable analysis. For these reasons, with abundant computational power and advanced whole boundary analysis methods, SV has become the first method of choice for the study of many interacting systems.

2.2.1.3   Heterogeneous and Mixed Associating Systems

Similar principles apply to heterogeneous and mixed associating systems as for self-associating systems. In 'ideal' sedimentation equilibrium governed by interactions through mass action law, each free and complex species will assume a Boltzmann distribution based on its respective buoyant molar mass. There may be many terms, but, in contrast to self-associating systems, it is usually possible to characterize the individual components separately, so as to determine the buoyant molar mass of each free species, from which follows the buoyant molar masses of all complexes of given stoichiometry. Another important opportunity arising in the study of multi-component mixtures is the potential to exploit differences in the signal contributions of the different macromolecular components by judicious choice of detection system and absorbance wavelength. This and other advanced modeling techniques make the analysis of interacting systems by SE quite tractable, despite the notoriously difficult problem of exponential decomposition of noisy data [83, 84, 160, 163–167]. However, much like self-associating systems, SV experiments are usually conducted first, or side-by-side, in order to provide a framework for the analysis with prior knowledge (or hypotheses) as to the number and stoichiometries of the complexes.

SV on heterogeneous reversibly associating systems can be extremely informa-tion rich. If the interconversion of associated and dissociated states is slow on the time-scale of sedimentation, hydrodynamic separation of the complexes can be ex-pected, which provides information on the number and sedimentation and diffusion coefficients of complex species. The opportunity to use different detection systems and detection wavelengths, as in the multi-signal sedimentation velocity (MSSV) approach, provides an additional data dimension, allowing for the discrimination of contributions from each macromolecular component to each sedimentation bound-ary [85, 91, 92]. Taken together, each complex species can be characterized by its stoichiometry, buoyant molar mass, and translational friction.

As in self-association, we can determine signal-weighted average sedimentation coefficients independent of the conversion kinetics.[21] These values can be deter-mined simultaneously at various signals that exploit differences in absorbance spec-tra or other signal contributions of the different components, from which informa-tion on the overall sedimentation of the whole system can be extracted. By varying the composition of samples, binding isotherms can be generated and globally fitted with appropriate interaction models. Notably, two-component systems (in contrast to one-component or self-associating systems) offer different ways to vary the com-ponent concentrations, some of which may lend themselves well to visually discern certain aspects of binding.

---

[21]In analogy to Eq. (2.15), for a simple bimolecular reaction it takes the form:

$$s_w = \frac{c_{A,\text{free}}\epsilon_A s_A + c_{B,\text{free}}\epsilon_B s_B + K c_{A,\text{free}} c_{B,\text{free}} (\epsilon_A + \epsilon_B) s_{AB}}{\epsilon_A c_{A,\text{tot}} + \epsilon_B c_{B,\text{tot}}}$$

assuming the absence of spectral changes upon association. In self-associating systems, the monomer extinction coefficients will cancel out and do not need to be known for s_w.

Multi-component systems forming complexes with short life-times that are in rapid association-dissociation equilibrium with the constituent free species exhibit characteristic multi-modal sedimentation patterns [34, 168, 169]. So far, only the two-component case has been studied in detail. At the root of the multi-modal boundary structure is the fact that the mass action law for the formation of mixed complexes between two macromolecular components now depends on the concentrations of two species that in general sediment at different rates. Let us refer to them as free species A and B, and, for the sake of conceptual simplicity, let us only consider the formation of a single 1:1 complex AB sedimenting with s_{AB}. We assume that the local chemical equilibrium governed by the mass action law $c_{AB} = K c_A c_B$ is always attained instantaneously everywhere during the entire sedimentation process. It immediately follows that there can be only two sedimentation boundaries, one containing all three species, and one containing either only free A or only free B (Fig. 2.3, bottom row). This is because there cannot be a region in the solution where free A and B coexist without forming complex; and there cannot be a region in the cell where there is complex without free constituents.[22]

Importantly, the sedimentation boundaries of interacting systems in SV are very different from those in size-exclusion chromatography due to the sample loading geometry. In size-exclusion chromatography a lamella of the mixture is injected on top of the column, and the slower migrating free species will eventually be separated such that little complex can be formed during migration on the column. However, in SV, the entire solution column is initially filled with the sample, such that the faster sedimenting complexes always remain in a bath of their slower-sedimenting constituents. Therefore, complexes can be more readily observed and studied, as they can re-associate after dissociation in a way that reflects their equilibrium and kinetic binding properties.

In SV of two-component interacting systems with fast kinetics, the boundary containing one of the free constituents is referred to as the 'undisturbed boundary,' as this corresponds to the sedimentation of the single, free species. The boundary containing a mixture of all species is referred to as the 'reaction boundary.' Just like rapidly interconverting self-associating systems, it has been long recognized that such a reaction boundary of a hetero-associating system must not be confused with that of a physical species [5, 168]. In fact, the properties of the reaction boundary can be quite counter-intuitive at first: its composition does not correspond to the stoichiometry of the complex; the velocity has an intermediate value between that of the larger free component and the true complex; increasing the concentration of one of the components may lead to an increase or a decrease of the reaction boundary s-value; and the component providing the undisturbed boundary is not necessarily the component in molar excess.

Simple physical rules were recently discovered for the propagation of reaction

---

[22]Similarly, a three-component mixture with rapid chemical equilibria can exhibit at most three boundaries, or generally, n-component mixtures in rapid chemical equilibria can exhibit only n boundaries.

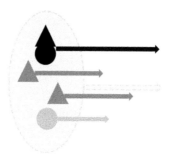

Figure 2.5 Illustration of a rapidly interacting system that can sediment jointly as an effective particle: A total concentration of two circles and three triangles, with equilibrium dissociation constant $K_D = 2$, will form one complex, one free circle, and two free triangles. The free circles sediment at a relative velocity of 2.0, the free triangles at a relative velocity of 2.5, and the complex at a relative velocity 4.0, as indicated by the length of the arrows attached. At these concentrations, binding constant, and relative velocities, the rapidly interconverting system can sediment jointly as an effective particle: The time-average velocity of circles is 3.0 (being half of the time free with relative velocity 2.0 and the other half in the complex with relative velocity 4.0), which matches the time-average relative velocity of the triangles (being two thirds of the time free at a relative velocity of 2.5 and only on third of the time in the fast-sedimenting complex at a relative velocity of 4.0). Therefore, they form a self-consistent effective particle with the relative velocity 3.0. The total composition of the effective particle is 3 triangles:2 circles, despite the reaction stoichiometry of 1:1, which keeps the fractional time circles spend free sufficiently short so that they do not fall behind the effective particle.

boundaries that naturally explain their properties, and treat the reaction boundary as being generated by an 'effective particle' [34,170]. Effective particle theory (EPT) leads to the fundamental insight that in a reaction boundary the time-average sedimentation velocity of molecules of all components must match, and be equal to that of the reaction boundary $s_{A...B}$ [34]. Considering that molecules of A and B will sediment at different rates when free, and assuming a nomenclature where free A sediments slower than free B, $s_A < s_B < s_{AB}$, it becomes obvious that there must be an asymmetry of composition in the reaction boundary since A has to be kept in the complex for a higher fractional time than B, thus there must be more total B than total A in the reaction boundary (Fig. 2.5). How much more will depend on the difference in the s-values of free A and B. If the composition of the total material loaded matches the condition for the reaction boundary, then there will be no undisturbed boundary.[23] However, all material that is present in excess of the necessary composition of the reaction boundary provides the undisturbed boundary. From this, it immediately follows that the transition from A providing the undisturbed boundary to B providing the undisturbed boundary, which is akin to a second order phase transition in concentration space, requires higher concentrations of B than the equimolar line, with the difference again depending on the difference in s-values of free A and B [34].

---

[23]This happens at the phase transition line Eq. (2.18), which describes the particular value of c_B for each value of c_A, dependent on the s-value and K, where only a reaction boundary exists [34].

SEDPHAT — The function Options ▷ Interaction Calculator ▷ Effective Particle Explorer Calculator provides, for simple heterogeneous interaction models, a movie depicting the coupled sedimentation and binding process faithful to given concentrations, sedimentation, and interaction parameters. It also provides a detailed display of various properties of the effective particle, including its velocity and composition.

Using the formalism obtained from EPT, it is possible to describe all properties of the undisturbed and reaction boundary in simple algebraic form:

$$s_{A\cdots B} = \begin{cases} (c_A s_A + c_{AB} s_{AB})/(c_A + c_{AB}) & \text{for} \quad c_B > c_B^*(c_A) \\ (c_B s_B + c_{AB} s_{AB})/(c_B + c_{AB}) & \text{else} \end{cases}$$

$$c_B^*(c_A) = \frac{K c_A (s_{AB} - s_A) + (s_B - s_A)}{K (s_{AB} - s_B)}$$

(2.18)

with c_A, c_B, and c_{AB}, the concentrations of free and complex species, respectively, as determined by mass action law, and $c_B^*(c_A)$ denoting the phase transition line between the concentration space where A or B is providing the undisturbed boundary. Similar expressions are available for the boundary amplitudes and compositions. In this manner, we can take advantage of the multi-modal boundary structure by calculating signal-weighted average sedimentation coefficients of the reaction boundary as a function of loading composition, which can be globally fitted along with s_w, and greatly aid in the determination of binding constants and hydrodynamic shape of the complex [92, 169, 171].

More detailed descriptions are available to describe the diffusion broadening of the reaction boundary in the framework of EPT [170]. The Gilbert–Jenkins theory developed in the 1950s provides an early iterative approach to describe boundary shapes in the limit of negligible diffusion (the 'asymptotic boundary' in the limit of infinite time) [168, 172], but did not find wide application [173]. Finally, for sufficiently monodisperse components, differential equations of sedimentation and chemical reactions can be fit directly to the observed sedimentation boundaries [174–176].

2.2.1.4 Solvent-Mediated Protein Interactions

Protein interactions may be more complicated than those described above, in that the energy of complex formation or conformational transition may be coupled to the concentration of a solvent component [124], such as a salt [123, 177], detergent [178–182], or other small ligand co-factor [147].[24] In principle, the sedimentation behavior of such interactions can be regarded and analyzed as a multi-component interaction where one of the binding partners happens to be small and

---

[24] A prototype for a salt-linked interaction is the self-association equilibrium of nucleophosmin (NPM1), which reversibly forms pentamers that increase in stability when salt ions screen the negative charges that cluster in the pentamer [183]. Lacking structural information, this linkage was initially misinterpreted as ion-binding induced conformational changes [184].

typically in high abundance. However, this may be simplified, and the co-factor for binding may be considered constant during the sedimentation process, provided it is present in large molar excess and does not exhibit significant gradients. If, in addition, the buoyant molar masses of the free and complex species are proportional to the complex stoichiometry, and/or negligible changes in the buoyant molar mass occur with co-factor binding, the sedimentation analysis becomes fairly straightforward. In such a case, the (partially) liganded species becomes the effective sedimenting particle for the interacting macromolecular system, and apparent binding constants for the macromolecular interactions may be obtained. Linkage relationships can account for the co-solute dependence of the apparent macromolecular binding constant [124]. These conditions can often be ensured, for example, for salt-dependent binding and for detergent-solubilized membrane proteins.

The study of membrane proteins solubilized in detergents, as pioneered by Tanford and Reynolds [178, 185, 186], or reconstituted in nanodiscs [187], is a highly developed field of research by AUC. While it was initially focused on SE, very powerful SV approaches have been developed recently by Ebel and co-workers [182, 188, 189]. Experimental strategies for working with such systems include the adjustment of solvent density to make the co-solute ligand neutrally buoyant, and density contrast variation techniques [179–182, 185].

2.2.2 Sedimentation of Concentrated Solutions

2.2.2.1 Sedimentation Equilibrium at High Concentrations

Even in the absence of chemical and charge interactions, particles always interact through the obligate requirement that they cannot penetrate, namely the volume of one particle is excluded from the solution volume available to another macromolecule. These repulsive interactions become relevant at high concentrations, and at small intermolecular distances resulting from high concentrations. Electrostatic forces can also contribute to repulsive forces.

This thermodynamic nonideality[25] becomes relevant at concentrations of a few mg/mL for folded proteins and tends to dominate the entire sedimentation process above 10–50 mg/mL, corresponding to a few percent in volume occupancy. However, the onset of nonideality may occur at significantly lower concentrations for charged polymers, or molecules with high frictional ratios (i.e., extended shapes that exacerbate volume exclusion).

The phenomenology of SE at high concentrations can be understood from the simple consideration that volume exclusion provides additional repulsive forces, growing with increasing concentration, which oppose the sedimentation and thus diminish the concentration gradients generated by the centrifugal field. Therefore, the concentration profiles will not be described by exponentials but will be increasingly more shallow than exponentials at higher concentrations at higher radii (Fig.

---

[25]This is to be distinguished from hydrodynamic nonideality occurring only in transport, which is introduced below.

2.6). Based on the local slope, it is still possible to calculate an apparent buoyant molar mass $M_{app,b}$:

$$M_{app,b}(r) = \frac{2RT}{\omega^2}\frac{d\ln(c)}{dr^2} \tag{2.19}$$

that can be related to the true buoyant molar mass using the thermodynamic relationship:

$$M_b\frac{\omega^2}{2}\left(r^2 - r_0^2\right) = \mu(r) - \mu(r_0) = RT\ln\frac{a(r)}{a(r_0)} \tag{2.20}$$

with the chemical potential μ and chemical activity a. Eq. (2.20) expresses the fact that the chemical activity still follows the familiar Boltzmann exponential, though not the concentration. A relationship between the apparent and true molar mass is then established through the activity coefficient γ, with $a = \gamma c$, as [190, 191]:

$$M_{app,b}(r) = \frac{M_b}{\left(1 + c\frac{d\ln\gamma}{dc}\right)} \tag{2.21}$$

In this manner, the sedimentation profiles in SE depend on the concentration dependence of the thermodynamic activity coefficient $\gamma(c)$, and the interaction free energy $RT\ln\gamma$. A similar, more general relationship exists for interacting multi-component systems, as described by Minton and colleagues [190].

In the concentration regime characterized by the onset of thermodynamic non-ideality, i.e., where three-body interactions can be neglected and electrostatic interactions are damped by the ionic strength of the buffer, the second virial coefficient may be determined and interpreted in the context of hard particle potentials [192]. This approach fails at higher concentrations where increasingly higher-order virial coefficients will need to be considered. Alternatively, the interaction free energy may be calculated in a closed form with scaled particle theory, which considers the work required to create sufficient space in a solution to accommodate a particle. Here, molecules can be represented by effective hard spherical particles with an effective volume as the sole parameter for solution of any concentration, including those approaching physiologically relevant conditions [193–195]. An example of this is shown in Fig. 2.7.

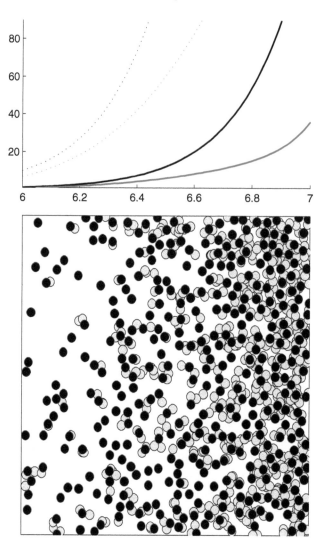

Figure 2.6 Principle of SE at high concentration illustrated by placement of particles of a finite size in a container in the presence of a gravitational potential. In this computer simulation, 1000 particles (solid circles) are sequentially added into an empty container at random positions. The probability of a particle being placed at a given radius in the container exponentially increases from left to right, as indicated by the black solid line in the top graph (with the dotted line showing a 10-fold magnification). This simulates the gravitational field, and the total density profile of particles represents a Boltzmann exponential in the absence of nonideality. Nonideality is caused by volume exclusion: Whenever a particle intersects with the occupied area of any of the previous particles, steric exclusion causes this location to be rejected (indicated as open circles, total n_{imp}). If it can be placed without intersecting any of the previous particles, it is added (indicated by solid circles, total n_{poss}). As the container fills up, increasingly more particle locations are excluded especially in the highly concentrated region. The density profile of placed particles (red solid and dotted lines) is very close to the total number of particles at low concentrations (left) and still monotonically increasing, but is shallower than the Boltzmann distribution and reaches close packing (right). This is a realistic simulation: According to scaled particle theory, the chemical activity coefficient is the inverse of the fractional available volume $\gamma = v_{tot}/v_{available}$, and to the extent that random placement is sampling available volumes, nonideality is related to the rate of placement rejection as $\gamma = (n_{imp} + n_{poss})/n_{poss}$, i.e., the ratio of total to black circles.

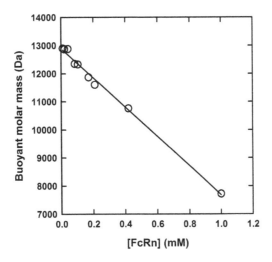

Figure 2.7 Buoyant molar mass of FcRn in 150 mM NaCl and 63 mM sodium phosphate (pH $= 6.0$) as a function of the neonatal Fc receptor FcRn concentration. The best-fit line is described by $M(1 - \bar{v}\rho) = 12,900 - 110 \times w_{FcRn}$, based on the experimental molar mass of 47,500 Da to convert the concentration scale from [FcRn] in mM to w_{FcRn} in g/L. The decrease of the apparent molar mass was interpreted in terms of volume exclusion [193, 195] — the values of the activity coefficient γ for FcRn were calculated using the best-fit above and equation 6 of Rivas et al. [195]. Treatment of the FcRn solution as a suspension of hard spherical particles (equation 10 in [195]), shows that the variation of $ln\gamma$ is best modeled in terms of FcRn having a radius of 2.8 nm.

2.2.2.2 Sedimentation Velocity at High Concentrations

The dynamics of the sedimentation process at high concentrations, where particles experience 'nonideality' from mutual interactions even in the absence of specific chemical reactions, is significantly more complex than SE, since in addition to the thermodynamic nonideality introduced above, hydrodynamic interactions (hydrodynamic nonideality) also occur. A major contributing factor to hydrodynamic nonideality is the finite size of the liquid column in which sedimentation is observed: this causes particle sedimentation to be accompanied by an equal volume flux of solvent in the opposite direction, which increases the sedimentation velocity (and friction) in the reference frame of the particle. This bulk effect is negligible if the molecules jointly occupy only a small fraction of the total solution volume, but will be significant if their volume fraction is high.[26] In addition, somewhat more locally, hydrodynamic nonideality also results from the coupling of each particle's motion

---

[26]This is a bulk solution effect. Simply, the molecules act in unison like a piston moving down in the fixed volume solution column, resulting in an upward return of the bulk solvent. In contrast to buoyancy, which is an obligate thermodynamic (hydrostatic) consequence of solvent displacement by each individual molecule independent of the others, this bulk displacement effect depends on the total number of particles and the velocity of particle movement and it disappears if there is no net transport. In contrast to hydrodynamic interactions, this effect does not depend on the inter-particle distances.

to solvent motion: particle sedimentation and diffusion will create a hydrodynamic flow-field that disturbs the sedimentation and diffusion of other particles. For highly concentrated solutions of polydisperse particles with various shapes, this process at present escapes full analytical description.

Only at the onset of significant hydrodynamic nonideality, which occurs at concentrations above ~1 mg/mL of globular proteins (or lower for particles with high frictional ratio), can we usefully describe hydrodynamic nonideality by a concentration dependent coefficient, k_s, in the form of a first order approximation in weight concentration:

$$s(w) = s_0 \left(1 - k_s w\right) \tag{2.22}$$

From statistical fluid mechanics, the expected value of k_s for a solution of uniform and spherical particles is $6.55\bar{v}$, as first shown by Batchelor [196]. This corresponds to a value of ~0.005 (mL/mg) for ideal spherical proteins (in the play dough formalism). Slightly larger values are usually measured, chiefly due to shape asymmetry of most proteins. For mixtures of different particles in this linear approximation cross-coefficients $k_{i,j}$ are required, in principle, but they are very difficult to determine in practice, and all nonideality coefficients are usually assumed in a first approximation to be identical.

Practical analytical ultracentrifugation analyses are restricted to this level of approximation. Macroscopically, the shape of the sedimentation boundaries can be extremely altered from the retardation of sedimentation at higher concentrations. Single boundaries of monodisperse particles will be sharpened, due to the stronger retardation of the leading edge at higher concentrations (Fig. 2.8). In mixtures of particles with different sedimentation coefficients and different boundary velocities, additional distortions in the boundary heights appear, owing to the Johnston–Ogston effect [197]. In the plateau region of the faster sedimenting species, the slower sedimenting species will experience stronger retardation (due to the higher total concentration) than in the plateau region of the slower boundary, causing an increase in the concentration and boundary amplitude of the slower species (Fig. 2.9). If the detection is specific to the slower sedimenting species, then the plateau region will be split in two levels, with a decrease in signal towards a lower plateau level at the position of the faster boundary.

If all particles present in solution lead to greater than ~5% volume occupancy (assuming relatively globular particles; a lower threshold will apply for more extended particles), more complex hydrodynamic interactions are to be expected and Eq. (2.22) will not hold. Inter-particle hydrodynamic interactions depend on the pair distribution function, i.e., the probability of finding one particle at a certain distance r to another. The flow-field of particles decays slowly like a long-range force with $1/r$, essentially coupling the sedimentation of all particles in solution. As a consequence, hydrodynamic interactions lead to alterations of the particle pair distribution function itself and to lateral structure formation in solution. For example, at high concentrations (above 10% volume occupancy) in mixtures of particles of different size or density, streaming effects can lead to lateral segregation where

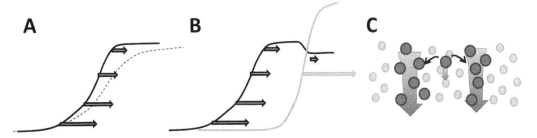

Figure 2.8 Different consequences of hydrodynamic interactions in SV. Lines depict concentration profiles and arrows depict the relative sedimentation velocity of the macromolecules at different points in the solution. *Panel A*: Boundary sharpening due to the increasing retardation at higher concentrations in the diffusionally spread boundary. The solid line indicates the boundary shape of a single class of macromolecules due to sharpening, the dotted line the boundary that would be observed ideally. *Panel B*: In a two-component system the slower component (concentration profile depicted as black line) experiences an additional retardation in the presence of a faster sedimenting component (its concentration depicted as gray line), leading to an increasing concentration of the slower component in the zone outside the boundary of the fast component. If the faster sedimenting component is optically undetected, signal profiles arise that highlight this accumulation of slower sedimenting material by a drop in signal at the point of the fast boundary. *Panel C*: At very high concentrations, streaming effects can lead to lateral structure formation, causing segregation of two components into columns of unmixed components that allows the faster sedimenting component to sediment collectively with higher velocity than individually in the mixed region.

columns of particles sediment faster collectively in their mutual flow fields. Such clustering effects will increase with increasing particle size and decreasing diffusion [198–200]. Other factors influencing the inter-particle distance distribution will, similarly, modulate the extent of hydrodynamic nonideality interactions, including repulsive or attractive reversible interactions [201, 202]. Furthermore, different collective sedimentation modes arise for semi-dilute polymers, which become entangled, such that the sedimentation coefficient becomes independent of chain-length and scales with concentration [203]. Tangling was also proposed as a contributing factor for the large sedimentation coefficients measured for amyloid fibrils [204].

2.3 SEDIMENTATION EFFECTS ON MACROMOLECULAR MIGRATION

Finally, we also need to consider whether the experimental conditions used for sedimentation will influence the measured macromolecular properties. There are various ways in which such effects may be conceived, and we will discuss, among others, the potential effects of hydrostatic pressure arising from the strong gravitational field on macromolecular conformation and interactions. Furthermore, we consider an interesting question: whether macromolecules bend, or align, due to the hydrodynamic forces arising from their migration in the sedimentation experiment, and thereby exhibit alterations in their translational friction.

More indirectly, sedimentation effects on the solvent or mediated by the solvent can be of significance. For example, as we have already discussed, solvent flow

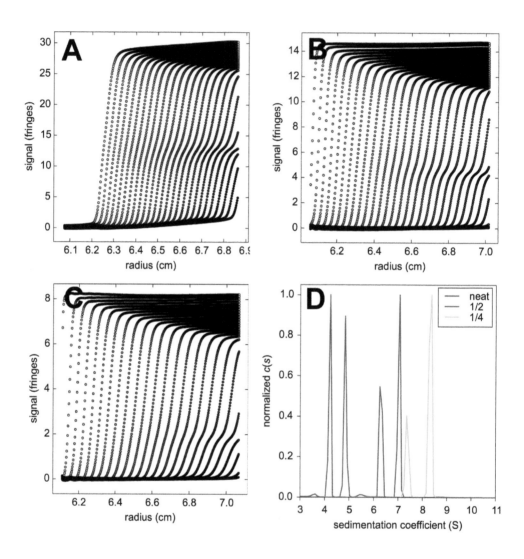

Figure 2.9   Illustration of SV at high concentrations: Shown are the sedimentation profiles acquired during the sedimentation of nucleosome core particles at 41 mg/ml (A), and a two-fold (B) and four-fold (C) dilution. The corresponding sedimentation coefficient distributions $c(s)$ are shown in D, illustrating the relative height of the boundaries. In addition to the significant retardation of sedimentation at higher concentration, the Johnston–Ogston effect causes a minor population of slower sedimenting particles to appear as a much higher population at higher concentrations.

exacerbates the consequences of volume exclusion and hydrodynamic interactions in sedimentation velocity. Furthermore, the sedimentation of solvent components, or pressure-effects on the solvent, create non-uniform bulk properties of the solvent, leading to anomalies in the observed macromolecular sedimentation.

2.3.1 Pressure

Ultracentrifugation can produce gravitational fields of up to ~300,000 g, which generate considerable hydrostatic pressure in the solution column. A typical 12 mm aqueous column spinning at 50,000 rpm in an analytical rotor will result in hydrostatic pressures of approximately 10 and 20 MPa in the middle and bottom of the cell, respectively. At a maximum rotor speed and solution column height the pressure differential will be as high as ~40 MPa across the cell. Although these pressures are an order of magnitude lower than those typically required for pressure-induced denaturation of proteins, and are in the range of average pressures experienced by organisms at the ocean floor [205], this hydrostatic pressure can still have profound consequences and influence the sedimentation process in various ways, such that this topic has received sustained attention over the years [206–210].

> SEDFIT – A function in the `Options ▷ Calculator` menu calculates the hydrostatic pressure at the bottom and middle of the solution column.

2.3.1.1 Compressibility of Solvent and Particles

Solvent compression is an immediate consequence of the applied gravitational field, and the pressure $p(r)$ at any given radius r is dependent on the centrifugal force, and the height and density of the solution column above. It follows that radial dependence of pressure:

$$p(r) = -\frac{1}{\kappa} \log \left[1 - \frac{1}{2} \rho_0 \omega^2 \kappa \left(r^2 - m^2 \right) \right] \tag{2.23}$$

results in a solvent density gradient:

$$\rho(r) = \rho_0 \left[1 - \frac{1}{2} \rho_0 \omega^2 \kappa \left(r^2 - m^2 \right) \right]^{-1} \tag{2.24}$$

where κ denotes the solvent compressibility [210]. Based on the compressibility of water, of $\kappa = 4.6 \times 10^{-4}/\text{MPa}$, density changes of around 1% would be expected at high rotor speeds and long (ca. 1.2 cm) aqueous columns [1]. Organic solvents have compressibility values that are 2- to 3-fold larger than water (see Appendix B), whereas mixtures of H_2O and DMSO ($< 70\%$) are less compressible than water [211].

The change in density across the solution column will alter the buoyancy of the sedimenting particle unless this is compensated for by the compressibility of the particle. Typically, proteins have a smaller compressibility than that of water [212, 213] (an estimate for average globular proteins is 2.5×10^{-4}/MPa [214]); accordingly, the change of water density will be amplified to a change of buoyancy across the cell, causing a reduction in sedimentation velocity at higher radii. This deceleration, will in turn, cause slightly increasing concentrations in the 'plateau' region [210]. Analytical expressions for this behavior have been derived, along with methods for analysis [210].

In the case of small proteins, due to solvent compressibility, deviations in the sedimentation coefficient and buoyant molar mass amounting to a few percent from ideal behavior may be observed, depending on the rotor speed and solution column height [210]. Pressure effects are exacerbated somewhat in organic solvents, due to their 2–3 fold higher compressibility, and, most importantly, for particles having densities closer to the density of the solvent. In fact, Cheng and Schachman have observed simultaneous sedimentation near the meniscus and flotation near the bottom of the solution column of polystyrene latex particles in H_2O/D_2O mixtures adjusted to an atmospheric pressure density slightly lower than that of the particles. This effect was attributed to the greater compressibility of the solvent than the particles, leading to negative buoyancy close to the cell bottom [206].[27]

The increasing pressure along the centrifugal solution column also increases the solvent viscosity. This can be very significant for organic solvents, [207], resulting in much stronger pressure effects than would be observed with a pressure-dependence of their density alone. For example, by increasing the pressure to 49 MPa at 30 °C, the viscosity of carbon tetrachloride and benzene increase by 21%, and 19%, respectively [215], whereas a 28% change in viscosity across a centrifugal solution column at 69,000 rpm was reported for acetone [216].

Water shows an anomalous pressure dependence of viscosity, due to the break-up of structured clusters at higher pressure. Above 33 °C, the viscosity monotonically increases with pressure; whereas at lower temperatures the viscosity decreases slightly with increasing pressure reaching a minimum at ∼100 MPa, before increasing slightly with increased pressures [217]. Data relevant for SV vary slightly: Harris and Woolf report that water at 25 °C and 31 MPa has a 0.4% lower viscosity than at atmospheric pressure (with about twice the effect for D_2O) [218], whereas Horne and Johnson only find a 0.18% decrease at 20.4 °C and 41 MPa [219]. In any event, the consequence of this for faster sedimentation will be partially compensated for by the pressure-dependent retardation of sedimentation from the increase in buoyancy due to water compressibility.

---

[27]This could be considered a special case of isopycnic banding, where the density gradient of the solvent is created by compressibility rather than co-solute concentration.

2.3.1.2 Pressure Dependence of Interactions

In any chemical reaction associated with a change in the volume between reactants and products, ΔV will lead to a pressure-dependence of the equilibrium constant K of the reaction:

$$\left(\frac{\partial \ln K}{\partial p}\right)_T = -\frac{\Delta V}{RT} \tag{2.25}$$

In fact, many reactions involve volume changes, including ion-pair formation and hydrophobic hydration [205, 220]. Attributed to electrostriction, the pH value of water increases by 0.3 units at 100 MPa [205].

Consequently, protein interactions can be affected by pressure, leading to pressure-induced dissociation of assemblies [205], although rarely at pressures achievable in the AUC. As reviewed by Harrington and Kegeles [209], polymerization reactions of proteins can be associated with positive volume changes (on the order of 10^{-3} mL/g), and therefore can be sensitive to elevated pressure. Examples include the monomer-polymer equilibrium of myosin [209], the depolymerization of tubulin [221], and dissociation of myelin proteolipid protein oligomers in Triton X-100 [222].

Straightforward experimental tests for pressure-dependent interactions are possible, provided sample of excellent purity and stability is available. These include the variation of the rotor speed, solution column height, and/or the increase in pressure through application of an oil layer on top of the aqueous solution column using synthetic boundary centerpieces (see below) [209]. Pressure chambers for filling conventional AUC cells under pressures up to 13.5 MPa have been described [223].

2.3.2 Hydrodynamic Drag Acting on the Particle — Do Molecules Bend or Align in Sedimentation?

Another important question is whether a macromolecular particle can be deformed or aligned by the frictional forces acting on it within the sedimentation flow. This will depend significantly on the particle size, its sedimentation velocity and frictional coefficient:

Particles with small sedimentation coefficients will achieve a small absolute sedimentation velocity, even at high rotor speeds (following Eq. (1.1)), and therefore experience only a small frictional force. Based on the spread of the sedimentation boundary (assuming monodisperse particles) with the distance of the boundary midpoint from the meniscus, it can be discerned for small particles that the rms displacement from diffusion is of the same order of magnitude or higher than the displacement from sedimentation. Therefore, the velocity due to sedimentation is on the same order as the mean velocity in diffusion, and the kinetic energy imparted by hydrodynamic drag from sedimentation is similar or lower than that from diffusion. We can therefore conclude that if any deformation were to occur from the frictional force, it would also likely occur from Brownian forces causing diffusion in the absence of sedimentation. It should be kept in mind that rotational diffusion occurs on a significantly faster time-scale than translational diffusion, and consequently anisotropy in the macroscopic translational diffusion is absent under these conditions. It can therefore be concluded that, for small molecules exhibiting diffusion-broadened boundaries, the sedimentation process cannot impose conformations on the macromolecule that are not sampled by ordinary Brownian motion, and that the ensemble of conformations cannot be significantly perturbed.

SEDFIT – A function in the Options ▷ Calculator menu can calculate the hydrodynamic frictional force for a particle of given sedimentation coefficient under various experimental conditions.

This empirical argument fails for very large particles that exhibit significantly lower diffusional spread under the conditions of the SV run. In this case, it is instructive to consider the forces involved. The sedimentation of the BSA monomer at 50,000 rpm will generate an overall frictional force of only ∼0.05 fN. For a flexible polymer of length L_0 behaving like a worm-like chain with persistence length P, the force F required to extend the end-to-end distance by z is $(3k_BTz)/(2PL_0)$ [224], or with k_BT being 4.1 pN×nm, $F \approx 6.15z/(PL_0)$ pN×nm (increasing with decreasing persistence length due to the higher chain entropy). Assuming an unfolded peptide with persistence length 0.4 nm [225], end-to-end extension by 1% of the contour length would require ∼0.15 pN, i.e., a force more than 3 orders of magnitude greater than the frictional force in SV.

In contrast, a 100 MDa nucleic acid sedimenting at 60 S at the same rotor speed will experience an overall frictional force of ∼100 fN. Based on a persistence length of ∼ 50 nm [224] we can calculate that the extension of the end-to-end distance by 1% would be achieved with a much smaller force of ∼1 fN. Despite this, it must be noted that the alignment or deformation requires differences in the frictional force at different locations within the particle, and such force differentials are certainly smaller than the overall frictional force from translation. Based on this back-of-the-envelope estimate, hydrodynamic friction during the sedimentation experiment may influence the macromolecular conformation and/or orientation of extremely large particles, such as the stretching of 100 MDa DNA strands.[28]

There is experimental evidence for very large macromolecules being subject to rotor-speed dependent deformation and/or alignment. In a more detailed analysis, Zimm has shown in a theoretical model for a flexible random coil (as a series of beads and springs with segmental hydrodynamic interactions) that the hydrodynamic drag at the ends is greater than in the middle, the latter experiencing greater hydrodynamic shielding [229]. This leads to the extension of the coil at the ends, and this distortion produces an increase in the overall friction, such that a decrease of the sedimentation coefficient with increasing rotor speed results. Quantitatively,

---

[28]For comparison, forces to break reversible molecular interactions are of the order of hundreds of pN [226], much like those required to elastically deform clathrin coated vesicles [227]. Covalent bonds rupture at several nN [228].

this effect was found to be significant for DNA molecules with molar masses >100 MDa [230]. This could explain the ~10% decrease of s for T2 DNA from 61.8 S at 30,000 rpm to 55.1 S at 60,000 rpm [229]. This theoretical model was extended more recently with simulations of the sedimentation of self-avoiding flexible polymers by Schlagberger and Netz [231, 232], who predicted compaction at low velocities due to recirculating flow fields, and a stretched trailing tail in the direction of the flow at higher velocities, ultimately resembling a tadpole [231, 232].

The rotational orientation of large macromolecules during sedimentation has been theoretically examined [233], and proposed as a possible mechanism, jointly with hydrodynamic inter-particle interactions, for the concentration-dependent increase in the sedimentation coefficient of some large DNA and tobacco mosaic virus [234]. Coupling of hydrodynamic interactions from concentration-dependence and rotational orientation of rods has been examined in theory and experiment by Dogic and colleagues [235] for bacteriophage fd. Again, these are comparably gigantic particles, resembling rods of 880 nm length and 6.6 nm diameter [235]. No evidence for such rotational alignment has been found for smaller rods in the size range from ~250–750 nm in a study with single-walled carbon nanotubes as model rods [236].

It should be noted that if directed hydrodynamic interactions from sedimentation cause changes in the macromolecular conformation, then the translational friction coefficient for sedimentation and diffusion will not be equal, and the Svedberg equation Eq. (1.8) will not apply.

2.3.3 Dynamic Hydrostatic and Viscous Interactions from Co-Solute Sedimentation

As sedimentation experiments are usually conducted using multi-component solvents, principally to screen charge effects, a secondary co-solute effect will occur due to the resulting dynamic concentration gradient of the co-solute.

Owing to the small sedimentation coefficient and large diffusion coefficient of small co-solute molecules, the concentration gradients are typically very shallow, and may be roughly approximated in terms of linear concentration increase from meniscus to bottom across the cell, with a slope increasing with time, and a hinge point close to the center of the solution column (somewhat displaced towards the bottom due to the sector-shaped configuration). In a standard long-column SV experiment, NaCl will, for example, exhibit an equilibrium concentration at the meniscus that is ~5% lower than the loading concentration, and at the bottom a concentration that is ~5% above the loading concentration [237]. Roughly twice this gradient is achieved with the heavier CsCl salt [238].[29] Typically, at NaCl loading concentrations of 150 mM, this difference does not generate significant enough changes in the solvent density and viscosity to affect sedimentation of proteins (although it will already generate signal offsets in the interference optical detection if

---

[29]CsCl will form steep SE gradients, and this has been classically used to band and purify DNA. This approach was used to demonstrate the semi-conservative nature of DNA replication [239].

co-solute sedimentation is not optically matched in the reference cell [237]). However, dynamic local changes in the solvent density and viscosity can significantly influence macromolecular sedimentation at higher co-solute concentrations, usually when close to the molar range or above.[30]

In this case, the relative time-course of macromolecular and co-solvent sedimentation, i.e., largely the macromolecular s-value, will govern where and when macromolecular sedimentation is altered. In fact, the decrease in solvent density and viscosity in the upper half of the solution column (i.e. at radii smaller than the hinge point) may not necessarily be affecting the macromolecule, since — dependent on its sedimentation coefficient — it may traverse this region prior to the time when dilution of co-solute occurs. If it sediments slowly (for example, below 10 S), then the drop in co-solute concentration in the upper half of the solution column will decrease the viscosity and buoyancy and locally increase the macromolecular sedimentation velocity. This can generate a shallow peak in the leading edge of the sedimentation boundary [238]. On the other hand, the increase in solvent density and viscosity in the lower half of the solution column cannot escape influencing macromolecular sedimentation, producing a local decrease in sedimentation velocity and, as a consequence, generating positively sloping plateaus. In extreme cases, for example, high concentrations of CsCl or sucrose, the solvent density gradients can be sufficient to exceed the density of proteins and to produce a flotation zone as in an isopycnic sedimentation experiment. Such anomalies are usually small, but if not computationally accounted for, can impair, for example, the quantitation of trace antibody oligomers in the presence of 5% sorbitol [240].

> Methods for the computational analysis of coupled co-solute and macromolecular sedimentation as described in [238] are available in SEDFIT.

---

[30]Such gradients would not be expected to be large enough to significantly perturb the solvation properties of the sedimenting macro-particles during their sedimentation process [238].

CHAPTER 3

Sample Preparation and Ancillary Characterization

HAVING discussed the fundamental principles and various aspects of sedimentation processes in AUC, we now shift the emphasis to suitable samples. In practice, this is intimately related to the question that is being addressed and the many experimental factors that govern data acquisition. Through careful, coordinated planning, it is possible to acquire data that optimize information content regarding the specific question under investigation. Crucial factors of the AUC experiment relate to the capabilities and limitations of the optical detection systems, and include buffer requirements for the system under study, the available amount of material, its concentration, purity, and stability over the course of the planned experiment. Additionally, important considerations are the determination of ancillary parameters required for the sedimentation analysis, such as buffer density and viscosity, as well as partial-specific volume of the particle of interest. A thorough understanding of all experimental details is critical for the extraction of reliable information and a comparison to theoretical predictions.

3.1 COMMON SAMPLE REQUIREMENTS

Considering the great variety of particles being studied in different fields, it is outside the scope of the current chapter to give detailed recommendations for sample preparation. Nonetheless, some common considerations are in order. Because the study of proteins is the most popular application of AUC, some of our comments will specifically apply to proteins and other biological macromolecules.

3.1.1 Purity

As with many biophysical techniques, AUC can tolerate impurities in the sample preparation if they are 'inert,' and 'silent' with regard to the method of detection. However, additional considerations are that impurities should not constitute a significant volume fraction of the solution, as this would disturb the sedimentation process via hydrodynamic interactions. Furthermore, high particle concentrations may create refractive index gradients that result in lensing artifacts when using the absorbance optical detection system.

Even though AUC provides the opportunity to carry out label-free experiments, detection may not necessarily be specific to the particle of interest. For example, UV absorbance at 280 nm will detect all proteins possessing aromatic amino acids. Increased specificity may be obtained, for example, through the absorbance optical detection of a chromophore absorbing in the visible region of the electromagnetic spectrum, where few proteins will contribute to the overall signal, or through use of the fluorescence optics (if an appropriate fluorophore is present). However, unintended signals may even arise in the latter, due to Raman scattering or the presence of adsorbed fluorophores bound to carrier protein (e.g., bilirubin on albumin) [99, 102]. Considering that many fluorophores are 'sticky,' it is advisable to take precautions against inadvertent cross-contamination in the sample purification, which may arise, for example, during chromatography. The interference optical detection method is the least specific, as it records refractive index differences between the sample and solvent, namely all unmatched components, including co-solutes of any size.[1] A detailed description of signal increments for the different detection methods follows in Section 4.3.2.

Depending on the method and nature of the impurity, 95% sample purity is usually a reasonable goal (matching the requirements of many other biophysical techniques). Considering that it is possible to detect and quantitate trace components at an abundance of 1% or less, some common sense is necessary to balance the requirements of sample purity and the level of detail in the analysis.

The nature of the impurities is more important than their actual quantity, specifically their size and whether they can be resolved from the molecules of interest in the sedimentation method that will be employed. Most SV analyses tolerate inert impurities that are more than twofold larger or smaller than the molecule of interest, as these will be resolved in a $c(s)$ analysis and thus excluded from further consideration. Even if they cannot be resolved, but are known to contribute to the signal-weighted s_w-value in a reproducible manner that can be measured separately, an approximate correction may be applied:

$$s_{w,adj} = \frac{s_{w,tot}c_{tot} - s_{w,imp}c_{imp}}{c_{tot} - c_{imp}} \tag{3.1}$$

where $s_{w,adj}$ is the adjusted s_w-value, $s_{w,tot}$ and $s_{w,imp}$ are the $c(s)$ integrated s_w-values for the total system and impurities, respectively, and c_{tot} and c_{imp} are the respective integrated signals [102].

In contrast to the high tolerance of sample heterogeneity in the $c(s)$ approach, purity requirements are more stringent when modeling the data directly in terms of discrete Lamm equation solutions that assume the presence of just one or a few

---

[1] An exotic exception are polymers with a refractive index equal to that of water [241]. These may even generate negative interference fringe shifts in salt solutions.

species. Only in exceptional cases is it possible to get a good fit with such discrete models, because even slight polydispersity, trace aggregates, buffer signals, or micro-heterogeneity (such as in protein glycosylation) lead to excess spread of the sedimentation boundary. In fact, the ability to achieve a good fit using a single discrete Lamm equation solution model while obtaining a best-fit diffusion coefficient that coincides with an independently measured diffusion coefficient from dynamic light scattering is a very rigorous criterion for purity [242].[2] This problem is also a concern when using Lamm equation modeling for interacting systems, as the method of data analysis is sensitive to micro-heterogeneity of binding properties [243].

SE has an intermediate susceptibility for impurities, which are usually not problematic if they are at least 10-fold larger in mass than the largest protein or complex of interest, as rotor speeds may be chosen to effectively pellet these very large particles. On the other hand, SE is particularly sensitive to low molar mass impurities, as these species distribute throughout the solution column and skew the interpretation of the concentration gradients from the molecules of interest, if not accounted for. Unfortunately, such impurities may not always be recognizable through quality control by SDS-PAGE, as they may elute out of the gel and/or not stain sufficiently. For this reason, it is highly recommended that the sample be purified by size-exclusion chromatography prior to SE. In addition, the elution profile may sometimes provide useful information on the sample purity and interaction properties. As will be discussed below, the use of size-exclusion chromatography also provides a method for obtaining a well-defined sample buffer through buffer exchange. For all AUC experiments, control of the sample purity and identity by mass spectrometry is highly useful, although usually not essential.

One final aspect of sample purity addresses the issue of whether the macromolecules of interest are intrinsically homogeneous or polydisperse with respect to size and mass. For example, biological macromolecules can be highly monodisperse, whereas synthetic polymers are usually not. This can make an important difference with regard to the data analysis models and interpretation, as exemplified by highly glycosylated proteins with an intrinsic size heterogeneity. Such size heterogeneity will hinder a quantitative analysis of the SV boundary broadening with a single diffusion coefficient model [244], and SV distribution approaches are preferred in this case. Similarly, the boundary broadening observed for nanoparticles is often a result of the intrinsic size dispersion, and it may be possible to use SV to characterize this distribution with very high resolution. It is important to establish the heterogeneity, if possible, as part of the information gathering for the experimental design.

---

[2]The successful application of this criterion depends critically on knowledge of the sedimentation and diffusion experimental temperatures.

3.1.2   Concentrations

Three separate requirements need to be balanced when choosing optimal sample concentrations:

1. Experiments need to be carried out at concentrations such that the phenomenon of interest will be observed. For example, concentrations favorable for the determination of equilibrium binding constants are usually those that result in significant populations of both bound and unbound states (although such studies will also benefit from concentrations far above and far below the equilibrium binding constants to create additional constraints). Similarly, studies with detergent micelles must be conducted above the critical micelle concentration. While this seems trivial, it sets constraints that need to be considered when planning experiments.

2. If possible, concentrations should be reasonably low, so as to avoid or minimize effects on the sedimentation process due to excluded volume, electrostatic, or hydrodynamic interactions.

3. Concentrations must be high enough so as to generate an observable signal with a sufficiently high signal-to-noise ratio, and low enough so as to remain within the linear range of detection for the optical detection chosen. In this manner, a determination of the parameters of interest occurs with the desired precision and accuracy.

3.1.2.1   Concentration Ranges for Different Detection Systems

Table 3.1 lists typical detection limits for the different optical systems. These limits translate into different molar or weight concentrations depending on the optical properties (i.e., extinction coefficients, refractive index increments, or fluorescence quantum yields) of the molecules or particles under study (see Section 4.3.2 on signal increments). Furthermore, these limits will depend on the type of sedimentation experiment, and the feature of the sedimentation data that carries the relevant information. For example, the determination of a signal weighted-average sedimentation coefficient s_w and the calculation of a $c(s)$ distribution from sedimentation boundaries in SV can be achieved with signal amplitudes smaller than the stochastic noise of data acquisition [80, 99]. In contrast, at least an order of magnitude higher signal is required for the analysis of a sedimentation boundary shape and its diffusional broadening, and even higher signals may be necessary when attempting to quantitate the presence of trace aggregates in the sample.

Obviously the nature of the detection system critically impacts the feasible concentration range. In principle, the size of the data set of a typical SV experiment is so big, $\sim 10^4$–10^5 data points, that macromolecular signal amplitudes well below the stochastic noise of a single data point can be utilized (Fig. 3.1). This allows for the analysis of fluorescence SV data collected for 4 pM EGFP [99](in the presence of carrier protein preventing surface adsorption, see below), as well as absorbance data acquired at optical density values of 0.005 OD or below, corresponding to ~10 nM of a 50 kDa protein with far UV (230 nm) detection [80]. Unfortunately, this strategy does not work well for interference data, which does not discriminate between solution components, and can therefore suffer significantly from signal offsets caused by imperfect match of co-solute sedimentation between the sample and the reference sector [237]. With relatively dilute co-solute, e.g., 150 mM NaCl or below, such imperfection will usually constrain the lower useful signal to ~10-fold the stochastic noise, or ~0.05 fringes (equivalent to ~15 μg/mL protein); higher minimum concentrations would be necessary with high co-solute concentrations (see also Section 4.3.4 for more information on such unwanted signal offsets from co-solutes).

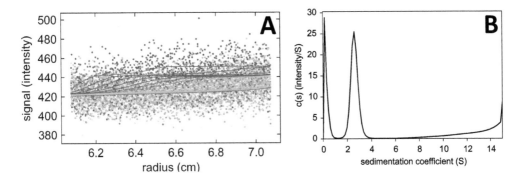

Figure 3.1 (*Panel A*) Fluorescence-detected SV data of 4 pM EGFP with 0.1 mg/mL κ-casein in PBS (circles, showing only every 2nd data point of every 20th scan) and best-fit sedimentation boundaries from a $c(s)$ distribution model (solid lines) [99]. Even though the sedimentation boundary is of similar height as the noise in the data, due to the entire data set being very large (~250,000 data points), a well-defined sedimentation coefficient distribution $c(s)$ can be determined. (*Panel B*) $c(s)$ distribution obtained from the analysis of the data in A.

The minimum signal for detailed boundary modeling of SV data is higher. In this category the minimum concentrations for typical proteins are usually lower in the interference optics than in the absorbance optics, around 0.1 fringes, mainly due to the higher signal increment in the latter system, and the better signal-to-noise ratio once the signal rises above that of the typical signal offsets from residual co-solute mismatch.

At the higher concentrations, absorbance data become nonlinear when optical densities exceed ~1.3 (see Section 4.3.3). However, the choice of a suitable wavelength can modulate the detectable concentration in the linear range: for proteins, detection at 230 nm offers a ~6-fold higher extinction (allowing for proportionally lower concentrations), and 250 nm offers a 2–3 fold lower extinction (allowing for proportionally higher concentrations). Another limitation presents itself in the absorbance optical system at high concentrations even for macromolecules that

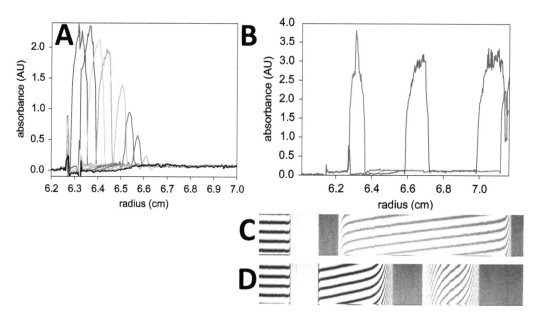

Figure 3.2 An example for boundaries with Wiener skewing. The strong refractive index gradient was created at a high protein concentration, scanned with absorbance at 560 nm where very small absorbance signals are observed (due to spectral cross-contamination, see Section 4.4.1). (*Panel A*) Due to the broadening of the sedimentation boundaries, the refractive index gradient is decreasing with time, which leads to a decrease in the peak of apparent absorbance. (*Panel B*) Same as (*A*) but at even higher concentrations leading to stronger apparent absorbance peaks. Scans at only two time-points are shown (red early, blue late), which correspond to the fringe pictures in (*Panel C*) and (*Panel D*), respectively, which show shadows (fringeless gray rectangular areas) at the points of the highest refractive index gradient.

do not absorb, as these can create refractive index gradients and corresponding aberrations in the absorbance scans due to lensing effects, referred to as Wiener skewing [245–247]. These usually take the form of artifactual peaks in their sedimentation boundary (Fig. 3.2). This artifact will depend on the sharpness of the concentration gradient in the boundary (i.e., increasing with concentration and molar mass) and thus impose an upper concentration limit in the absorbance system even for particles with very little absorbance. Interestingly, observing the same solution column using the Rayleigh interference system will produce different artifacts, depending on the focal plane: when the laser beam is focused at the midpoint of the cell, fringe blurring may be observed, but this is not accompanied by artifactual fringe displacement; whereas with a focus at the 2/3 plane of the cell, artifactual fringe displacement will occur without fringe blurring [246].

Fluorescence optical data are readily adjusted to a wide concentration range by variation of the laser power, photomultiplier gain, and focal depth. An upper concentration limit occurs due to the inner filter effect, expected at μM concentrations of fluorescein or EGFP. This can be avoided, in part, with a reduced focal depth and a correction of the ensuing radial signal intensity gradients in the data analysis [100] (Section 4.1.5). Even higher concentrations can be studied by addition of unlabeled material in a tracer experiment.

Interference data are, in principle, infinitely linear with regard to concentration, and therefore the Rayleigh interferometer is the detection system of choice for protein concentrations above 10 mg/mL. However, a limit on the concentration gradients observed may arise from practical requirements in reconstructing a faithful fringe shift profile. Limits of signal linearity and radial resolution will be described in more detail below in Sections 4.1.3 and 4.3.3.

One excellent strategy for modulating the useful concentration ranges of the absorbance and interference optical systems is through the choice of centerpieces with different optical pathlengths. While the data presented in Table 3.1 are for standard 12 mm pathlength centerpieces, 20 mm, 3 mm and 1.5 mm pathlength cells are also commercially available. An even shorter pathlength, achieved by a thin gasket between two windows, has been used by Lewis and Minton in the sedimentation equilibrium study of highly concentrated hemoglobin solutions using absorbance optics in the visible range [248]. The combination of optical pathlengths (3 mm and 12 mm) and detection wavelengths (230, 250 and 280 nm) often allows for the study of a 50–100 fold concentration range of samples producing the same absorbance.

When carrying out SE experiments, a distinction needs to be made between the loading concentration and the concentrations observed once equilibrium is attained. As SE generates a concentration gradient that will depend on the rotor speed, the loading concentration is not always so critical. Part of the solution column will be much more dilute, especially in regions close to meniscus at higher rotor speeds, while the part close to the cell bottom will be much higher in concentration than the loading concentration.[3] Thus, if only little prior knowledge about the size of the molecules is available, an appropriate range for detection can often be found only after SE data are acquired. For this reason, it is highly useful to acquire absorbance data simultaneously at different wavelengths, as they can report with good signal-to-noise and linear detection on different regions in the solution column.

3.1.2.2 Concentration Considerations for Ideal Sedimentation

Hydrodynamic nonideality will arise at high sample concentrations based on the finite volume occupancy of the particles in solution. Typically, the onset of significant hydrodynamic nonideality for globular proteins occurs above 1 mg/mL (assuming a precision of ~1% and a nonideality coefficient of ~0.01 mL/mg). However, the nonideality coefficient increases strongly with the frictional ratio, such that more

---

[3]This observation can feed back to questions of sample preparation. Initial sample concentration may not be necessary in SE, since SE itself will generate a concentration gradient. This effect can be enhanced with a long solution column, although it will take longer time to reach sedimentation equilibrium; this, in turn, may be compensated for by using an optimized time-varying centrifugal field [249].

extended molecules will generate hydrodynamic nonideality interactions at much lower concentrations.[4]

Thermodynamic nonideality also arises from the finite volume occupancy and possible electrostatic interactions, as expressed in the second virial coefficient. For uncharged, spherical proteins, a $\sim$1% effect can be expected at concentrations of $\sim$1 mg/mL. Predictions of the effective excluded volume for ellipsoidal and/or charged particles, which will exhibit stronger nonideality, were implemented in the software program COVOL by Harding and co-workers [251].

A concentration series can reveal where, in individual cases, observable nonideality sets in. If it is possible to carry out experiments under conditions where nonideality is negligible this will greatly simplify data interpretation.

3.1.2.3 System-Dependent Concentration Choices

Interacting systems should be studied over a concentration range encompassing the value of K_d, ideally covering a range at least as large as 0.1 to 10-fold K_D. Concentration series are also important for systems that are expected to behave as non-interacting, ideally sedimenting particles. It is only through a concentration series that one can ensure that the particles are indeed non-interacting, and that the sedimentation properties are those of the macromolecules of interest. If concentration-dependent nonideality is observed, extrapolation to infinite dilution may be carried out.

Thus, in the vast majority of studies it will be necessary to conduct experiments over a wide concentration range, and unless the sample is limiting, it is a good practice to do so in a single run to minimize systematic error.

In the case of a two-component interacting system, the choice of concentration can be imagined as sampling a trajectory in the plane of component concentrations. Traditionally these trajectories have been chosen either parallel to one component's concentration axis, generating titrations in which the concentration of one component is kept constant and the other increased; along a diagonal, generating a dilution series with constant molar ratio; or along another diagonal with the concentration of one component increasing and the other decreasing so as to keep the total concentration constant, as in a Job plot [252]. These trajectories aid in the visual inspection of the results, but are by no means required for a modern nonlinear regression of the data.

When studying multi-component interactions, it is indispensable that each component be studied individually over a large and relevant concentration range. This will aid in the verification and quality control of the sample stock concentrations, but also establish the self-association properties of each component.

---

[4]A classification of solution behavior of various classes of macromolecules on the basis of their hydrodynamic nonideality, termed 'conformation zoning' was described by Pavlov, Rowe, and Harding [250]. The nonideality of sedimentation increases from globular or branched, to random coil, semi-flexible, rigid, and the extra rigid regime.

An experiment planner is available in SEDPHAT (Generate functions; and Options ▷ Interaction Calculator ▷ Effective Particle Explorer Calculator). This function can visualize the binding isotherm for a given two-component interaction in the concentration plane (or for three-component interactions in the three-dimensional concentration space), and allows the user to apply constraints for available volume, maximum concentrations and signals. A trajectory in the concentration plane can then be graphically entered across the feasible concentration space, whereupon SEDPHAT calculates a plan for mixing a desired number of samples along this trajectory from given stock solutions. The program then reports the amount of material needed for conducting such an experiment.

3.1.2.4 Considerations for Sample Preparation

The final step of sample preparation, prior to AUC, requires a measurement of the concentration. Even though sample concentrations will be determined experimentally on the AUC, knowledge of the expected signal and concentration is extremely helpful for setting up the experiment. For example, in the case of interacting systems, this will allow for the preparation of mixtures with the desired molar ratios. Furthermore, some configurations (such as interactions with molecules that do not provide a suitable optical signal) may require one to rely on *a priori* knowledge of the loading concentration in the analysis.

In the case of samples that absorb in the UV-VIS, a measurement of sample concentration using a benchtop spectrophotometer is the method of choice, assuming that an extinction coefficient is known (see Section 4.3.2 on signal increments). Careful dry weight determination of concentration [253] would be the gold standard, but this is usually not feasible. Sample concentrations measured utilizing colorimetric assays are usually not accurate and should be avoided [254]. When comparing signals measured on a benchtop spectrophotometer with those of the AUC absorbance detector, small differences (typically 10% or less) should be expected. These differences are due to the limited photometric accuracy of spectrophotometers (discussed in Section 6.1.5), the depletion of aggregates and scattering particles in the centrifugal field, and possible effects from wavelength inaccuracies (Section 6.1.4).

When preparing mixed samples for AUC it can be highly beneficial to adopt a pipetting scheme that generates constraints that can be applied during data analysis. For example, repeat pipetting of the same volume with the same pipette tip can ensure sufficiently accurate volumes for computational constraints that the volumes (concentrations) be equal. However, the volume adjustment of a pipettor is not sufficiently accurate and reproducible to be relied on as a constraint in AUC data analysis. As a second example, the dilution of a pre-mixed stock of two components will ensure identical molar ratios in all samples, whereas the repeated mixing of different volumes of the individual stocks in the same volume ratios will generally reproduce the same molar ratio less accurately.

Another important consideration for sample preparation is the need for chemical equilibrium prior to SV experiments. This becomes particularly important for interacting systems with slow exchange kinetics, as equilibrium may not be attained during the time required to load samples into the cell assembly and subsequent temperature equilibration in the rotor chamber. Slow equilibration may be encountered for interacting systems with high affinity: this problem becomes especially pronounced when preparing samples by diluting high concentration stock mixtures for which the dissociation rate constant may be rate limiting in achieving the new equilibrium. Slow equilibration may also occur in systems with moderate affinity but large conformational changes. Unfortunately, the reaction kinetics may only become known after the SV analysis. For the analysis of equilibrium binding constants of slow dissociating systems, it may be necessary to conduct control experiments with different mixing schemes [99].

Finally, it is important to prepare the samples in a well-defined buffer suitable for the specific question being addressed. This can be achieved primarily through size-exclusion chromatography or exhaustive dialysis. This is important because the buffer properties, namely density and viscosity, impact the data analysis, as does the buffer dependent effective partial-specific volume of the sample under study. The importance of a well-defined buffer will be a recurring theme in the sections that follow.

3.1.3 Loading Volumes

In SV, the standard sample volume in a 12 mm centerpiece is 400 μL. Optimal information content will be achieved using the maximum fill volume of $\sim$450 μL, but such volumes are practically more difficult to achieve. In any case, to create a well-defined meniscus, the centerpiece should not be completely filled. Smaller sample volumes can be utilized, usually to conduct SV experiments at high concentration using limited sample; in such cases good data can be obtained using solution columns of 300–350 μL. SV data obtained from as little as 150–250 μL can also be very informative. All sample volumes are reduced by a factor of 4 when using 3 mm pathlength centerpieces. For a relationship between solution column heights and approximate filling volumes for standard double sector centerpieces, see Appendix D.

Sample volume requirements for SV can be reduced further by analytical zone centrifugation, in which specialized band-forming centerpieces (Section 5.2.1) are used to layer a lamella of sample over a more dense solvent column at start of during centrifugation [255, 256]. Unfortunately, a drawback of this configuration is that initial sample distribution is not as well defined, and therefore the analysis is not quite as detailed as the standard uniform loading configuration.

In the case of SE experiments, volume requirements are balanced by the time required to reach sedimentation equilibrium. For small molecules <10 kDa, which usually have a high diffusion coefficient, 300–400 μL samples can be used, whereas for larger molecules 130–150 μL samples are common. Smaller volumes of $\sim$100 μL can be used for very large molecules, but the drawback of shorter solution columns is their limited information content due to fewer data points and less curvature, which often only allows for the determination of an average molar mass, and their requirement for higher rotor speeds to generate sufficient concentration gradients. Their resulting higher concentration gradients report on a given concentration with fewer data points, and are closer to the limit imposed by the optical radial resolution, which depends on the optical system used (see Section 5.5.2 below). The benefit of short solution columns is their much more rapid attainment of sedimentation equilibrium (Section 5.5.2), which can be extremely useful for the study of unstable proteins, as demonstrated by the characterization of the tubulin monomer-dimer equilibrium by SE using 50 μL samples [257]. The longer time requirement at larger solution columns can be counteracted partially using time-varying rotor speeds following an optimized initial overspeeding schedule [249]. The relationship between solution volume in SE and time requirements is discussed in more detail in Section 5.7.2.

A typical experiment would necessitate a sample with a stock solution that can be conveniently studied in 3 mm centerpieces at the highest concentration, and in a series of 3-fold dilutions in 12 mm centerpieces. This configuration would require a total of ~300 μL of the concentrated stock sample in standard SV experiments, and ~120 μL total in a standard SE experiment.

As will be described in more detail in Section 4.3.4 on buffer signal offsets, when using the interference optics in SE or SV it is imperative that the sample and reference volumes are precisely matched. This is usually not essential when using the absorbance optics.

3.1.4 Stability

An important requirement is that the samples are stable for the duration of the AUC experiments, including the time required to fill the cell assemblies and to temperature equilibrate the rotor. Typical times for standard SV experiments on medium sized proteins include 2–3 hours for setup (although this could potentially be reduced to 1 hour) and 3–4 hours of run time. The time required for SE is traditionally much longer, typically on the order of days, but strongly dependent on solution column height and diffusion coefficient, as discussed above and in more detail in Section 5.7.2. Archibald experiments, which evaluate the initial depletion of material near the meniscus and accumulation near the bottom, report the average buoyant molar mass of the sedimenting particles [258, 259], and therefore are an alternative approach to short-column SE. Similarly, time-modulated over-speeding schedules can significantly educe the time for SE [249].

Sample stability is a major concern for proteins, which may have limited conformational stability, be subject to proteolytic activity, and/or undergo irreversible aggregation. It is therefore a good practice, when working with proteins, to schedule the last step of sample purification just prior to the start of the AUC experiment. It is common experience that protein samples are somewhat more stable in the centrifugal field than when kept at the same temperature on the laboratory bench.

In the absence of other considerations, SE experiments can be carried out at low temperatures, such as 4 °C, in order to minimize degradation.

Under these conditions, the vast majority of proteins studied in the authors' laboratories have been sufficiently stable for SE to be attained, sequentially, at three different rotor speeds over a few days. Such a 'multi-speed' SE experiment creates constraints that greatly facilitate the analysis of both non-interacting and interacting systems [160]. A detailed study on the measurement of the equilibrium constant by SE has shown that low levels of degradation (resulting in the formation of smaller sized impurities) have the greatest impact on the study of high-affinity interactions [80].[5] Conversely, it can be expected that aggregation processes will have a larger impact on the estimates of binding constants for low-affinity interactions, due to the formation of artifactual oligomers. Importantly, if sample degradation occurs, this cannot be assumed to be reproducible from cell to cell in the same run.

Protein stability is less critical for SV experiments, due to the shorter experimental time-scale. For this reason, it is advantageous to carry out the SV experiments at 20 °C to avoid large temperature corrections for the buffer viscosity (and to avoid long temperature equilibration times). In exceptional cases where the protein is unstable in the buffer conditions of interest, it is possible to keep the sample in a buffer in which it is stable, and use analytical zone centrifugation to overlay the sample on the buffer of interest (Section 5.2.1). This effectively results in a rapid buffer exchange due to co-solute diffusion and migration of the protein into the solvent column during the rotor acceleration phase.

Following SV experiments, samples can be resuspended by taking the cell assemblies out of the rotor and gently turning them over repeatedly to remix the contents (Section 5.2.4). Often a repeat SV experiment on resuspended samples will produce almost identical results.[6] Failure to reproduce the same result is usually a sign of irreversible sample aggregation or pelleting at the extremely high concentrations encountered near the bottom of the solution column during the first SV run (see [260] for the properties of the sediment), and not a sign of degradation impacting the analysis of the sedimentation boundary of the first run.

Some systems require special considerations, such as RNA, which can be particularly susceptible to degradation by ubiquitous RNAses. AUC studies are possible

---

[5]Sample stability following SE may be tested by resuspending the sample and carrying out an SV experiment in the same cell. Due to the shorter column it is advisable to use a lower rotor speed than would be used for a standard long-column SV experiment, in order to obtain a sufficient number of scans. Even though this will provide a low-resolution size distribution, it should expose the unexpected presence of low molar mass species. Alternatively, it is possible to visualize low molar mass species by SE at higher rotor speeds.

In cases where there is significant degradation, SE may never be reached due to continual degradation. Also, one may only reach the desired curvature at higher rotor speeds than expected. It may be helpful to recover the samples post SE and test by SDS-PAGE [80].

[6]This is the case, for example, for bovine serum albumin (BSA) after sedimentation at 20 °C and 50,000 rpm. In fact, BSA can be re-run many times, which is one reason why this protein provides an excellent reference material for testing instrumental calibrations (see Section 6.2).

provided suitable precautions are taken, including the inactivation of RNAses by treatment of the cell windows and centerpieces during cell cleaning, the possible addition of small amount of RNAse inhibitors to the sample buffer and the compulsory use of gloved hands and RNAse free pipettors and loading tips.[7]

The requirement for an oxygen-free environment, which is necessary for the study of hemoglobins and related systems, can be accommodated by incubation of the cell components in the required atmosphere, followed by cell assembly and sample loading in a glove box. Alternatively, a desiccator may be used to store and equilibrate the cells, with subsequent cell loading under a nitrogen tent.

3.2   CHOICE OF SOLVENT AND BUFFER

The most commonly used solvent for AUC is water. In many respects it is an ideal solvent when compared to organic solvents, which have a 2–3 fold higher compressibility and exhibit a very significant pressure-dependence of the viscosity. These effects can be compensated for by use of lower rotor speeds and shorter solution columns. Notwithstanding these complications, organic solvents are also commonly used, and have a long history of applications in AUC.

Due to its chemical similarity to H_2O, deuterated or heavy water, D_2O, is often used in density contrast or density matching experiments with aqueous buffers. However, it should be noted that this substitution may cause subtle changes in hydration, protein binding affinity, and allostery, among others [262–265], in addition to a different refractive index and solvent viscosity (see Tables in the Appendix). Importantly, the use of D_2O can also produce changes in protein partial-specific volumes and masses through substitution of deuterium for exchangeable hydrogens [266]. This drawback does not usually occur when using $H_2\,^{18}O$ [106, 267–269] instead. The even heavier water $D_2\,^{18}O$ has likewise been shown to be very useful as a solvent in AUC [266].

A broad range of co-solutes and buffers can be used for AUC. Besides the chemical compatibility with the centerpiece material (which may be charcoal-filled Epon, aluminum, or titanium, see Section 5.2.1), the main considerations are that the co-solutes are (1) well-defined, (2) provide sufficient ionic strength to screen macromolecular charges and (3) are not present at high concentration, if possible, to avoid the generation of either refractive index gradients or density gradients.

While the first point may seem trivial, it is not always fulfilled in practice. For example, when working with proteins it is common practice to add high concentrations of kosmotropes such as glycerol. The presence of glycerol will have a substantial impact on the solvent density and viscosity, as well as the apparent partial-specific volume. Pipetting an exact volume of glycerol is very difficult due to its high viscosity; to achieve a reproducible and well-defined condition, addition by weight is preferred. If possible, buffer exchange by size-exclusion chromatography

---

[7]Not all RNAs require such stringent treatment. If the RNA has significant tertiary structure, it is usually quite stable. For example, no special treatment was necessary in [261].

or exhaustive dialysis to equilibrium with the goal of removing the glycerol is desirable for the AUC experiments, unless this would lead to a significant loss or destabilization of the protein.

A final buffer exchange is also recommended when using interference optics, because co-solutes sediment and unmatched components could result in significant refractive index gradients and corresponding interference signal contributions. The strategy is therefore to exactly match the chemical composition and sample geometry of the reference sector, which will ideally compensate for all co-solute signals in the sample. Such a perfect match in chemical composition can be reliably achieved by dialysis or size-exclusion chromatography, but not, for example, by using separate batches of identically prepared buffer. This becomes a problem when the co-solute concentrations are very high, usually several hundred millimolar or more, as even a perfect chemical match will still result in co-solute signal contributions, due to residual imperfections in the matched geometries of the sample and reference solution columns. In such cases, additional signal offsets can be accounted for in the data analysis [237]. Similarly, if a final sample dialysis or chromatographic buffer exchange is not possible, and high concentrations of co-solutes cannot be avoided, a good strategy may be to load pure solvent without the co-solute in the reference sector and subsequently model the sedimentation of the co-solutes in the data analysis. This would usually be represented as an additional slowly sedimenting and strongly diffusing component [237].

Stringent requirements for refractive index matching do not apply to the absorbance or fluorescence optical detection due to signal selectivity. However, a well-defined buffer composition is still necessary. Furthermore, the sedimentation of the macromolecule can be substantially altered by preferential binding or exclusion of co-solutes, as discussed in Section 2.1 above. This is true for most proteins, in particular, if the density of the solution (including co-solutes) is significantly different from the solvent alone (without co-solutes). Thus, when using high concentrations of co-solutes in density matching or density contrast experiments [270,271], allowance must be made for changes in the protein $\bar{v}$. As outlined above (Section 2.1.2), such considerations are essential, for example, when working with polyelectrolytes, such as nucleic acids.

The use of salts to screen long-range electrostatic interactions and essentially eliminate charge effects can be usually met, when studying most proteins, with the addition of at least 10 mM of a supporting electrolyte. The addition of 130–150 mM NaCl, as found in phosphate buffered saline, will suffice to screen charge interactions without creating complications with regard to refractive index gradients.

The presence of co-solutes that absorb light at the detected wavelength will impact the choice of the optical detection system. For example, nucleotide co-factors absorb at both 260 and 280 nm,[8] and require the use of the interference detection

---

[8]ATP, for example, has an extinction coefficient of 15,400 $M^{-1}cm^{-1}$ at 260 nm, and approximately one tenth of that at 280 nm. A 100 μM solution will therefore have an absorbance of 1.5 at 260 nm (in a 12 mm column), which rules out the use of absorbance optics at 260 nm.

optics dependent on their concentration. Other common co-solutes that will impair, to a lesser extent, the absorbance detection are buffers such as TRIS and HEPES, which at concentrations above 20 mM may be problematic at 230 nm. A check of the buffer spectrum on a benchtop spectrophotometer (blanked against water) is recommended when using absorbance detection and as a rule, the total absorbance of the buffer and the protein should not exceed 1.3 OD (taking into consideration the optical pathlength of the AUC centerpiece). Accordingly, the useful protein concentration is lower when using buffers with higher absorbance.

Special care needs to be taken with reducing agents such as DTT and β-mercaptoethanol, whose absorbance properties change with oxidation state. These substances are best used at low to moderate concentrations (e.g., 1 mM DTT or below), provided the data analysis allows for the fact that the baseline absorbance will increase with time, even though the reference buffer is matched [272]. To monitor the magnitude of buffer absorption, it can be advantageous to use pure solvent in the optical reference sector, and to model the buffer absorption strictly as a separate sedimenting species, usually characterized by very low s and high D values [237].

Significantly higher buffer absorbance can occur when nucleotides form part of the buffer, with their strong UV absorbance limiting light transmission and the utility of the absorbance system.[9] In this case, modification of the protein absorbance properties, either by genetic inclusion of tryptophan analogs, such as 5-hydroxy-tryptophan, which extends the UV extinction slightly to ~310 nm [273, 274], fusion with a fluorescent protein [275], or extrinsic labeling with a chromophore in the visible spectral range, all allow it to be studied selectively in a complex mixture. Any fluorescent label, for example, will have a significant extinction at the visible excitation wavelength, as well.

With regard to the study of protein in an active state that requires and processes GTP, Rivas and colleagues have shown how it is possible to conduct SV experiments of FtsZ fibrils in the presence of an enzymatic GTP regeneration system (acetate kinase and acetyl phosphate) to keep the GTP concentration constant [276, 277].

Finally, when working at very low protein concentrations, typically nanomolar or below, it may be necessary to prevent sample losses due to surface adsorption by adding an inert 'carrier' or 'blocking' protein that will adsorb to the surfaces of the centerpiece and windows. Choices documented in the AUC literature include 0.1 mg/ml of BSA, lysozyme, and κ-casein [99] and consideration for this choice is based on their overall signal contribution, which is significant for BSA [99, 102], and their inertness with regard to the protein or system of interest. This needs to be established in control experiments that ascertain whether the sedimentation properties of the proteins to be studied are altered in the presence of the carrier

---

[9]For example, experiments can be carried out with a buffer having an absorption of 0.5 OD, provided the protein does not contribute more than 0.5–0.8 OD.

protein. This may not always be trivial.[10]

3.2.1 Measuring Solvent Density

For most AUC experiments it is necessary to know the precise solvent density and viscosity to determine molar masses and derive hydrodynamic shapes. Aqueous buffer densities and viscosities far from the standard values of water at 20 °C will be encountered, especially when working at high or low temperature, or when buffer components such as sucrose or glycerol are present. Very large deviations from standard conditions are observed with non-aqueous systems, such as organic solvents.

The density and viscosity of aqueous buffer solutions can be calculated with sufficient accuracy based on their composition using tabulated data, assuming that the effects of temperature and co-solute are independent and additive (which holds for dilute solutions, see Appendix B). The software package SEDNTERP (freeware provided by Dr. John S. Philo, with the latest version implemented at sednterp.unh.edu) [278] is one tool frequently used for such calculations. However, an experimental determination of these parameters is often preferred, at least to confirm the buffer composition, and mandatory if tabulated density and viscosity data are not available for some of the buffer components. Densimetry for this application is usually straightforward, for example with a Kratky balance [279], as a far lower precision is required ($\sim$0.0001 g/mL) than is necessary for the determination of partial-specific volumes discussed below [279].[11]

3.2.2 Measuring Solvent Viscosity

Knowledge of the solvent viscosity is important for any hydrodynamic interpretation of the sedimentation transport process, including the absolute values of s and hydrodynamic friction. Frictional ratios f/f_0 can be compared to theoretical predictions only after they are converted to standard conditions of water at 20 °C. On the other hand, thermodynamic quantities, such as the (buoyant) molar mass, along with several other quantities of potential interest, such as percent oligomer in protein preparations, or the composition of multi-protein complexes, will be independent of the solvent viscosity. Likewise, the solvent viscosity is irrelevant in SE experiments, other than influencing the time required to attain equilibrium.

For cases where viscosity is important, it is prudent to carry out the SV exper-

---

[10]While this control experiment can often be carried out at higher concentrations, it may be complicated by the existence of association equilibria, if subunits dissociating at lower concentration preferentially interact with the carrier.

[11]There has been a slight controversy in the field whether solvent density or solution density should be used. At higher macromolecular concentrations, the chosen density will determine the magnitude of the resulting nonideality coefficients; a discussion of this topic can be found in the work of Harding and Johnson [280]. However, the difference will vanish at low concentrations of macromolecules. Most commonly the solvent density, not solution density is used, and this framework is adopted in the present book.

iment as close as possible to standard conditions. It is particularly advantageous to avoid large temperature corrections as the viscosity of water changes from 1.3 cP at 10 °C to 1.002 cP at 20 °C, and 0.798 cP at 30 °C (see Appendix B for more details). In the absence of other considerations, SV experiments are usually conducted at 20 °C, noting that it is important to regularly check the instrument's temperature calibration, which may be subject to drifts or sudden changes (see also Section 5.3.1 below).

Similarly, when measuring the solvent viscosity it is important to verify the accuracy of the viscometer by determining a viscosity of 1.002 cP for water at 20 °C. Since the accuracy of the s-values is typically better than 1%, any viscometer should offer at least three significant decimal places. In many AUC laboratories, rolling ball viscometers with a temperature accuracy of 0.02 °C are used for routine viscometry. In exceptional cases, when dealing with solvents containing a high concentration of surface-active co-solutes, concerns regarding the accuracy of the viscometer may arise due to surface films. For this reason, it is sometimes preferable to use the sedimentation coefficient of a molecule with well-known s-value, such as BSA monomer, as a standard and compare this in a standard buffer to the solvent of interest, in the same SV run as a validation of the measured viscosity. However, assumptions with regard to the inertness of the reference molecule and with regard to preferential solvation and buoyancy differences may be required in this approach.

3.3 DETERMINING THE PARTIAL-SPECIFIC VOLUME

The determination of the partial-specific volume is important for any AUC experiment in which absolute knowledge of the molar mass and/or sedimentation coefficient is sought. This is because the sedimentation analysis reports on the buoyant molar mass $M_b = M(1 - \bar{v}\rho)$ of the sedimenting particle. However, there are situations in which the determination of the buoyant molar mass is sufficient to solve the problem at hand, for instance, in the study of heterogeneous interactions. Specifically, SE experiments designed to determine binding constants (in particular, for heterogeneous interactions), or assess the purity of a sample, can frequently be conducted without prior knowledge of the precise mass or partial-specific volume of the sedimenting particle, and, similarly, without an exact knowledge of the buffer density. Instead, the analysis can proceed by setting the partial-specific volume to zero, such that the molar mass estimates become identical to the buoyant molar mass M_b. Alternatively, if any estimate of $\bar{v}$ is available (such as the value of 0.735 mL/g experimentally observed for an average unmodified protein [281] and predicted from a compositional analysis of the human proteome [95]), say $\bar{v}'$, this value may be used to produce an apparent molar mass M', which will be not too far from the true value if the partial-specific volume is a good approximation. Effectively, any operational estimate $\bar{v}'$ provides a particular scale producing apparent molar mass values M' that are related to the true molar mass by:

$$M = M' \frac{(1 - \bar{v}'\rho')}{(1 - \bar{v}\rho)} \tag{3.2}$$

with ρ' the operational value for the solvent density. Like buoyant molar masses, apparent molar masses on this scale are additive for assembly processes, with the usual caveats, namely the absence of partial-specific volume changes upon assembly.

In hydrodynamic experiments, the frictional ratio f/f_0 also depends on the value of $\bar{v}$, since it involves the calculation of a molecular volume. In analogy with Eq. (3.2), an operational $\bar{v}'$ introduces a $(f/f_0)'$-scale, which via Eq. (1.4) relates to the true f/f_0 according to:

$$(f/f_0)' = \frac{R_0}{R_0'}(f/f_0) = \left(\frac{\bar{v}}{\bar{v}'}\right)^{\frac{1}{3}}(f/f_0) \tag{3.3}$$

Even though the introduction of such an operational $\bar{v}'$ (and related ρ'- and η'-values) leads to different M- and f/f_0-scales for expression of the mass and friction-related parameters, this does not change anything in the equations governing sedimentation, nor does it introduce any approximations in the data analysis of the sedimentation process. Only the interpretation of absolute numerical values would require a later interpretative step of the numerical results using the transforms for M' and $(f/f_0)'$.

If the partial-specific volume is required for data interpretation, or — equivalently, if a true molar mass and friction scale are required — the three most common approaches to determine the partial-specific volume are AUC itself, densimetry, and predictions based on the composition.

3.3.1   Ultracentrifugal Approaches

The simplest and most powerful approach to determine $\bar{v}$ is by use of the AUC itself, using either SE or SV. Both approaches will require independent thermodynamic or hydrodynamic prior knowledge of the molecule of interest, or the use of density contrast experiments. Ultracentrifugal approaches have the virtue of requiring very little material (just enough to carry out the AUC experiment) and not requiring an independent measure of the sample concentration.

3.3.1.1   Prior Knowledge of Molar Mass or Hydrodynamic Shape

The easiest and most common situation is one in which the molar mass M_a of the macromolecule under study is known through mass spectrometry, or composition. In this case, the buoyant molar mass determined in the AUC experiment can be used to calculate the density increment $d\rho/dw$ Eq. (2.3), or alternatively, the apparent partial-specific volume ϕ' as $\phi' = (1 - M_b/M_a)/\rho$. If an operational $\bar{v}'$ value was used to preliminarily analyze the AUC experiment, this becomes $\phi' = (1 - M'(1 - \bar{v}'\rho')/M_a)/\rho$. In changing the operational $\bar{v}'$ value to the ϕ'

so determined, one obtains a new molar mass scale on which the true molar mass appears.

This approach is very common, for example, in the study of heterogeneous protein interactions by SE, where prior experiments with each protein component alone will provide their respective buoyant molar masses, and by inference the partial-specific volumes. However, the analysis of protein interactions can continue to be carried out entirely in the buoyant molar mass or $\bar{v}'$-scale, which for heterogeneous interactions may be chosen conveniently to equal ϕ' of one of the macromolecular components, so as to provide a molar mass scale that gives the true macromolecular mass for one of the components.

As discussed above (Section 2.1.3), the apparent partial-specific volume ϕ' is equal to the partial-specific volume $\bar{v}$ of the macromolecule, or the sedimenting particle, only under conditions in which preferential solvation is absent and the hydration shell is neutrally buoyant. However, if we have an estimate for the degree of hydration, B_1, and thus the mass of the complete sedimenting particle, $M_{SP} = M_a(1 + B_1)$, then the experimentally measured buoyant molar mass leads to the partial-specific volume of the complete sedimenting particle including hydration, $\bar{v}_{SP} = (1 - M_b/M_{SP})/\rho$, and from here to the macromolecular $\bar{v}_a$ via Eq. (2.10). Again, this transformation is often not essential, for example, when studying heterogeneous protein interactions where each M_b can be determined independently.

The situation is more complicated in the case of self-associating systems, since it may not be possible to accurately measure the buoyant molar mass of the monomer if the accessible concentration range does not lead to complete dissociation. On the other hand, in this case perhaps conditions can be established where stable oligomeric assemblies (say n-mers) are observed, such that ϕ' can be determined from their buoyant molar masses as $\phi' = (1 - M_b/(nM_a))/\rho$. This requires knowledge of the oligomer size n, and in the case where n is not too large it may be possible to obtain this number simply based on a prior expectation for a reasonable value of ϕ'. However, higher oligomers may not be distinguishable with confidence.[12] For reversible self-associations it is best practice to acquire and globally model an experiment series studying the concentration dependence of the observed buoyant molar masses.

Unfortunately, when studying reversible self-association processes by SE, in contrast to purely heterogeneous associations, it will be very important to determine the protein partial-specific volume or buoyant molar mass of the monomer. The apparent $\bar{v}$ scale cannot be maintained due to a correlation of the measured molar

---

[12] A 1% uncertainty in the value of $\bar{v}$ leads to a $\sim$4% uncertainty in the apparent molar mass of proteins. Such a mass precision is clearly sufficient for discriminating between a monomer and a dimer, but would make the distinction of an undecamer from a dodecamer very difficult.

mass with the binding constant(s).[13]

An alternative strategy for estimating the partial-specific volume from SV data is possible, although more rarely applied, with prior knowledge of the macromolecular frictional ratio, along with a diffusion coefficient. The hydrodynamic shape of the sedimenting particles may be known, for example, from electron microscopy, and the diffusion coefficients may be the result of modeling of the sedimentation boundary in SV, the result of dynamic light scattering experiments, or size-exclusion chromatography. Jointly, f/f_0 and D lead to the radius R_0 of the compact sphere of the same volume as the particle, and therefore to the volume and mass of the displaced solvent. On the other hand, the measured s leads to the buoyant molar mass $M_b = sRT/D$, via the Svedberg equation Eq. (1.8). The buoyant molar mass and the mass of the displaced solvent provide the partial-specific volume, because M_a may be considered the sum of M_b and M_1: $\bar{v}_{SP} = (1 - (M_b/(M_b + M_1)))/\rho$. This is particularly useful for characterizing particles with known shape, which may include detergent-solubilized membrane proteins [188], and nanoparticles.

Frequently, such centrifugal determination of $\bar{v}_{SP}$ can provide more detailed information about the composition of the sedimenting particle (e.g. [284]).

> A model for the $c(s)$ distribution of particles with known shape but unknown $\bar{v}$ is implemented in SEDFIT. It exploits the relationships between the diffusion and sedimentation coefficients and the frictional ratio and translates the information of the diffusion coefficient extracted from the sedimentation boundary shapes into best-fit partial-specific volume values.

3.3.1.2 Density Contrast by Sedimentation Equilibrium

The method of density contrast, based on a variation of the solvent density ρ, is a powerful approach for an independent determination of M and $\bar{v}$. It was first described in detail by Edelstein and Schachman for the characterization of proteins using SE [266, 285, 286]. A plot of the experimental buoyant molar mass measured in SE as a function of solvent density should yield a straight line where the ordinate intercept (i.e., extrapolation to $\rho = 0$) gives the protein molar mass, and the abscissa intercept (i.e., extrapolation to $M_b = 0$) determines the point of neutral buoyancy where $\bar{v} = \rho^{-1}$ (Fig. 3.3).

---

[13]It is highly advantageous to study self-association processes by SV alongside SE [161, 282, 283]. In the case of SV, there exists an analogous correlation between the end-points of the isotherm of s_w-values. However, the s_w-values do not need to be buffer corrected to standard conditions for the isotherm analysis, and their analysis is therefore independent of the protein $\bar{v}$. Thus, the global analysis of SE and SV can help to elucidate the protein $\bar{v}$ (or ϕ', respectively).

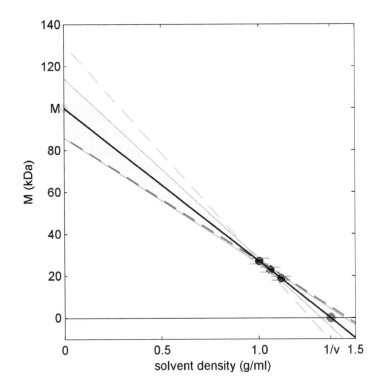

Figure 3.3 Schematics of a density contrast SE experiment and analysis. The black circles represent the measured buoyant molar mass of a 100 kDa protein with partial-specific volume $\bar{v}$ of 0.73 mL/g, in buffers at densities of 1.000 g/ml, 1.056 g/ml, and 1.111 g/ml, corresponding to aqueous buffers with approximately 100% H_2O, 50% H_2O/50% $H_2{}^{18}O$, and 100% $H_2{}^{18}O$. The solid black line depicts the best-fit and extrapolation required to determine the molar mass M from the intercept with the ordinate axis (cyan circle), and the inverse partial-specific volume $1/\bar{v}$ from the intercept with the abscissa (red circle). Error bars indicate a typical 5% error in the statistical precision of the measurement of M_b. They propagate to uncertainties in the values for M and $1/\bar{v}$ indicated by the gray area. If the experiment were performed with D_2O instead of $H_2{}^{18}O$, then H/D exchange would increase the measured buoyant molar mass by a factor k of ~1.0155 (for most proteins) in 100% D_2O [266]. Unless accounted for, this will lead to the regression and extrapolation shown as a red dashed line; introducing errors in the analysis. If, on the other hand, the same solvent densities were achieved with a co-solute such as sucrose, preferential hydration of 0.3 g/g water bound to the protein will diminish the M_b values measured in the experiment, and the resulting extrapolation will follow the dashed green line.

An important question is which molar mass and partial-specific volume are obtained. As explained below, the answer will depend on the nature of the density contrast agent and how the different densities are achieved.[14]

In the original Edelstein–Schachman approach buffers made with different mixtures of H_2O and D_2O or $D_2\,{}^{18}O$ were used [266, 285, 286]. A table of density and viscosity values for various water isotopes is found in the Appendix B. Since D_2O is expected to partition into the hydration shell at the same proportions as in the bulk, the hydration shell remains neutrally buoyant; we will consequently measure the molar mass and partial-specific volume of the macromolecule, namely M_a and $\bar{v}_a$ (or the apparent partial-specific volume ϕ' if co-solutes are present [266]). However, a complication arises when the exchangeable hydrogen atoms on the protein surface are replaced with deuterium, as this leads to a small increase in the molar mass as well as a decrease of the partial-specific volume dependent on the $D/(D+H)$ molar ratio, r_{DH} [266]. Based on data for many proteins, an average increase in mass due to H-D exchange of $k=1.0155$ was reported (mainly one replaceable amide hydrogen in the polypeptide backbone per amino acid residue) [266]. For mixtures with different H_2O/D_2O ratio, the value reduces proportionally to $r_{DH} \times k$, leading to expressions for the measured buoyant molar mass of $M_b = M_a(r_{DH} \times k)\left(1 - \bar{v}_a\,(r_{DH} \times k)^{-1}\,\rho\right)$ (i.e., the mass increases proportionally with $r_{DH} \times k$, but not the volume), and the extrapolation shown in the red dashed line in Fig. 3.3. Even though the value of k is small, its effect is amplified by the long extrapolation to zero density; consequently this correction needs to be considered.

Fortunately, the need to account for H/D exchange can be avoided by the use of heavy oxygen water $H_2\,{}^{18}O$ [106, 267–269], which has a density very similar to D_2O, and has now become more readily available. This is especially true considering that this application poses no requirements for extreme isotopic purity, mainly just the need to know the heavy isotope fraction. Unfortunately, the same is not the case for other heavy water isotopes, such as $D_2\,{}^{18}O$, which cause H/D exchange but would be highly desirable to improve the experimental solvent density range.

In contrast, if the same solvent is used throughout (e.g., H_2O) and the density is raised with different co-solute concentrations, one may assume that the co-solute

---

[14]In addition, higher-order density variation approaches are possible. For example, the Edelstein–Schachman method was extended to determine the partial-specific volume and molar mass of a pigment-containing detergent-solubilized intrinsic membrane protein [287]. The difficulty in such systems is that the pigment contributes to the protein buoyant molar mass, in addition to the solubilizing detergent, such that the oligomeric state of the protein is not trivial to predict. In this work, the protein was brought into a series of different detergents, chosen to have different partial-specific volumes. Measurement of the buoyant molar masses of the protein/detergent complexes in densities providing neutral buoyancy for the respective detergent (or extrapolation to this density) provided data for the relationship of $M_b(\rho)$ of the protein/pigment only, from which the protein partial-specific volume and molar mass could be obtained simultaneously. In conjunction with the theoretical mass of the conjugated peptide, this enabled the determination of the protein oligomeric state.

is excluded from the hydration shell of the macromolecule.[15] For example, a density of 1.10 g/mL can be achieved by addition of either 2.3 M KCl, 2.5 M NaCl, 1.4 M $(NH_4)_2SO_4$, 25% (w/v) sucrose, or 40% (w/v) glycerol [269]. If the invariant particle model [106] applies and $M_b(\rho)$ is linear with density, then the properties observed are those of the complete sedimenting particle (M_{SP} and $\bar{v}_{SP}$), including hydration (green dashed line in Fig. 3.3). In fact, such an approach has been used to obtain a thermodynamic measure of protein hydration [24, 120]. However, if interpreted as purely the macromolecular mass M_a and $\bar{v}_a$, then both values will be overestimated [270]. Nevertheless, after accounting for macromolecular hydration, as shown by Lustig and colleagues [271], an advantage of the co-solute approach is the higher range of densities that can be reached (at least when compared to H_2O/D_2O density contrast), which can be particularly attractive when working with membrane proteins in detergents of high density [270, 271] (although protein contrast will also be diminished).

Whichever method is applied, it becomes apparent from the long extrapolation illustrated in Fig. 3.3 that the density range that can be covered experimentally is critical for the precision of this approach [266]. This problem is exacerbated when studying particles of higher density, such as metal nanoparticles.

A practical problem that sometimes arises relates to the transfer of an aqueous stock sample at moderate concentration to a higher density solvent, especially if the sample cannot be easily dialyzed without risking unacceptable losses. Three important considerations are (1) that the sample ultimately be studied in a well-defined solvent;[16] (2) that the sample will not be diluted to an extent such that large errors in M_b occur due to a poor signal-to-noise ratio; and (3) that the required final accuracy in the parameters of interest can be achieved (for example, this would be different depending on whether one wants to distinguish a trimer from a dimer, or a dodecamer from an undecamer). In some cases, if the stock sample is not sufficiently concentrated, modest volumetric dilution of the sample may be the only option for buffer exchange, using a solvent of the highest possible density as diluent. Considering all factors, this may include co-solutes and/or heavy water [288].

For the ultracentrifugal approach, it is important to achieve the highest possible precision for the values of M_b because of the large error amplification [266]. As described by Edelstein and Schachman, under ideal circumstances, a precision of $\bar{v}$ of ± 0.003 mL/g can be achieved [286]. Statistically, slightly better estimates can be achieved in the modern implementation of a direct global fit to all SE traces at different densities and rotor speeds [289], including the traditional low-speed and high-speed conditions. Similarly, the statistics of the linear regression can be improved by including more data points from more solvent densities. However, it should be noted that this does not eliminate the fundamental problems of error

---

[15]This assumption will not necessarily hold when using salts to raise the density, especially with polyelectrolytes, due to preferential salt binding.

[16]This is especially important if densimetry cannot be performed on the entire sample.

amplification caused by experimental solvent densities being far from the density of particles under study.

In an implementation of the Edelstein–Schachman method, SEDPHAT allows for the global analysis of SE and/or SV experiments in solvents with different densities. Importantly, this includes the consideration of H/D exchange in the different experimental buffers containing a specified H/D ratio, with a user-provided proton exchange coefficient k (defaulting to the value of 1.0155).

Sample purity is of critical importance in the application of this method. This is due to the fact that species of different buoyant molar mass will be present in different proportions within the radial window of observation (which does not encompass the entire solution column). The different species' contributions to the recorded signal will depend on the steepness of the Boltzmann distribution, which, in turn, will change with buffer density. Thus, larger molecules that may be selectively partitioned into the 'shadow' of observation close to the bottom of the solution column at low solvent density, will present a more shallow exponential at higher solvent density and thereby contribute more to the buoyant molar mass. Hence, in the presence of macromolecular heterogeneity, rather than representing a density-invariant average over all macromolecular components, the M_b value will be biased toward higher values at higher density. These errors will be magnified in the extrapolation and lead to under-estimates of both the average M_a and the average $\bar{v}_a$. Density contrast experiments are therefore ill-suited for the study of intrinsically mono-disperse samples in preparations that are not highly pure, as well as macromolecules that are polydisperse, such as synthetic polymers.

For similar reasons, the application of the density contrast SE method by Edelstein and Schachman is not possible under conditions where reversible self-associations are present [266, 269, 286]. In this case, a very complex situation arises due to the re-distribution of monomeric and oligomeric species in the different concentration gradients arising in solvents of different density. If unrecognized, this can lead to a qualitative error in $\bar{v}$ [269].[17]

Therefore, a method that does not rest on the analysis of concentration gradients should conceptually be more robust for density contrast analysis. This is possible in SV.

---

[17]To illustrate the importance of this point, let us consider a 50 kDa, $\bar{v} = 0.73$ mL/g protein in a monomer-dimer equilibrium studied at a loading concentration of $3 \times K_D$, where the monomer and dimer are equally populated. In a 4 mm solution column spanning 6.8 cm to 7.2 cm, useable data are usually obtained from 6.81 cm to 7.13 cm. At 15,000 rpm, the apparent 'cell-average' buoyant molar mass would be 20,581 Da in H_2O and 15,574 Da in $H_2{}^{18}O$, implying an apparent $\bar{v}$ of 0.685 mL/g using the density contrast analysis. This error originates from a shift in the monomer-dimer ratio across the visible cell corresponding to an increase in average molar mass of 6,400 Da. If a 5% error in protein concentration is allowed for (assuming a higher concentration in the higher density buffer), the error in the apparent $\bar{v}$ is slightly worse, with a value of 0.681 mL/g [269].

3.3.1.3 Density Contrast by Sedimentation Velocity

Whereas the Edelstein–Schachman approach relies on the measurement of the buoyant molar mass in SE, a time-honored density variation approach is based on the change in sedimentation coefficient at different solution densities [21, 290–292]. It is based on the fact that the experimentally measured s-value depends on solvent density, viscosity, and particle partial-specific volume as described in Eq. (1.5). Therefore conducting SV experiments on the identical sample in buffers of different densities allows for a determination of the particle density. In fact, if the experimentally measured s-values are just viscosity corrected, a functional form equivalent to the dependence of the buoyant molar mass on solution density appears, and a plot of $(s_{xp} \times \eta_{xp})(\rho)$ will look virtually identical to that of $M_b(\rho)$ in Fig. 3.3. The advent of diffusion deconvoluted sedimentation coefficient distributions $c(s)$ has provided the s-resolution that is necessary for the successful implementation of SV density contrast analysis [269, 293]. This is further leveraged in the framework of global modeling of a single $c(s)$ distribution simultaneously to SV data from multiple experimental conditions [269]. An example of a density contrast SV experiment is shown in Fig. 3.4.

SEDPHAT has a flexible model to carry out analyses of density contrast SV experiments termed Hybrid Global Continuous Distribution and Global Discrete Species, where a single $c(s)$ distribution can be fitted to sedimentation boundaries from multiple experiments of different buffer density (allowing for a scaling factor to accommodate concentration errors). Different globally constrained $\bar{v}$ values can be applied to species within certain s-ranges.

It is clear that density contrast SV comes at the expense of more material than SE, and it has the downside of also needing reliable solvent viscosity information in addition to the density. Also, the method assumes that the different solvent conditions do not influence the macromolecular frictional coefficients (and, like in SE, the oligomeric states).

Nonetheless, density contrast SV offers several important advantages. As the number of data points is an order of magnitude greater than in an SE experiment, and since boundary positions are easier to measure than exponents of the Boltzmann distributions, s-values can be determined much more precisely than M-values. Typically, the precision of s-values from samples run side-by-side in the same experiment is between 0.1–1%, dependent on the signal-to-noise of the data, whereas errors in M_b from SE are typically 5%. Furthermore, the high hydrodynamic resolution afforded by the diffusion-deconvoluted sedimentation coefficient distributions $c(s)$, allows for a much greater tolerance in sample impurity or intrinsic sample heterogeneity. Based on the framework of sedimentation coefficient distributions, it is quite straightforward to restrict the density contrast analysis to the species of interest sedimenting within a defined range of sedimentation coefficients, in units

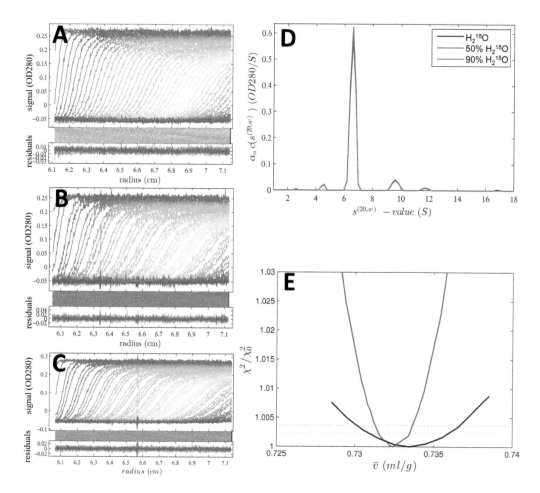

Figure 3.4 Density contrast SV for an IgG sample in solvents containing 100% H_2O, 50% $H_2{}^{18}O$, and 90% $H_2{}^{18}O$ (*Panels A–C*) [269]. Absorbance profiles at 280 nm were recorded side-by-side in the same run at 50,000 rpm. The sedimentation coefficient distribution $c(s_{20,w})$, determined in the global analysis on the basis of given solvent densities and viscosities, reveals a major species containing some trace aggregates and breakdown products (*Panel D*). The partial-specific volume is a global parameter, iteratively adjusted so as to yield the minimum global χ^2 of the fit. The projection of the error surface at different $\bar{v}$-values is shown in *Panel E* for IgG (black), and BSA (blue) in a similar density contrast SV analysis [269], yielding values highly consistent with results from dry weight measurement and densimetry [294]. BSA data (not shown) were collected at a higher signal-to-noise leading to narrower confidence intervals. The two gray lines highlight the 68% and 95% confidence levels.

of $s_{20,w}$, while excluding all other species (these can be assigned a different $\bar{v}$). This allows for the study of imperfectly pure samples and polydisperse macromolecules, without inflating errors in the resulting $\bar{v}$.[18] If density contrast SV is carried out using direct boundary modeling, globally fit to all data sets, a dual source of information can be exploited, since the modeling of the diffusional boundary spread produces information on the ratio s/D, which is proportional to M_b. This provides the potential for a higher precision in the resulting $\bar{v}$ estimate.

Unavoidably, the precision of the approach will still be limited by the available density contrast: If we denote the ratio of buoyant molar masses, or viscosity-corrected sedimentation coefficients, as R^*, with:

$$R^* = \frac{M_{b,2}}{M_{b,1}} = \frac{s_2 \eta_2}{s_1 \eta_1} \tag{3.4}$$

then the relative error in R^* will be amplified to a relative error in $\bar{v}$ by [269]:[19]

$$\frac{\sigma_{\bar{v}}}{\bar{v}} = \frac{\sigma_{R^*}}{R^*} \times \frac{R^*(\rho_1 - \rho_2)}{(R^* - 1)(R^*\rho_1 - \rho_2)} \tag{3.5}$$

For example, for a protein with $\bar{v}$ of 0.73 mL/g in an aqueous solvent with $\rho_1 = 1.006$ g/mL and in a buffered solution of 90% $H_2{}^{18}O$ with $\rho_2 = 1.103$ g/mL, the amplification of errors in R^* is by a factor of 0.73. Therefore, in order to achieve a precision of $\bar{v}$ of ~1%, one would need to determine R^* with a precision of ~1.4% or better, which may be easily achieved experimentally [295–297]. By contrast, metal particles with a of $\bar{v}$ of 0.15 mL/g would, in the same buffers have an R^*-value of 0.98, and an amplification factor of ~42 [269], which is unacceptably high.

The SV approach to density contrast naturally permits the study of reversibly self-associating molecules, if identical macromolecular concentrations are loaded in all samples such that their association state remains the same.[20] Even though concentration gradients do appear in the region of the sedimentation boundary, which could potentially have different shapes and different local chemical equilibria in solvents of different density and viscosity, the relevant concentration for the overall

---

[18]In addition, the danger of low-level protein degradation during the long time of SE biasing M_b and the resulting $\bar{v}$ is eliminated in SV.

[19]If a factor k for H-D exchange is introduced, a slightly different expression is obtained:

$$\frac{\sigma_{\bar{v}}}{\bar{v}} = \frac{\sigma_{R^*}}{R^*} \times \frac{R^*(\rho_H - \rho_D)}{(R^* - k)(R^*\rho_H - \rho_D)}$$

as described in [269].

[20]It is interesting to examine again the susceptibility to concentration errors. Let us assume again the same monomer-dimer system described above, with s-values of 3.5 S and 5.5 S for monomer and dimer, studied at a concentration of $3 \times K_D$, where the dependence of the relative populations of monomer and dimer is steepest. In pure H_2O we would observe a weighted-average s_w-value of 4.833 S and in $D_2{}^{18}O$ it would be 3.177 S. If the latter sample was 5% more concentrated, the new chemical equilibrium would shift the s_w-value to 3.185 S. This propagates to an error of only 0.0012 mL/g in the density contrast analysis of $\bar{v}$, which is within the uncertainty of the method.

transport in the cell is the average concentration in the plateau region, which alone will determine the weighted-average sedimentation coefficient s_w (compare Section 2.2.1 and Fig. 2.4) [154]. In fact, it is not necessary for the association scheme to be known if samples contain equal macromolecular loading concentrations. This is, of course, with the caveat that the variation of the solvent itself does not cause changes in the self-association state, which should be considered when using either D_2O or other co-solutes. $H_2\,^{18}O$ appears to be the safest choice in this regard.

Finally, density contrast SV is the natural choice for large particles [298,299] that cannot be studied by SE due to the excessive time required to attain equilibrium in solution columns of reasonable length, and lower limits of the rotor speeds that can be applied.

In the global modeling framework of SEDPHAT, the density contrast SV analysis may also naturally be combined with SE data, to exploit the virtues of both methods, to further enhance the precision, and to provide an internal check for consistency of solvent viscosity-dependent and viscosity-independent approaches.

> Density contrast modeling in SEDPHAT provides for the global direct boundary analysis of SV data acquired at different densities (with/without deuterium exchange dependent on the solvent D/H ratio), in the framework of a sedimentation coefficient distribution. It can naturally be combined and extended to global SE/SV analysis with direct modeling of all data.

3.3.1.4 The Neutral Buoyancy Method

For particles that exhibit densities $\bar{v}^{-1}$ within the range of experimentally accessible solvent densities, a straightforward approach to determine $\bar{v}^{-1}$ is a search for the matching solvent density in which the sedimenting particle is neutrally buoyant, ρ_n. It follows that $\bar{v} = \rho_n^{-1}$. It should be noted that the partial-specific volume provided by this method is naturally that of the complete sedimenting particle, $\bar{v}_{SP}$ for the conditions in which the experiment is carried out. However, as we have previously seen in Section 2.1.3, the apparent partial-specific volume ϕ' is identical to $\bar{v}_{SP}$ at the matching point. ϕ' will include a hydration contribution to the partial-specific volume if density matching is achieved through the use of co-solutes. On the other hand, if density matching is achieved using H_2O/D_2O, hydration contributions will disappear, but H-D exchange may take place (as noted above). Consequently, different matching densities may be observed depending on the contrast agent. This distinction will need to be consistently maintained if the partial-specific volume determined is to be used in further experiments.[21]

---

[21] A well-documented example is that of detergent micelles that show a lower matching density in the presence of sucrose or glycerol when compared to H_2O/D_2O mixtures [270]. This results from the effects of preferential hydration that has a lower density than the bulk sucrose or glycerol solution. D_2O partitions to the same extent as bulk solvent within this hydration shell.

One traditional way to determine the matching density is to generate a density gradient and observe macromolecular banding resulting from sedimentation in regions of lower density and flotation in regions of higher density — this is isopycnic density gradient centrifugation [300, 301]. There are only a limited number of co-solutes that generate a sufficient density gradient. An alternative approach is to use a solvent that approximates the condition of neutral buoyancy. The correct condition can be verified experimentally quite accurately by SV through the absence of either sedimentation or flotation.[22] Finding the density matching point through trial-and-error adjustments may be more tedious than in isopycnic sedimentation. Even though simple density matching does not provide the additional information on the sample's density, heterogeneity, and molar mass afforded by isopycnic sedimentation, it can be carried out for a greater variety of solvent and sedimentation conditions. Naturally, this method is independent of molar mass or sedimentation coefficient, and may be applied in either SV or SE mode.

The neutral buoyancy approach can be particularly valuable for systems in which it is challenging to determine accurate values of M_b or s; this may include, for example, highly heterogeneous systems. Similarly, the method can be useful when chemically dissimilar solvents have to be used for density contrast [302, 303], and/or in cases which may render the linear extrapolation questionable.

Finally, conditions of neutral buoyancy can be usefully applied in the study of multi-component systems, such as the protein/detergent systems. Matching the solvent density to one of the components (e.g., the detergent), results in neutral buoyancy for this component — accordingly this will not contribute to the buoyant molar mass of the sedimenting complexes. In this context, the neutral buoyancy method serves as a control experiment to establish correct density matching conditions.

3.3.2 Densimetry

Densimetry provides a direct measurement of the density of a solution. Density measurements at different macromolecular concentrations allows for a determination of the macromolecular density increment and partial-specific volume in a thermodynamically well-defined manner. This comes at the price of requiring an independent measurement of the weight-concentration of the sample, the need for dialysis, and considerable quantities of material. However, in different settings, densimetry is quite versatile.

---

[22]Strictly, unless the neutral buoyancy is performed with uniform simple solvents, any solvent mixture or co-solute will necessarily generate a small composition gradient. Even the use of a pure solvent will generate a density gradient in the centrifugal field arising from the intrinsic solvent compressibility. This can lead to sedimentation close to the meniscus and flotation close to the bottom of the cell. A detailed presentation and analysis of such a case with the goal to measure the compressibility of polystyrene latex particles was reported by Cheng and Schachman [206]. Such effects, considered negligible in the present context, would ultimately limit the precision of this approach.

The instrument used exclusively in conjunction with AUC is a Kratky oscillator, in which a U-shaped glass tube is filled with the solution of interest, and the mass contained in the fixed volume is measured from the vibrational frequency of the tube, as detailed by Kratky, Leopold and Stabinger [279] and Elder [304]. The mass contained within the fixed volume is that of the solvent and dissolved macromolecule, and for the interpretation of the measured density in terms of macromolecular mass and volume, we need to account for the volume of solvent displaced by the macromolecule. With w denoting the weight concentration of the macromolecule, and ρ_0 the density of the solvent, the solution ρ density becomes:

$$\rho(w) = \rho_0 + w - w\bar{v}\rho_0 \tag{3.6}$$

The slope of this relationship provides the density increment:

$$\frac{d\rho}{dw} = (1 - \bar{v}\rho_0) \tag{3.7}$$

and the partial-specific volume [266]:

$$\bar{v} = \rho_0^{-1}\left(1 - \frac{d\rho}{dw}\right) \tag{3.8}$$

In addition to volume exclusion by the macromolecule, other factors contributing to the partial-specific volume will include any disturbance in the structure of the solvent around the macromolecule, such as electrostriction and preferential binding or exclusion of co-solutes.

Therefore, to obtain a well-defined partial-specific volume consistent with thermodynamic definition, all solution components other than the macromolecule must be held at constant chemical potential. This can be achieved by dialysis equilibrium, using the dialysate to dilute the macromolecular stock solution to different final concentrations for densimetry, as for AUC [305]. Establishing a well-defined buffer by dialysis equilibrium is particularly important when studying polyelectrolytes, as they contain counterions when weighed. Similarly, protein samples may contain significant amounts of buffer salts when resuspended from lyophilized powders. In cases where no slow solvation effects are expected, dialysis can be replaced by chromatographic buffer exchange *via* size-exclusion or affinity chromatography. Dilution of a stock of macromolecular solution in different buffers is not suitable and will lead to incorrect results. Also, when re-dissolving the sample from a crystalline state, it should be ensured that the samples are properly and uniformly rehydrated, since aggregates or micro-crystals may exhibit a different partial-specific volume. Samples that tend to coat the surface of the oscillator or those that exist in a mixture of different phases, as may be the case with some detergents, present special problems, which may be addressed by confining the analysis of slope in $\rho(w)$ to an appropriate concentration range. However, detergent solutions cannot be properly dialyzed, rendering densimetry results unreliable. Furthermore, some proteins may not be sufficiently stable during dialysis [306]. Another practical consideration when using buffers of high co-solute concentration is the need to carry out the dialysis in a closed system to prevent errors from the evaporation of water [307], and for the same reason, it may be advisable to retrieve the sample from the dialysis bag without removal from the bath [117]. Finally, accurate temperature adjustment of the sample in the densimeter is essential, due to significant dependence of aqueous density with temperature [111], and care must be taken that no air-bubbles are present in the sample. A detailed account of densimetry experiments with many useful hints can be found in [306].

In contrast to the AUC-based methods described above, densimetry requires a precise knowledge of the macromolecular concentration in units of mg/mL. To know the concentration in weight units is, in fact, key in densimetry: if the macromolecular concentration is known only in molar units, densimetry will only provide the buoyant molar mass, just like the AUC experiments![23] In this case, the only possible information gained from the combination of AUC and densimetry would be that of using M_b from AUC to provide an estimate for the molar concentration of the macromolecule,[24] but unfortunately this will not help in determining either the molar mass or partial-specific volume. Thus, measuring molar masses of microscopic particles by AUC really ultimately hinges on weighing the material on a macroscopic scale for densimetry.[25]

It is important to be clear about the definition of the particle that is weighed, or similarly, the weight-based extinction coefficient used for spectroscopic concentration measurements, for the purpose of the w concentration scale of densimetry. Usually, it cannot be the weight of the complete sedimenting particle, as hydration contributions are excluded from the dry weight and spectroscopic extinction contributions. For this reason, the $\bar{v}$ obtained from densimetry is not that of the complete sedimenting particle, $\bar{v}_{SP}$. Instead, it is the apparent partial-specific volume ϕ', as shown in Section 2.1.3, accounting implicitly for all solvation and preferential binding effects for the given solvent conditions Eq. (2.7).

For illustrative purposes, we consider polyelectrolytes. Because of electro-neutrality requirements, dry material will always contain counterions, which will be replaced in dialysis by buffer ions. As outlined in detail by Eisenberg [305], irrespective of this, the molar mass measured in AUC will remain that molar mass considered for the weight measurement, even if it is 'off' relative to another defini-

---

[23]With the macromolecule in molar concentration units c, Eq. (3.6) becomes $\rho(c) = \rho_0 + Mc - Mc\bar{v}\rho_0$, with the slope $d\rho/dc = M(1 - \bar{v}\rho_0)$.

[24]To obtain the concentration in molar units from density experiments, we can rearrange to $c = (\rho(c) - \rho_0)/M_b$.

[25]This is true even if we use theoretical-compositional approaches to estimating $\bar{v}$ (Section 3.3.3, below), as these rely on tabulated values that were determined previously by knowing the weight concentrations of the constituents on a macroscopic scale, or calculated themselves based on previous experimental data. In the case of macroscopic particles of known volume, the partial-specific volume is not required, but densimetry is still necessary for the solvent density. See Section 6.1.7 for a further discussion of this topic.

tion of the particle by an arbitrary factor x, since:[26]

$$M_b = M \frac{d\rho}{dw} = (xM) \frac{d\rho}{d(xw)} \tag{3.9}$$

For example, if the added weight to the solution considers only the weight of the co-valent macromolecule without counterions, then bound ions will contribute equally to the densimetry after dialysis equilibrium as to AUC experiments, and — as long as the same equilibrated buffers are used — cancel out, such that the measured buoyant molar mass from AUC jointly with the measured density increment will lead to the covalent macromolecular molar mass without counterions. This is very convenient, since it elegantly eliminates the need to know any of the hydration and solvation properties of the macromolecules, as long as equilibrium dialysis is carried out for both AUC and densimetry.

On the other hand, the fact that the resulting molar masses from AUC will be those used in the definition of the weight concentration for densimetry can also be a disadvantage. For example, if the weight concentration of a glycoprotein were to be measured spectrophotometrically using the extinction of the peptide bond, the sugar moieties would not be accounted for in either the weight concentration or in the final molar mass from AUC, and may therefore not be recognized. Likewise, if a molar extinction coefficient ε_{molar} is known (perhaps estimated from the known amino acid sequence) and then transformed into a weight-based extinction coefficient by $\varepsilon_w = \varepsilon_{molar} \times M'$ using the sequence molar mass M', then any errors in the assignment of M' will re-appear in the M-values resulting from the AUC analysis, which can mask errors in the assumption of M'. In any event, gross errors could be caught in a critical inspection of the $\bar{v}$ value arising from densimetry, and it is a good practice to compare the densimetry values with theoretical expectations (see below).[27]

To avoid concentration and densimetry errors due to contamination, it is advantageous to carry out a SV run with $c(s)$ analysis as a control. If the sample is pure, refractometric detection using the interference optics may be useful to determine the sample weight concentration.

In addition to questions related to the definition of the macromolecule, an important consideration in densimetry is the statistical error arising both in the

---

[26] As an extreme choice, one could pick $x = 1/M$, in which case the combination of densimetry and determination of M_b via AUC will again lead to the molar concentration, $c = (\rho(c) - \rho_0)/M_b$. In a reversal of the usual work-flow, that may help, for example, to determine a spectroscopic molar extinction coefficient ϵ_{molar}.

[27] A method has been described by Stothart [308] that applies density contrast in densimetry of macromolecular solutions to simultaneously determine ϕ' and an absolute concentration scale w. Similar to density contrast AUC, density contrast densimetry exploits different solvent densities of heavy water to generate an experimental variation of the density increment as an additional source of information, from which ϕ' can be determined. This approach has not found many applications in the literature, perhaps due to added complexities of H-D exchange when using H_2O/D_2O mixtures, which adds another unknown parameter in the analysis [308].

concentration measurement as well as the densimetry itself. Based on Eq. (3.9), we can see that concentration errors in densimetry will propagate proportionally into the resulting molar mass in AUC. If the concentration measurements are based on compositionally predicted UV extinction coefficients, errors are on average already 4% [82], corresponding to an error in $\bar{v}$ of 0.029 mL/g. Errors introduced into $\bar{v}$ due to the finite statistical precision of the densimetry can be minimized by using appropriate amounts of material. A simple error analysis can be used to determine the concentration range necessary, if we consider only one measurement at the highest macromolecular concentration [279]. In this case, $\bar{v} = \rho_0^{-1}(1 - (\rho_1 - \rho_0)/w_1)$, and the error in the partial-specific volume estimate is on the order $\Delta\bar{v} \approx (d\bar{v}/d\rho_1)\Delta\rho = -(\rho_0 w_1)^{-1}\Delta\rho$, neglecting all other error terms in comparison with the error of the densimetry, which we may assess as $\Delta\rho$ $\sim 3\times10^{-6}$ g/mL. With a target precision of $\Delta\bar{v}$ of 0.001 mL/g, one would require a highest weight concentration of $w_1 = \Delta\rho/(\rho_0\Delta\bar{v})$, or ~ 3 mg/mL. The volume required for densimetry is typically 1.5 mL, and as a consequence, at least ~ 5 mg of material would be required [115, 306, 309]. Although the material is recoverable, the required amount is prohibitive in most cases [115].

In summary, densimetry is an elegant method for determining the protein partial-specific volume, or density increment, but it is not routinely applied in practice due to the requirement for large amounts of material, the need to know the absolute weight concentration, and the requirement for dialysis.

3.3.3 Theoretical Predictions

The calculation of the partial-specific volume based on the partial-specific volume of the macromolecular components or atomic molar volumes is a highly developed field (e.g. [111, 281, 309–313]). In the case of unmodified proteins, values can be reliably predicted based on the amino acid composition with an accuracy of approximately 1%, depending on the method used [111]. Such accuracy is usually sufficient for the determination of the molar mass from the buoyant molar mass with an error of 5% or better. This error is small enough for the assignment of stable oligomeric states using AUC (except for the most difficult cases of oligomers containing ten or more protomers). The accuracy in $\bar{v}$ is more critical when studying weak self-associations using SE; here, additional error correlation and amplification occur. The accuracy of these theoretical predictions far surpasses that of the densitometric method when applied given the amounts of material typically available in practice (see above). Therefore, along with the determination of ϕ' by AUC itself via M_b and the usually known molar mass of the macromolecule (see above), use of the calculated partial-specific volume is the most common approach in the modern literature of protein studies.

What is determined from the volumetric prediction is largely $\bar{v}_a$, and therefore this approach is most useful for dilute solutions with a density close to that of water, such that significant preferential hydration or solvation effects are not observed. However, to the extent that the predictions combine experimentally derived values, some solvent interactions may be already accounted for [111,310,311]. Nevertheless, this approach should be used with caution, for instance in the case of highly charged macromolecules, such as nucleic acids, unless it is based on tabulated data valid for the specific buffer conditions to be used in the experiments. Similarly, preferential interactions with co-solutes require caution, and methods for approximating preferential hydration effects to arrive at a better estimate of the experimentally relevant ϕ' in the presence of osmolytes or salts have been developed [313,314]. More recently, molecular dynamics simulations have been employed to calculate the partial molar volume of small compounds and macromolecules, in a way that captures the effects of solvent rearrangement [315–317].

3.3.3.1 Using Traube's Rules for the Prediction of Partial Molar Volumes from Chemical Structures

Traube showed that molecular volumes can be calculated as an addition of elemental atomic volumes plus a 'molecular co-volume' that provides corrections for interactions [318]. The underlying atomic and molecular partial molar volumes are generally different from crystallographic or van der Waals volumes [319,320]. Høiland has compiled partial molar volumes of a large number of small chemical and biochemical compounds, as well as partial molar volumes for different chemical groups, which may be combined to calculate partial-specific volumes of a macromolecule of given structure [319]. Durchschlag and Zipper described a more direct *ab initio* approach to calculate partial molar volumes as:

$$\bar{V}_a = \sum V_i + V_{CV} - \sum V_{RF} - \sum V_{ES} \tag{3.10}$$

where V_i are the volume increments for atoms (or atomic groups), V_{CV} is the correction for co-volume (usually 12.4 mL/mol), V_{RF} the correction for ring formation, and V_{ES} the correction for electrostriction and ionization [320,321]. Additional terms occur with specific classes of compounds, accounting, for example, for charges of ions, or for micellization of surfactants [320]. Tables of atomic volumes and ions are provided by Durchschlag and Zipper, determined so as to provide best match with experimental partial volumes of different compounds, together with detailed step-by-step examples. These ab initio partial molar volumes may be combined with tabulated experimental values of known compounds. The accuracy of this approach is estimated to be ~2-3% compared to experimental values [321]. This has, for example, been applied to peptide-nucleic acid (PNA) moieties [283].

3.3.3.2 Prediction of Protein Partial-Specific Volumes from Amino Acid Composition

Amino acid residue partial-specific volumes $\bar{v}_i$ exhibit values ranging from 0.60 mL/g for aspartic acid to 0.90 mL/g for leucine, and the relative content of different amino acids will be the determinant for the protein $\bar{v}_a$. As shown in a genome-wide computational analysis, the contributions of different amino acids average out across the known human proteins to a Gaussian distribution with mean of

0.735 mL/g $\pm$ 0.010 mL/g [95], but with a significant fraction of proteins, especially small proteins and peptides, falling outside this range. A large compilation of experimental values for different biological macromolecules was assembled by Durchschlag [309]. The compositional prediction offers a convenient approach to estimate the protein $\bar{v}_a$ with sufficient accuracy for most AUC applications for proteins with known amino acid composition.

A concise overview of this topic was provided by Perkins [111]. In pioneering work on compositional prediction based on the principle of molar volume additivity, Cohn and Edsall have compiled partial volumes of all amino acid residues based either on molar group summation or experimental measurements, and showed that the protein partial-specific volume can be determined as a sum:

$$\bar{v} = \frac{\sum \bar{v}_i w_i}{\sum w_i} = \frac{\sum N_i V_i}{\sum N_i M_i} \tag{3.11}$$

with $\bar{v}_i$, N_i, M_i, and w_i denoting each residue partial-specific volume, copy number in the protein, molar mass, and weight fraction, respectively [311, 312, 322].[28] Alternative amino acid tables were later produced solely on the basis of densimetry by Zamyatnin [310], or on the basis of crystallographic volumes by Chothia and colleagues [323], and consensus values derived as average molar volumes from the different approaches were proposed by Perkins [111], which are close to the Cohn–Edsall values [311]. The latter are the most commonly used data, and tables of Perkins consensus values and Cohn–Edsall values are provided in Appendix A.

There is a SEDFIT function OPTIONS ▷ CALCULATOR ▷ calculate vbar, dn/dc and extinction280 from protein sequence) to predict partial-specific volume, molar and weight-based extinction coefficient, (along with the refractive index increment) for given polypeptides on the basis of Cohn–Edsall and Perkins consensus values for the volumes and standard or per residue temperature corrections, respectively.

It should be noted that the values are usually tabulated for a temperature of 25 °C. There is a small but significant linear temperature dependence of the partial-specific volume, below temperatures at which thermal transitions of heat denaturations occur [309]. The temperature dependence of $\bar{v}$ impacts the transformation of experimental s-values to $s_{20,w}$-values Eq. (1.5), as illustrated in Fig. 3.5. As reported by Durchschlag [309], the mean value of published temperature increments $\Delta\bar{v}/\Delta T$ is 4.5×10^{-4} (mL/g)/K, with most values ranging between 3.5–5×10^{-4} (mL/g)/K. Partial molar volumes of different amino acid residues as a function of temperature have been measured by Makhatadze et al. [324] and Hedwig and Hinz [325], and are listed in Appendix A.

---

[28]The calculation of the partial-specific volume should be made with a sequence that corresponds exactly to the macromolecule in solution, i.e., including tags, loops that might not be resolved and missing in a crystal structure, and include all modifications.

3.3.3.3 *Partial-Specific Volumes of Glycoproteins and Other Protein Conjugations*

The same principle of additivity in molar volumes can be applied to account for protein decoration with modifications, such as carbohydrates, PEG, or lipid chains. In this case, we have:

$$\bar{v}_a = \frac{\bar{v}_{pp}M_{pp} + \bar{v}_{np}M_{np}}{M_{pp} + M_{np}} \tag{3.12}$$

with the indices 'pp' and 'np' labeling 'polypeptide' and 'non-protein' values.

It can be particularly valuable to apply mass spectrometry for the determination of the total mass $M_a = M_{pp} + M_{np}$, so that the amount of non-peptide mass M_{np} can be calculated, knowing the amino acid contribution M_{pp} [326]. Alternatively, an optical approach can often be used based on the combination of UV absorbance and refractometric IF signals, as the carbohydrates do not usually absorb at 280 nm.[29] It should be noted that less quantitative approaches, such as SDS-PAGE or size-exclusion chromatography are not suitably accurate for this purpose. For monomeric proteins, the experimental buoyant molar mass can be used to determine the mass of non-protein contribution if the partial-specific volume of the latter is known or can be estimated [327].

In principle, the partial-specific volume for any non-peptide compound may be determined through chemical group summation using Traube's rules, or from tabulated values of compounds. Of particular interest are carbohydrates, for which tables of partial-specific volumes have been compiled by Durchschlag [309], and some examples for n-linked oligosaccharides are provided in [328]. A likely range of partial-specific volumes of carbohydrate moieties occurring in glycoproteins is 0.602–0.642 mL/g [329]. For cases where the non-protein weight fraction is small, using an average value, such as 0.61 mL/g as an estimate may yield adequate precision for the AUC analysis [281].

---

[29] In order to eliminate contributions from impurities and obtain the highest precision, this is best carried out with an SV experiment in which both IF and ABS optics are employed, followed by a $c(s)$ analysis. Integration over the same $c(s)$ peak in the absorbance data and in the interference data will provide equivalent loading signals of the main boundary. Assuming the non-protein part does not contribute to the UV signal at 280 nm, the total absorbance signal A will be

$$A = \varepsilon_{PP}^{(w)} \times c \times d \times M_{PP}$$

(where $\varepsilon_{PP}^{(w)}$ is the weight-based extinction coefficient of the polypeptide at 280 nm in units of OD(mg/mL)$^{-1}$, c molar concentration, d optical pathlength), and the total fringe signal F will be

$$F = \varepsilon_{F,PP}^{(w)} \times c \times d \times M_{PP} + \varepsilon_{F,NP}^{(w)} \times c \times d \times M_{NP}$$

(where $\varepsilon_{F,PP}^{(w)}$ and $\varepsilon_{F,NP}^{(w)}$ are the weight-based signal coefficients of the polypeptide and the non-protein part in units of fringes(mg/ml)$^{-1}$). The ratio of non-protein to polypeptide mass follows direction from the signal ratios as

$$M_{NP}/M_{PP} = \left(\varepsilon_{PP}^{(w)}/\varepsilon_{F,NP}^{(w)} \right) \times (F/A) - \left(\varepsilon_{F,PP}^{(w)}/\varepsilon_{F,NP}^{(w)} \right).$$

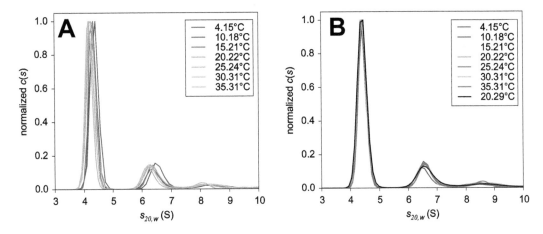

Figure 3.5 Comparison of $c(s)$ profiles from a SV run at 50,000 rpm of BSA in PBS at a range of temperatures. In *Panel A*, data are corrected only for the temperature dependence of solvent density and viscosity, in *Panel B* the temperature-dependence of the partial-specific volume of BSA is additionally accounted for *via* Eq. (1.5). Rotor temperatures were measured during the sedimentation run with a calibrated iButton® as described in Section 5.3.1.

3.3.3.4   Prediction of Solvent Effects on Protein Partial-Specific Volumes

The methods described above are applicable for dilute solutions in the absence of significant preferential solvation or hydration. For proteins, this excludes conditions of aqueous buffer with high co-solute concentrations, proteins with substantial net charge (like halophilic enzymes), and proteins in buffers with strongly binding co-solutes such as detergents and denaturants. Under these conditions, the apparent partial-specific volume ϕ' is significantly different from the macromolecular partial-specific volume $\bar{v}_a$.

The magnitude of preferential solvation will depend very much on the co-solute, as well as on the protein studied. Examples of co-solutes not interacting with the protein other than being excluded from the hydration shell have been examined in detail by Ebel et al. [120] by measuring the buoyant molar mass of aldolase in solutions with different sugars. It was found that the assumptions of the invariant particle model held, with the exception of glycerol and trehalose. The preferential hydration parameter B_1 increases slightly with increasing sugar size, with values ranging from 0.17 g/g for glucose to 0.29 g/g for α-cyclodextrin. More complicated, nonlinear behavior was observed for glycerol and trehalose [120]. From the magnitude of measured hydration, Ebel et al. concluded that the protein may not be covered simply by a uniform hydration layer, but instead form a surface-residue-dependent patchwork of hydration, consistent with the notion introduced by Kuntz of the residue-dependent non-freezable hydration [330] and with modern views of residue-dependent local hydration from current molecular dynamics simulations of thermodynamic hydration [107, 108, 132].[30]

This makes it non-trivial to precisely assign the degree of bound water in the case of non-interacting co-solutes. However, as a first approximation for the case of not too high co-solute concentrations (solution density not too different from water), and for 'non-interacting' and purely excluded co-solutes, where the invariant particle model can be assumed to apply (and in the absence of other information), it may be useful to substitute the $\bar{v}_a$ value with a hydration-corrected estimate of

$$\phi' = \bar{v}_a - \rho^{-1}B_1\left(1 - \bar{v}_1\rho\right) \tag{3.13}$$

which is a special case of Eq. (2.7) with B_1 spanning a range of 0.15 – 0.5 g/g [22–24]. As we have seen in Section 2.1.3, this value is solvent density dependent. A solvent density-independent expression can be achieved in the framework of the complete sedimenting particle, which would require counting the hydration towards the molar mass. A method for the more detailed prediction of hydration exploiting additivity principles and the experimental amino acid hydration data from Kuntz was presented by Timasheff [331].

For the case of co-solutes that specifically interact with proteins, the work of Timasheff describes a method for estimating ϕ' for a variety of co-solute conditions, including GuHCl [332], urea [331], salt and amino acids [313], and glycerol [119] (and references cited therein). For GuHCl and urea, it is based on the amino acid composition dependent preferential binding of both water and co-solute [331], whereas for salts, sugars, and amino acids they are based on the observation that preferential interactions were found to be proportional to the protein surface area [313, 314].

3.3.3.5 Prediction of the Partial-Specific Volume of Nucleic Acids

Different nucleotides and base pairs exhibit different $\bar{v}_a$, as compiled by Durchschlag [309, 320] and reproduced in Appendix A, with values significantly lower than those of most amino acids. The partial-specific volume of nucleic acids of known composition can be estimated using the same additivity principles.[31] However, the prediction of apparent partial-specific volumes ϕ' of nucleic acids requires special precautions due to the strong ion binding of these polyelectrolytes, and the dependence of ϕ' on the identity of the counter-ions [141]. This was particularly dramatic in the traditional analysis of molar masses from CsCl density gradients, or when studying the limiting ϕ' of NaDNA and CsDNA in water [141].

Experimental determination is possible, for example, via the measured buoyant molar mass in SE given the known molar mass, or via density contrast in solutions

---

[30]Such residue-dependent thermodynamic hydration should not be confused with hydrodynamic hydration operationally assigned to explain macromolecular hydrodynamic properties [131], which is very different in origin and may be very different in magnitude.

[31]This is the basis of the use of AUC as an efficient method to determine the base composition of genomes, complementing results from cloning and sequencing [333, 334].

of high salt concentrations [335]. Using the former approach, recently, Hatters and co-workers have systematically determined ϕ' of homo-oligonucleotides in 100 mM sodium phosphate, pH 7.4, 20 °C, and obtained values of 0.56, 0.58, 0.61, and 0.54 mL/g for dA, dC, dT, and dG, respectively [336]. These values are consistent with those determined by Bonifacio et al. [337] by SE for double-stranded short oligonucleotides of mixed base composition in 100 mM KCl, 20 mM sodium phosphate, pH 7.0, ranging from 0.538 to 0.578 mL/g with an average of 0.56 mL/g. This value is slightly lower than the average value recommended by Durchschlag in a solution of 200 mM NaCl [309]. Under the conditions of 100 mM KCl, 20 mM NaPi, pH 7.0 in the experiments of Bonifacio, RNA exhibited a ϕ' of 0.508 mL/g [337], significantly lower than that of DNA.[32]

In this context, it should be noted that hydration values for DNA are very different from those of proteins, with values of 0.8–1.0 g/g reported for short dsDNA [337].

---

[32]This value is distinctly lower than the value of 0.54 mL/g suggested by Durchschlag for RNA [309]. However, the lower value is consistent with an overestimate of the apparent molar mass of 40 bp dsRNA in similar salt conditions compared to the sequence mass, when assuming *ad hoc* a higher ϕ' of 0.55 mL/g [338].

TABLE 3.1   Typical Signal and Concentration Ranges for SE and SV Experiments with Different Optical Systems.

optical system	minimum signal/concentration			maximum signal/concentration	
	SV for $c(s)$, s_w	SV for detailed boundary modeling	SE 4-5 mm column, average concentration[5]	SV for $c(s)$, s_w or detailed boundary modeling	SE 4-5 mm column, average concentration[5]
absorbance[1]	0.005 OD [80]	0.1 OD; ~80 µg/mL @280nm[3]; ~10 µg/mL@230 nm[4]	0.03 OD[8]	1.3 OD[11]; ~2 mg/mL@250nm	1.0 OD[9]
interference[1]	~0.05 fringes[6]; ~15 µg/ml protein	0.1 fringes	0.05 fringes[8]	unlimited[10]	unlimited[10]
fluorescence	1 pM EGFP[2]		0.1 nM EGFP[2]	µM EGFP[7]	unknown[12]

These benchmarks assume that the instrument optical detection systems perform within the manufacturer's specifications, and assume average extinction or signal increment values when considering protein concentrations in mg/mL. [1]Assuming 12 mm centerpieces, values will be 4-fold higher with 3 mm centerpieces, and 1.7-fold lower with 20 mm centerpieces. [2]Data are for EGFP in PBS [99], values will depend on the fluorophore and solvent. Parameters assume a 10 mW laser power, high photomultiplier voltage, and typical focusing depth. [3]Assuming an average protein extinction coefficient at 280 nm of 1.0 OD/(mg/mL). [4]Assuming a 6–7 fold higher protein extinction at 230 nm when compared to 280 nm; this will depend on the ratio of aromatic amino acids. [5]This refers to the average concentration in the analyzable radial range. [6]Due to the sensitivity to buffer signal imperfections, it is harder to go to very low signal-to-noise levels. [7]Upper limit will depend on the fluorophore and magnitude of the inner filter effect. [8]At relatively high rotor speeds lower concentrations are possible with longer solution columns, at the cost of longer equilibration times. [9]Highest average signal at which an average buoyant molar mass can be determined; loading concentration may be higher depending on the rotor speed. [10]Unlimited, but gradients cannot exceed 70 fringes/mm (see Section 4.1.3). [11]Assuming a twofold lower protein extinction at 250 nm when compared to 280 nm (usually a factor $0.3 - 0.5$ is found for $\epsilon_{250}/\epsilon_{280}$ depending on the tryptophan to tyrosine ratio). [12]Virtually all fluorescence optical detection data in the literature utilize sedimentation velocity.

CHAPTER 4

Data Acquisition

T HE SUCCESSFUL accomplishment of an AUC experiment depends critically on properly acquired data and a detailed understanding of the optical detection systems. This chapter will discuss different features of the optical detection systems most commonly used in analytical ultracentrifugation. The optimal design of signal acquisition requires a balance between the experimental concentration requirements and the optical properties of the macromolecules, solvent and co-solutes. In addition, the signal-to-noise ratio and different degrees of freedom in the acquired data for each optical system need to be considered, in the context of the methods used for the computational data analysis and the ability to extract the desired macromolecular parameters from the experiment.

Historically, the goal of AUC data acquisition was to obtain a signal that is proportional to the local concentration at a specific point in space and time, such that theoretical models describing the evolution of the macromolecular concentration $c(r, t)$ can be simply applied to the measured data $a(r, t)$ through a proportionality constant such as the extinction coefficient. This is intuitively straightforward and allows for the direct application of the theory of macromolecular motion to data analysis. It must be recognized, however, that this goal is strictly never attainable for various reasons: the finite optical resolution will always result in a convolution of the signal across a radial range; a scanner with finite scan velocity will report on different regions in the cell at slightly different times; and unknown or at least poorly defined baselines and radial- or time-dependent baseline profiles may occur. These issues can be highly relevant for a detailed analysis [339]. In addition, due to stray light, and becuase absorbance signals are acquired with a finite wavelength bandwidth, they will at some level become nonlinear. Signals may also be subject to a time-varying or spatially varying signal magnification, but these are more commonly observed with fluorescence data [100, 340].

Fortunately, modern data analysis permits a more comprehensive consideration of optical signals, in which these effects can be computationally accounted for in a mathematical model with experimentally well-defined parameters. While the practically unattainable ideal of an optical signal solely reflecting the local concentration at a given time was essential in the era of pre-computer data analysis, it is not necessary anymore. We only require data acquisition that reports on macromolecular sedimentation in a well-defined and quantitatively well-understood manner.[1]

The following sections discuss specific aspects of sedimentation data and its relationship to macromolecular concentration, many of which are common to all optical systems. A comprehensive and detailed review of the design and calibration of different traditional optical systems has been written by Lloyd [72].

4.1 RADIAL DIMENSION

4.1.1 Radius Calibration

The goal of virtually all AUC detector configurations is to report on the radial concentration distribution across the cell as a function of time. For this purpose, the optical systems need to be calibrated in the radial direction.

4.1.1.1 Reference Positions in the Counterbalance of a Reference Cell

The standard counterbalances provided by Beckman–Coulter have openings with edges at well-defined distances from the center of rotation, designed for radial calibration of the absorbance and interference optical systems, as described in the manufacturer's instructions (Fig. 4.1 Panel A).[2] Detailed step-by-step screenshots and instructions for radial calibration of the interference optical system can be found in the protocols provided at `https://sedfitsedphat.nibib.nih.gov`. This calibration must take place with a spinning rotor, but should be done at a sufficiently low rotor speed (usually 3,000 rpm) so that centrifugal stretching of the rotor is negligible and does not lead to displacement of the reference marks.[3] Rotor stretching occurs as a function of rotor speed, as described in Section 5.2.3, but is not a problem, as long as we correctly measure the actual radial position from the center of rotation during the sedimentation experiment. Rotor stretching can be easily observed by imaging the counterbalance with the interference optics, which after radial calibration at low rotor speeds, will show a shift of the reference edge

---

[1]An example of non-traditional data acquisition is the fixed-radius signal in some custom-build detectors, which report on the time-course of sedimentation in a way that is suitable for sizing of big, non-diffusing particles [53, 341]. An adaptation of this detection can be implemented with the absorbance scanner repeatedly covering only a very short radial range (unpublished).

[2]For this to work properly, the mask insert for the counter balance that contains the reference holes must be oriented correctly (as described in the manual) since the reference edges are at different locations for the two sectors and the masks have different widths for the 'inner' and 'outer' reference holes.

[3]Furthermore, unavoidable systematic errors in the radial calibration may arise from the fact that the rotor is spinning around its center of gravity, which may not coincide with the geometric axis if imbalanced; however this error is likely negligible [72]. Further, there is little evidence for rotor precession in modern instruments, which would manifest itself as a slow periodic oscillation of the meniscus position [342] or baseline interference patterns [58].

image to higher radii at higher rotor speeds, by a magnitude of up to 0.02–0.03 cm (Section 5.2.3).

Figure 4.1 Radial calibration masks in the standard counterbalance (A), custom steel masks [296] (B), and a pattern deposited onto windows by lithography(unpublished)(C).

In principle, a radial calibration does not need to be carried out prior to each run, unless modifications or service on the optical system or centrifuge drive took place since the last calibration, because once the calibration is correct, it does not depend on what is imaged in the light-path. However, changes in the performance of the mechanical components with time warrant a regular radial calibration (Chapter 6).

The radial positions of the two reference positions in the counter balance may not be highly accurate and cannot be perfectly imaged, which can significantly limit the possible accuracy of the measured quantities. Let us assume that the radial calibration is skewed such that the innermost radius r_1 appears at $r_1' = r_1 + \delta r_1$, and the outermost radius r_2 appears at $r_2' = r_2 + \delta r_2$. We can describe this as a magnification error $\delta m = \delta r_2 - \delta r_1$ and a displacement error $\delta d = (\delta r_1 + \delta r_2)/2$. Interference experiments imaging several counterbalances side-by-side in the same rotor and comparing the apparent radial positions of their reference marks as they appear in the interference optics allow one to assess the maximally possible accuracy of the standard radial calibration.[4] In this way, the standard deviation δr was found to be 0.003–0.004 cm [339].

This has consequences for the accuracy of the measured s-values. The determination of the sedimentation coefficient essentially measures a linear velocity, $v = (r_2 - r_1)/\Delta t$ at a given centrifugal field, which has an average magnitude of $g = \omega^2(r_1 + r_2)/2$ in the cell. These will now appear to be $v' = v\,(1 + (\delta r_2 - \delta r_1)/(r_2 - r_1))$ and $g' = g\,(1 + (\delta r_2 + \delta r_1)/(r_1 + r_2))$. This re-

---

[4]Unfortunately the same experiment cannot be easily done using the absorbance optics. Radial calibration of the absorbance optics results in a relation between fixed distances on the counterbalance and voltages that drive the stepper motor on the absorbance scanner. The analog-to-digital (A/D) converter on the motor has a maximum of 4,096 values, and calibration radii need to be such 5.85 cm and 7.15 cm correspond to A/D values of ∼600 and ∼3,700, respectively. This allows the stepper motor to cover the entire radial range with a maximal A/D range and avoids issues with A/D values close to the limits of 0 and 4,096. Setup should be carried out by a Beckman Coulter service engineer.

sults in a relative error of $s'/s = (1 + \delta m/(r_2 - r_1)) \times (1 + 2\delta d/(r_1 + r_2))^{-1}$. Using the reference positions in the standard counterbalance located at 5.85 cm and 7.15 cm, and the measured errors in δr, the error in s'/s will have contributions of $\sim 0.39\%$ from the magnification error, and $\sim 0.04\%$ from the displacement error. In contrast, the variation of s-values in replicate samples run side-by-side in the same rotor with the same radial calibration can be as good as 0.1% [295, 296]. This suggests that the magnification error, which results in errors in the distance travelled, is far more important than the error in the absolute position, which results in centrifugal field errors. In SE, radial calibration errors play a much smaller role, as the magnification errors only translate to errors in M_b of $\sim 0.1\%$, and statistical errors in M_b from noise in the concentration signals are usually much larger.

A very useful test for the accuracy of the radial calibration with different optical systems is an inspection of the apparent meniscus position of a solution column at high speed in the same run (see below). Obviously, the meniscus position should be observed at the same radius, and any mismatch would indicate the presence of calibration errors. However, different optical systems may well exhibit different absolute displacement errors, and that this may not translate into significant errors in relevant macromolecular parameters such as the buoyant molar mass or sedimentation coefficient. This is due to the relatively small impact of displacement errors.[5] Conversely, distinct optical systems may project the meniscus to the same apparent radius, but each exhibit different magnification errors leading to significantly different sedimentation parameters. Therefore, the meniscus position alone, as any single point measurement, does not suffice as a criterion for correct calibration.

The issue of radial displacement and magnification errors was addressed in a large benchmark study including >129 data sets from 79 different instruments measuring the same reference sample cell assemblies [297]. The standard deviation of the sample meniscus position was 0.015 cm for the absorbance system and 0.037 cm for the interference system. These values may be taken as an estimate for the statistical variability of the displacement error, translating to errors in s-values by 0.23% and 0.57%, respectively. The magnification error, as determined from the scans of a periodic precision mask (see below), were found in the absorbance system to range from -6.5% to $+3.1\%$ with a mean of -0.43% and a standard deviation of 1.36%, and for the interference system with a mean of 1.85% and a standard deviation of 4.14%. As expected, no correlation between sample meniscus position error and ratios of s-values was observed, consistent with the notion that displacement and magnification errors are largely independent.

4.1.1.2 Imaging of Periodic Precision Masks

The use of external controls is highly useful to correct for errors in counterbalance masks, the automated calibration in the absorbance system, as well as errors in

---

[5]For this reason it is generally not advisable to constrain the meniscus position for the analysis of one optical system to the meniscus position measured in another optical system.

the manual radial calibration procedures made during the radial adjustment of the Rayleigh interference optics. A precision fabricated steel mask with a set of seven holes, each precisely 1 mm wide and 1 mm apart (Fig. 4.1 Panel B), has been designed for use as an external control to obtain an accurate measure of the radial magnification [296]. Sandwiched between two optical windows in the middle of the AUC cell assembly, it can be applied for the absorbance optical system (preferably in intensity mode), the Rayleigh interference system, and after coating with FITC, the fluorescence detection system. Ideally, the mask is imaged in a separate rotor hole alongside the samples of interest.[6] From the periodic pattern of signal/shadow with edges at well-defined intervals of 1.0011 mm (accounting for the angle of the radial scan in the sample sector relative to the mid-point of the cell) (Fig. 4.2 Panel B), a magnification correction factor can be calculated, independent of the instrument internal calibration (Section 6.1.3). Even though the mask does not measure an absolute displacement or address its error, it can help correct for the magnification error which, as we have seen above, is the aspect of radial calibration with the greatest impact on the accuracy of the s-values.

Software MARC (Mask Analysis for Radial Calibration) is available for the analysis of the edges of the scans. The analysis provides a correction that can be applied to the measured sedimentation coefficient [296, 297] (and unpublished).

More recently, an improved method was developed where a mask is deposited directly on the window by lithography (Fig. 4.1 Panel C), which can be done with better accuracy and higher precision than the fabrication of steel mask (unpublished). In addition to the higher density of edges, this mask also includes a second set of edges with the same constant pitch for the reference sector, symmetrically offset relative to that of the sample sector, increasing the number of data points for radial calibration (Fig. 4.2 Panel C). Furthermore, the joint analysis of the two patterns allows the user to correct for the rotational misalignment of the calibration window (unpublished). Ultimately, the highest accuracy will be achieved by direct image analysis of the fringe pattern in different illuminations through the lithographically patterned mask (Fig. 4.2 Panel D).

4.1.1.3 Linearity of Radial Position

An additional benefit of the precision patterned reference mask over the default two-point calibration is the opportunity to assess the linearity of the radius measurement across the cell [296]. If nonlinearity is detected, a polynomial regression of the measured apparent radial intervals of edges vs their true radial intervals allows a re-mapping of the radius data from the SV experiments collected in the same

---

[6]It was shown that the high centrifugal fields lead to a constant displacement, but not to stretching of the mask [296].

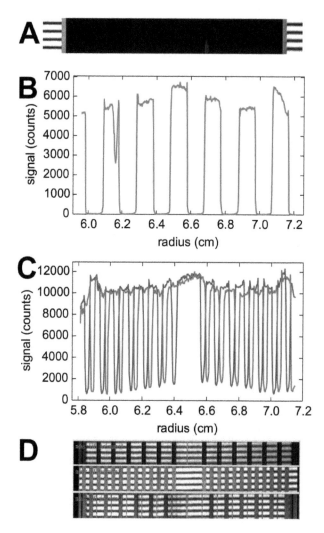

Figure 4.2 (*Panel A*) Fringe patterns measured with the Rayleigh interference optics for the standard counterbalance; (*Panel B*) Pattern of transmitted intensity with the custom steel masks [296]; (*Panel C*) The transmitted intensity patterns from the sample (blue) and reference sectors (red) obtained from the lithographically patterned window (as shown in Figure 4.1C) using the absorbance system, (*Panel D*) Composition of camera images of interference fringe patterns at different illumination delay times observed for the lithographically patterned window (unpublished) .

run to remove the nonlinearity [296]. In the multi-laboratory benchmark study a minority of instruments exhibited significant nonlinearity, and re-mapping the radial dimension typically led to slight improvement of the fit quality and the *s*-values [297].

The software MARC can rewrite experimental radial scan files taking into account a polynomial relationship between true and apparent radii measured by the patterned masks.

A second cause for nonlinear errors in the measured radial positions can be Wiener skewing of light traversing sample solutions with high refractive index gradients. Lensing effects in the gradient can cause a nearly complete loss of light at regions of the highest gradients (Fig. 3.2) and radial shifts or distortions at lower gradients. In this case, recorded radial positions of the solution column projected in the plane of the detector differ from the true radii [245]. The artifacts at highest gradients can readily be observed in the form of 'black bands' in SV sedimentation boundaries, whereas the more subtle effects from lower concentration gradients will impact SV and SE similarly. The latter are usually small in effect and negligible under most conditions[7], but will be exacerbated at very high macromolecular concentrations and long optical pathlengths, especially if the optics are not properly aligned or focused [246, 343].

4.1.2   Meniscus and Bottom

The meniscus and bottom of the solution column are notoriously difficult to image, principally because these interfaces create optical artifacts. At the same time, these regions present important landmarks in the sedimentation experiment, and play important roles in the data analysis, especially the meniscus in SV and the bottom in SE, which will usually be computationally refined.

In the case of the meniscus, it is important to recognize the two bounds for its position: the lowest radius safely identified as located in the airspace above the solution column, as well as the highest radius adjacent to the meniscus where light is entirely traversing the solution column. These should be identified and used as constraints for the computational optimization of the meniscus position in the data analysis. These bounds are also crucial in the error analysis. A similar approach for the cell bottom position is generally not feasible due to the opaqueness of the centerpiece, further obscured by the accumulation of material close to the bottom,

---

[7]Such effects result in errors of the order of a few percent with BSA at concentrations of 50 mg/ml in 3 mm pathlength cells [343].

and possible pelleting and surface film formation.[8] This leaves only a one-sided limit for observation, i.e., that of a minimum radius of the bottom position. Fortunately, computational approaches for the determination of the bottom from the analysis of back-diffusion in scans at different points in time or SE at different rotor speeds usually provide satisfactory results.

The location of the 'true' meniscus position cannot be determined [339].[9] At very low rotor speeds the meniscus artifact may have a finite 'width' due to contact (surface tension) with the windows [342, 349]. Furthermore, the appearance of the meniscus and bottom will depend very much on the optical detection system (Fig. 4.3), and at rotor speeds above a few thousand rpm, the image will be dominated by optical effects, such as the refraction of light at the interface [342], and the focal plane of the optics [350].

It seems, at first, that the bottom position should be experimentally better defined since it is not dependent on the height of the solution column and instead fixed by the rotor and centerpiece geometry. However, in practice this assumption does not hold true. Different molds for centerpieces introduce variation in the bottom position, even though most should be expected at roughly 7.2 cm. Furthermore, pelleting of material can shift the effective position to lower radii and the high concentration gradients of accumulating particles can create optical artifacts at radii smaller than the bottom position. Finally, rotor stretching will shift the bottom position to slightly higher radii depending on the rotor speed.

---

[8]In order to obtain a better optical definition of the bottom location, a small amount of transparent and immiscible liquid of higher density may be added to raise the solution column [344]. Liquids used include silicone [344], Kel-F polymer oil [345], and perfluorotributylamine (FC-43) [346] (with the latter being the most common choice). Caution is advised due to concerns regarding possible degradation of macromolecules at the interface [93]. Raising the bottom of the solution column was historically of interest for improving selected analysis approaches in SE, for example those based on mass conservation of individual scans, Archibald analyses of the approach to equilibrium, or the study of very short solution columns. None of these methods is widely used anymore, and there is little benefit to raising the bottom position with current analysis methods. These artificial bottom positions also addressed a problem with old style wide window gaskets that obscured part of the solution column close to the bottom of the centerpiece; current gaskets have an improved design that does not obscure any part of the solution column. Immiscible liquids are very rarely added in current AUC practice, as the packing of proteins against the interface can cause sample heterogeneity [347]. In a study of the interaction of IgG1-Fc and sCD16 [327], sedimentation equilibrium experiments were carried out multiple times both in the presence and absence of added FC-43. Use of the FC-43 led to the reproducible loss of material and ΔG values that did not show a parabolic dependence with temperature, but rather an undefined scatter.

[9]This is true despite the fact that some AUC data analysis software is equipped with algorithms for automated meniscus recognition. This should be regarded as a convenient method for obtaining an initial estimate of the meniscus position, which will have to be optimized and/or adjusted to explore error propagation at limits of the optical artifacts. Automated meniscus recognition algorithms have no scientific basis beyond rough localization of the artifact region (even though they may procedurally be very well defined, for example, by recognizing the maximum of a spike or similar feature) [348].

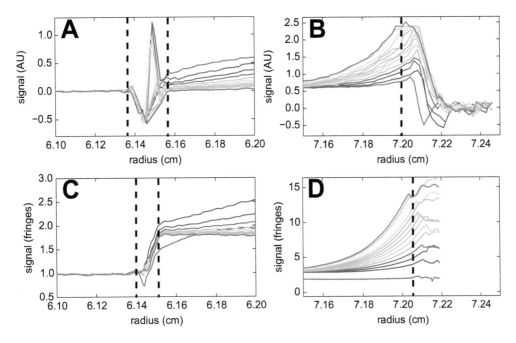

Figure 4.3 Meniscus (*Panels A and C*) and bottom (*Panels B and D*) artifacts in absorbance (*A and B*) and Rayleigh interference (*C and D*). Data are from the same cell assembly acquired during the same run. Upper and lower bounds of possible meniscus and lower bound of the bottom position are indicated as black dotted vertical lines.

4.1.2.1 Absorbance

The dominant feature of the meniscus in absorbance optics is caused by refraction at the interface, which results in a dip of the transmitted intensity, leading to an artificial spike in the absorbance trace with a positive orientation for the sample meniscus and a negative orientation for the reference meniscus (Fig. 4.3 Panel A). The meniscus artifact sometimes exhibits a more complicated fine structure (e.g., see the meniscus artifacts in the intensity profiles of sample and reference sector shown in Fig. 4.4).

It was customary, particularly when using the absorbance optics, to load a slightly larger volume into the reference sector, in order to clearly identify the sample meniscus and use its apparent radial position as a fixed parameter in data analysis. However, when using the interference optics, this would result in a signal mismatch due to differential distribution of small co-solutes (e.g., buffers and salts), which is undesirable (see Section 4.3.4). The advantage of sample meniscus visualization, with the exact position now a fitted parameter in modern data analysis, is significantly outweighed by the advantages of flexibility in the optical systems used and the ability to include detection by interferometry to better observe unexpected features. It is therefore prudent to routinely aim for matching reference and sample volumes and matching menisci with combined optical features that can largely cancel out or result in a more complex combined appearance.

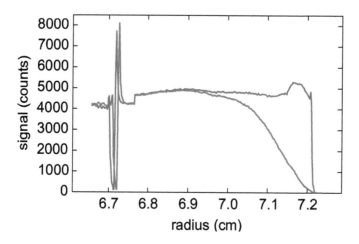

Figure 4.4 Radial intensity scan showing the transmitted light intensity in sample (red) and reference sectors (purple).

Limitations in the absorbance optical system with regard to signal linearity, signal gradients, and radial resolution do not allow for an accurate determination of the bottom position, especially in the steep gradient of accumulating material (Fig. 4.3 Panel B). However, it is possible to acquire data in 'intensity mode,' where the transmitted light intensity for the sample and reference sectors ($I_S(r)$ and $I_R(r)$, respectively) are reported separately. As shown in Fig. 4.4, this allows for a better definition of the lower limit in the bottom position from the transmitted light in the reference sector, unimpaired by accumulating macromolecules. The bottom positions in the sample and reference sectors can be taken as identical. The intensity data can later be transformed into absorbance values $A(r)$ using the transformation: $A(r) = log_{10}(I_R(r)/I_S(r))$.

4.1.2.2 Interference

The meniscus artifact generally appears sharper when using interference optics (Fig. 4.3 Panel C). The most obvious effect is often a lack of fringe count continuity at the meniscus, arising from very steep fringe shift gradients (in excess of 0.5 fringes per pixel) close to the interface and its associated optical aberrations. In the case of inadvertent volume mismatches between the sample and reference sectors (Fig. 4.5), the two menisci can be distinguished from the optically unmatched refractive index change associated with solvent and air compression. This leads to a positive signal gradient with the sample meniscus at lower radii than that of the reference meniscus and a negative signal gradient when the menisci are in the opposite orientation.

Due to the greater tolerance for concentration gradients, its higher radial resolution, and unlimited linearity, it is often possible to image considerably closer to the bottom position than in the absorbance optics (Fig. 4.3 Panel D). However,

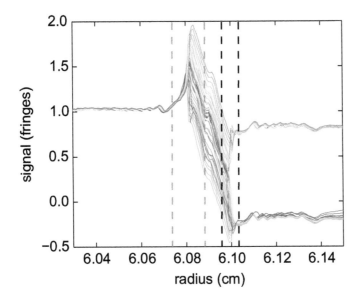

Figure 4.5 Close-up of the meniscus region with an unmatched solution column imaged using interference optics. The lack of continuity in fringe counts at the meniscus region results in different baselines in the solution column. This integral fringe shift can be accounted for easily in the data analysis. Visually estimated bounds for the reference (gray) and sample meniscus (black) are indicated as dotted lines.

due to the presence of reflections and shadow effects, it is doubtful that the bottom position can be imaged with great accuracy.

4.1.2.3 Fluorescence

For several reasons, the appearance of the meniscus and bottom positions is very different in fluorescence scans (Fig. 4.6) as compared to absorbance or interference data.

In the absence of any dissolved fluorophores, at low laser power and photomultiplier voltage there may not be any signal associated with the meniscus.[10] However, when the photomultiplier is set to very high sensitivity and the optics are well aligned, it is possible to see an increase in the baseline signal from air to water due to Raman scattering of water (using a 10 mW laser) [99]. While it may not be desirable to raise the photomultiplier voltage and sensitivity to high values in the course of the sedimentation experiment (dependent on the fluorophore concentration), this procedure can be carried out following the SV experiment, prior to rotor deceleration so the meniscus can be visually identified.

---

[10]To better define the meniscus, the use of an immiscible liquid of lower density than water has been proposed. This liquid, which floats on top of the solution column, is spiked with a fluorophore for detection [351]. However, the creation of an additional interface raises the same concerns raised for the addition of heavy liquids to better define the bottom of the solution column (see above).

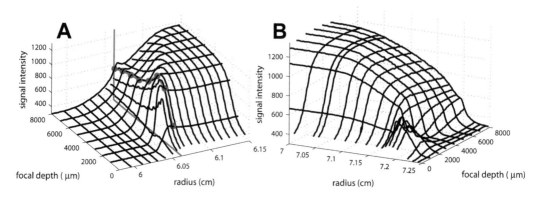

Figure 4.6 Meniscus (*A*) and bottom (*B*) region as imaged by fluorescence [100]. Due to the finite numerical aperture of the confocal path, the width of the artifacts will depend on the focal depth. For reference, the location of the computationally best-fit meniscus position is indicated (purple lines and blue dots, and projection on the axis in red, A). Data are recorded at very low centrifugal fields such that no sedimentation is observed; therefore these solely represent the optical properties of the system.

In the presence of dissolved fluorophores (i.e., when fluorescent macromolecules are loaded into the cell and scanning commences prior to their concentration depletion at the meniscus), the meniscus position is clearly visible through an increase in the signal at the air-liquid interface.[11] (Fig. 4.6 Panel A) [100]. Depending on the focal depth of the FDS, it may even be possible to record a small spike in the fluorescence signal, reminiscent of absorbance meniscus artifact. However, due to the lower optical radial resolution, the meniscus will not be a sharp transition, but rather a gradual increase, superficially resembling a convolution of an edge with a Gaussian beam.[12] The half-width of this transition will depend on the focal depth of the optics (increasing with increasing depth into the cell) (Fig. 4.6). This gradual transition may, at first glance, be mistaken for a sedimentation boundary, which may lead to further uncertainty of the meniscus position. This effect is particularly pronounced when using medium to large focal depths, as may be discerned from Fig. 4.6. Furthermore, when estimating lower and upper bounds of the true meniscus position, it is important to note that the tail of the optical convolution produces some fluorescence signal at radii outside the solution column close to the interface (e.g., regions in Fig. 4.6 Panel A at radii smaller than the red lines).

Similarly, the bottom position will appear unfamiliar to an experimenter accustomed only to the absorbance or interference systems (Fig. 4.6 Panel B). Prior to significant accumulation of fluorescent material, at the bottom of the solution column the signal drops continuously to a baseline value defined by the dark count

---

[11] In the current FDS scans proceed from the bottom of the cell to the inside, unlike the absorbance scanner, but this is irrelevant in the present context.

[12] Refraction effects of the confocal conical beams add more complicated physical effects than simple convolution.

of the photomultiplier and the fluorescence of the centerpiece itself. The limited optical resolution and finite angle of the excitation and detection cone lead to a gradual drop of signal near the cell bottom, an effect exacerbated at larger focal depths. However, the detailed shape of the artifact at the bottom is very different from that at the meniscus due to clipping of excitation and/or detection cone by the edge of the centerpiece. This does not happen at the meniscus, where refraction at the interface is expected. A model that accounts for the observed signal decrease is, to a first-order approximation, that of circular beam with radius δ_{beam} centered at r with a clipped and obscured segment close to the bottom b, leading to a remaining intensity given by the remaining area fraction:

$$B(b, \delta, r) = \frac{1}{\pi} \arccos\left(\frac{b-r}{\delta_{\text{beam}}}\right) + \frac{(b-r)}{2\pi\delta_{\text{beam}}} \sqrt{1 - (b-r)^2 \delta_{\text{beam}}^{-2}} \qquad (4.1)$$

for $b - r < \delta_{\text{beam}}$ [100]. Clearly the signals do not exhibit a discontinuous slope at $r = b - \delta_{\text{beam}}$ predicted by Eq. (4.1), due to optical convolution. Therefore, this model should be used in conjunction with a model for the radial convolution, for example that of a Gaussian beam with full width at half maximum σ:

$$s(r, t) = \frac{\int e^{-(r'-r)^2/\sigma^2} c(r', t) dr'}{\int e^{-(r'-r)^2/\sigma^2} dr'} \qquad (4.2)$$

where $s(r, t)$ denotes the measured signal, and $c(r, t)$ the local concentration in fluorescence signal units; σ may be taken from the optical resolution determined when carrying out the radial calibration. Combined, this often allows the user to model a portion of the data affected by the bottom artifact [100]. Unfortunately, inasmuch as the effective beam radius is usually not known, this does not lead to reliable estimates for the bottom position. Nevertheless, the time-course of signal accumulation in this range can add valuable information on the sedimentation process under study.

The distance of how much the optical artifact at the bottom extends into the solution column (corresponding to δ_{beam}) can be used as a simple graphical estimate of an effective cone diameter of the excitation and/or detection beams. Since their geometry is constant across the scan of the solution column, meniscus artifacts will start to arise at radii at and below a similar distance from the meniscus. Therefore, approximately the same distance should be left between the meniscus and the smaller fitting limit. This exclusion of data close to the meniscus is necessary due to a lack of an appropriate model for the more complicated refraction processes taking place at the meniscus, which often significantly influence the shape of the signal profile and usually cannot be properly modeled with sedimentation equations despite the smooth appearance with similarity to sedimentation boundaries (Fig. 4.6 Panel A). The loss of data can be minimized, however, with a choice of small focal depth, which leads to a small effective beam diameter and concomitant low emission signal [100] (see also Section 4.1.5 below).

4.1.2.4 Computational Determination

In most cases, when using modern SV data analysis, a computational approach to determine the meniscus is the method of choice. This relies solely on the progression of properly imaged sedimentation boundaries, which implicitly define their starting position at time zero, a value equivalent to the meniscus. Any fixed, graphical determination of the meniscus position (other than those of the lower and upper bounds) can significantly constrain the analysis, worsen the fit, and bias the result [339]. Usually the precision of the computationally defined meniscus in SV exceeds the optical resolution, and in a standard control experiment (as described in Section 6.2), the statistical error of the meniscus position was estimated to be ~ 0.002 cm or better, less than the nominal distance between data points acquired by the absorbance optics in the same experiments [297].

The fact that a zero-time boundary position coincides with a parameter that can be optically observed, though not precisely localized, is not just a convenient test for internal consistency in the analysis, it can also be used to diagnose experimental imperfections: if the calculated meniscus position is outside the graphically plausible zone, this can be taken as an indication of nonideality or, more commonly, convection. The best course of action is usually to take note of the experimental imperfection but leave the meniscus unconstrained [339] to obtain the least biased result.

The computational determination of the meniscus position is in stark contrast to all traditional, pre-computer approaches to SV analysis, which have had to rely on a graphical determination as a pre-condition for data analysis. There are rare cases of sedimentation velocity data with very broad boundaries, or extremely low signal-to-noise ratio, in which the meniscus position is not sufficiently well-defined by the data. In such cases, the optical bounds for the meniscus may be narrower than the statistical limits for the meniscus parameter, and the optically discerned constraints should be used. For these instances, two data analyses can be carried out by fixing the meniscus at the lower and upper optical limit, respectively, which will define the error propagation of the meniscus uncertainty on the parameters of interest.

> SEDFIT allows for a graphic determination of the lower and upper bounds to both the meniscus and bottom positions, within which the meniscus and bottom parameter may be adjusted during experimental data modeling. It is possible, and advisable in special cases, to fix either the meniscus or bottom position to pre-determined values.

The bottom position is usually of no interest in typical high-speed SV experiments and can be excluded from the data analysis. The only feature containing information on the bottom position is the back-diffusion region. If any back-diffusion is to be included in the analysis, then the bottom position should allowed to refine in the nonlinear regression of the data. The same holds true if significant signal is contributed by very small particles, which quickly assume an equilibrium distribution where the back-diffusion extends over essentially the whole solution column. If, on the other hand, the back-diffusion is from large particles, it will be compressed in a very narrow radial range (the last mm of the solution column), and exhibit a very steep signal increase at radii > 7.1 cm. Due to limitations in the performance of the optical detection systems in imaging steep concentration gradients, as well as nonideality effects setting in at the high local concentrations, this region should be excluded from the data analysis. In this case, the bottom position is irrelevant, and should simply be constrained to an expected value (e.g., 7.2 cm).[13]

As indicated above, a slightly different situation arises in the modeling of fluorescence SV data, wherein, dependent on the focal depth, optical artifacts can extend further into the solution column than back-diffusion. In this case, modeling of the shadow from the end of the solution column requires the bottom position to be computationally refined in the data analysis.

When determining the best-fit bottom position from SV data analysis (i.e., when including part of the back-diffusion region into the analysis), there is only a lower bound from optical detection to corroborate the computational result. Furthermore, dependent on the highest concentration in the immediate vicinity of the bottom, significant thermodynamic nonideality may be present, in addition to pelleting and formation of films or gels. Therefore, for large particles under conditions where they exhibit steep concentration gradients the computational bottom should be regarded as an 'effective' bottom position. Because the objective of the sedimentation experiment is not the determination of the bottom of the solution column, this seems to be of little relevance.

SE experiments, on the other hand, present a very different problem. The meniscus position is not very important, since concentrations there are comparatively low, and a graphical determination from inspection of the optical traces is sufficient. In fact, a computational optimization of the meniscus parameters from SE analysis models would typically yield very poor results. However, the bottom position is highly important for any analysis requiring an assessment of the total mass of sedimenting material. A model determining the bottom position from the invariance of the total mass in SE attained in the same sample at a range of rotor speeds usually yields excellent results [160, 166, 167].

---

[13]Since steep back-diffusion regions of large particles cannot be included in the data analysis, it is ill-conceived to spend significant computational effort in trying to solve the partial-differential equation for sedimentation in this region, as suggested by Cao and Demeler [352], at the expense of poor precision in modeling the sedimentation boundary [353]. Rather, for data not including back-diffusion, it is more practical to use a computational model for free sedimentation in a semi-infinite solution column [33] where neither the bottom position nor the back-diffusion region play any role.

4.1.3 Radial Resolution

The radial resolution has two components — the resolution of data points, which is largely dictated by the absorbance scanner movement (and analog to digital converter) or pixel density in the camera of the interference system; and the optical resolution, i.e., the radial region of light that will be sampled for any reported radial point. Both should be compared with the characteristic radial dimension of features in the concentration distribution, as well as their gradients.

The resolution of data points is usually higher than the optical resolution, and therefore not limiting.[14] In the absorbance system, a target value is entered by the user, typically 0.003 cm for SV experiments and 0.001 cm for SE [167], but the spacing and density of data points reported in the scan files will vary slightly. This is largely inconsequential for the data analysis.[15] Currently, data point resolution for the interference and FDS systems are approximately 0.0007 and 0.002 cm, respectively. Even though the difference in the data point density alone does not impact the sedimentation analysis beyond a statistical improvement, it is one of the factors be considered when weighting the different data sets in a global analysis [354].

In the global models in **SEDPHAT** a weighting factor can be introduced by raising or lowering the nominal error of data acquisition [354], or by automatically rescaling the weighting factor to compensate for the number of data points [354] *via* the check box in the upper left corner of the xp parameters window.

The optical resolution is more critical, and this is dictated by the optical path, the alignment and focusing of the optical system, and/or the pinhole size. It can be estimated as the signal response to a sharp edge.

In the fluorescence system, the optical resolution is estimated from the width of the signal transition in and out of the reference solution during radial calibration. Typical values are ~0.016 cm at average focus depths when measured with radial resolution of 0.002 cm, leading to a convolution of the signal as described in Eq. (4.2) assuming a Gaussian beam intensity; this can be accounted for directly in the data analysis [100].[16]

---

[14]In fact, the radial density of data points may be down-sampled to achieve improved computational efficiency. In **SEDFIT** and **SEDPHAT**, this is possible but usually not necessary (and switched off by default). But it may be advantageous when analyzing data on a computer platform with low computing power.

[15]For the purpose of TI noise calculation, which requires the same radial grid points, **SEDFIT** linearly interpolates the scans onto the same radial grid. However, the best-fit rmsd reported is that for the actual data.

[16]A physical limit is the pinhole size, which is usually 50 μm.

The additional fitting parameters in SEDFIT for the analysis of fluorescence data can be found in the menu Options ▷ Fluorescence Tools. The parameter accounting for optical resolution is termed Signal Radial Convolution, and should be fixed in the data analysis, since there is little opportunity for computational refinement during a fit of sedimentation data. (Since convolution is computationally intensive, it can be intermittently fixed to zero, to speed up preliminary non-linear regression of other parameters, and then be restored for a second-stage refinement.) Controls for parameters specific to fluorescence data can also be invoked also using the ALT-F key shortcut.

The absorbance system appears to have a higher optical resolution, with a value of 0.008 cm judged from imaging of the edges [339]. The finite optical resolution has no significant bearing on the accuracy of the analysis of absorbance profiles even without computational compensation, except for extremely high gradients [339]. In this regard, it is interesting to analyze the effect of a finite radial bandwidth Δr on the measured apparent absorbance a^* at a given radius r. We can roughly illustrate the situation by averaging over the transmitted light intensity measured within a radial region:

$$a^*(r) = -\log_{10}\left\{ \frac{1}{\Delta r} \int_{r-\Delta r/2}^{r+\Delta r/2} 10^{-a(r')} dr' \right\} \tag{4.3}$$

with a 'true' radial absorbance distribution $a(r)$. If the latter is not constant but has a slope, $a(r') \approx a(r) + (r'-r)(da/dr)$, this generates the leading error term:

$$a^*(r) - a(r) = -\log_{10}\left\{ \frac{1}{\Delta r} \int_{r-\Delta r/2}^{r+\Delta r/2} 10^{-(r'-r)(da/dr)} dr' \right\} \tag{4.4}$$

and an underestimate of the absorbance by a magnitude increasing with increasing slope da/dr. This results from the fundamental nonlinearity of absorbance, derived from the measurement of the transmitted light; the effect is similar to that for a finite wavelength bandwidth. However, this effect is small: with an optical resolution of ∼0.01 cm, the error would be larger than the typical noise in the data acquisition only for gradients greater than 2.5 OD/mm. It is rare to measure such sharp sedimentation boundaries, but this can affect the detection of the back-diffusion region.

This situation is much improved with the interference optics, because the intrinsic radial resolution is higher, and the error term resembles a symmetric box average, which only produces errors in second derivative terms.[17] However, there is a theoretical maximum gradient imposed by the manner in which the interference

---

[17]Similarly, the error term in Eq. (4.4) caused by the nonliearity in the absorbance data would disappear if one were to directly fit the incident reference and sample intensity data with models incorporating convolution integrals [355].

pattern is imaged and transformed. Interference fringe shift traces are constructed by comparing the phase of the light/dark pattern in neighboring columns of pixels in the camera (compare Fig. 1.11). Since no absolute fringe shift value exists at each column of pixels in the camera, relative fringe shifts can only be measured by establishing a continuous fringe displacement trace through neighboring pixels. The simplest algorithm to ensure continuity would be to assign neighboring fringe shifts an integer offset so that they differ by less than half a fringe. At the given radial density of pixels in the current commercial instrument, this continuity algorithm imposes an upper limit on the measurable slope of ~70 fringes/mm. For proteins, this would correspond to a limiting concentration gradient of ~20 (mg/mL)/mm in 12 mm pathlength centerpieces, or ~70 (mg/mL)/mm in 3 mm, respectively. Stronger gradients can be readily observed, for example, in the SV boundaries of samples at very high concentrations, leading to integral fringe 'skipping' in neighboring radial points (Fig. 4.7). To a significant degree, this can be addressed by post-centrifugal data processing, for example, by also requiring continuity in the derivatives of the fringe shift trace.[18]

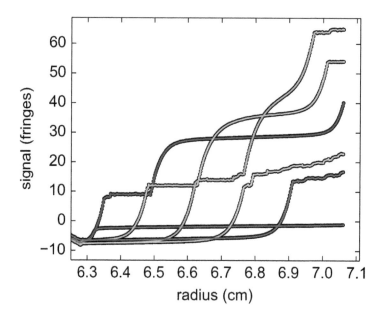

Figure 4.7 Fringe skipping in interference data at high concentration gradients. Data are from the sedimentation of 64 mg/mL maltose binding protein at 50,000 rpm. It may be discerned that where the steepest part of the boundaries would be expected, horizontal lines are recorded, due to incorrect assignment in the fringe shifts from neighboring pixels in the camera picture (compare Fig. 1.11) when the signal gradient exceeds approximately 500 fringes/cm.

---

[18]It should be noted that the true measured data at each pixel column consists only of a fringe phase to which an arbitrary offset can be added across all pixels, and/or an integer offset to each pixel column, without changing the true experimental information content of the raw data.

SEDFIT can optionally check and flag data exhibiting very strong gradients. The control in the Options ▷ Loading Options and Tools menu allows for this check to be switched on or off, and, when toggling, allows the user to input the desired threshold. By default, this is switched on, with a threshold of 500 fringes/cm (i.e., 70% of the theoretical maximum for fringe skipping in interference data).

In addition, SEDFIT has functions for post-processing (prior to data analysis) of interference scans with very high gradients showing 'fringe skipping.' Processing is based on continuity requirements of second- and higher-order derivatives of the radial fringe displacement trace. In an iterative fashion, integral fringe shift offsets are added so as to achieve continuity; the resulting scans can be saved as new data files.

4.1.4 Time-Invariant (TI) Noise

TI noise is a radially dependent, but temporally constant baseline offset. The measured signal takes the form:

$$a^*(r,t) = a(r,t) + b(r) \tag{4.5}$$

where a^* and a represent the time- and radial-dependent signals in the presence and absence of TI noise, respectively, and $b(r)$ the TI noise profile. TI noise may occur for different reasons, and is observed in all detection systems. Fortunately, these noise features can be easily calculated and added to the model; they can also oftentimes be visually recognized. Their calculation only requires a model for the experimentally measured signal distributions, and multiple data sets as a function of time (SV) or rotor speed (SE). Consideration of TI noise greatly improves the quality of fit and may provide better parameter uncertainties obtained from a statistical analysis of the fit, despite the introduction of a significant number of additional fitting parameters (with TI noise constituting a set of arbitrary offsets, one for each radial grid point).[19]

In SEDFIT and SEDPHAT, the calculation of TI noise can be switched on by checking the respective checkbox in the parameter menus. This will render the radially constant baseline offset switch inactive (as this is a special case of TI noise).

4.1.4.1 Physical Origin

In the absorbance optics, TI noise arises from ubiquitous tiny scratches and imperfections, which can sometimes be observed upon inspection of the window with the unaided eye (Fig. 4.8). Visibility notwithstanding, after data acquisition and sedimentation analysis these can be easily discerned in most absorbance SV data

---

[19]To test whether the inclusion of TI noise parameter has a significant influence on the confidence intervals of a parameter of interest, an error analysis based on covariance matrix (available in the Statistics menu of SEDPHAT) can be carried out both with and without TI noise.

sets, either directly from correlated deviations in the data at the same radius in different scans, indirectly from correlated features in the residuals, or from vertical stripes in the residuals bitmap (which is described in Section 4.3.5) (Fig. 4.8). As demonstrated in Fig. 4.8, absorbance TI noise is not expected to exhibit any low-spatial-frequency structure.

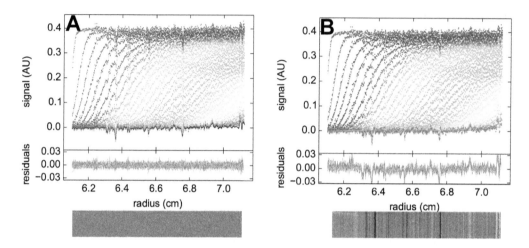

Figure 4.8 Comparison of the analysis of absorbance data accounting for TI noise (*Panel A*) and without consideration of TI noise (*Panel B*). Both are based on the same absorbance data from the sedimentation of BSA acquired at a rotor speed of 50,000 rpm and at a wavelength of 280 nm. Top: Experimental absorbance data (points), best-fits (colored solid lines), and calculated TI noise (black line in *A*); Middle: residuals of the fit; and Bottom: Residuals bitmap. The analysis accounting for TI noise in *Panel A* results in an rmsd of 0.0048 OD, whereas the same analysis without TI noise consideration in *Panel B* results in highly correlated residuals with an rmsd of 0.0088 OD. The residuals incorporate many of the radially invariant TI noise features.

The ubiquitous presence of TI noise in SV data suggests that it will always be a very significant contributor to the noise signal in SE scans, as well, even though it may not be as readily identified in the absence of obvious spikes. Clearly, a single SE scan alone cannot provide sufficient information to distinguish a macromolecular signal from TI noise and statistical noise, but it can be unraveled from multi-speed SE data sets. The latter may require minor modifications in Eq. (4.5) to account for the effect of rotor stretching causing a (predictable) translation of the noise features [160].[20] Since multi-speed SE analysis is highly advantageous also for the reason of allowing implicit mass conservation analysis, TI noise is routinely considered in SE analyses [166, 167].

---

[20] As described in Section 5.2.4, corrections in TI noise for rotor stretching are recommended for absorbance data, but not for interference data.

In SEDPHAT, the checkbox `TI noise with rotor stretch` in the experimental parameter window of SE data will account for a rotor-speed dependent shift in TI noise following predetermined stretch moduli for the different rotors. This requires the rotor type to be specified. This correction is recommended for absorbance data, but not for interference data, where significant contributions to TI noise may not be rotor speed dependent (Section 5.2.4).

Shortly after the development of TI noise analysis, our laboratory has introduced an alternate mode for the utilization of the absorbance optical system, whereby the use of a reference sector is abandoned and each solution column is filled with a different sample (see Section 4.3.2 below), referred to as pseudo-absorbance [356]. Consider the definition of the absorbance as the decadic logarithm of the ratio of transmitted light in the absence (I_0) and presence (I_S) of a sample as a difference from separate contributions. In the pseudo-absorbance mode we do not separately measure I_0, but assume instead that it can be factored into a radial dependent component, which may be modulated by an uncorrelated temporal component allowing for scan-to-scan drift:

$$A(r,t) = \log_{10}\left(\frac{I_0(r,t)}{I_S(r,t)}\right) = \log_{10}\left(I_0(r) \times 10^{\beta(t)}\right) - \log_{10}\left(I_S(r,t)\right)$$

$$\hspace{3cm} (4.6)$$

$$= b(r) + \beta(t) - \log_{10}\left(I_S(r,t)\right) = A_{pseudo}(r,t) + b(r) + \beta(t)$$

We can recognize that it is identical to the form of Eq. (4.5) if the signal in the absence of sample is described as a TI offset, in combination with an additional orthogonal term for time-dependent — but not radially dependent — drifts in light intensity (termed RI noise, see Section 4.2.5 below).[21,22] Indeed, the major change in the recorded light intensity for pure solvent is the position-dependent sensitivity of the photocathode in the absorbance system. In contrast to standard absorbance data, TI noise for pseudo-absorbance data may have significant low spatial-frequency components (Fig. 4.9 Panel B). Thus, the negative base-10 logarithm of the transmitted intensity, if combined with TI and RI noise, can play the same role as absorbance data without the need of a reference [356, 357]. When data are acquired in the 'intensity mode', separate transmitted intensities for the sample and reference sector are available, and each can separately be converted to pseudo-absorbance data. Practical drawbacks of this approach are discussed below

---

[21]In the absorbance system of the Optima XL-A, the intensity from the reference sector is not acquired truly simultaneously to the sample sector. But fluctuations in the power of the lamp flashes are accounted for using a separate incident light detector, and remaining variation is absent or sufficiently slow, such that the requirements in pseudo-absorbance for stability during a single scan are usually met.

[22]Thus, AUC differs from bench-top spectrophotometers in that the additional radial and temporal dimension allow discarding the reference measurement, as long as potential I_0 variations are uncorrelated with sedimentation signals.

(Section 4.3.2). An example of pseudo-absorbance data is shown in Fig. 4.9 Panel A, for the same experimental data as shown in the absorbance analysis of Fig. 4.8, but prior to the conversion of transmitted intensities of reference and sample sector to regular absorbance (Fig. 4.9).

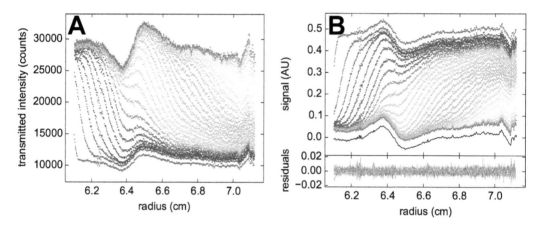

Figure 4.9 The same experiment as shown in Fig. 4.8, with raw transmitted intensity profiles in the sample sector (A) and pseudo-absorbance data (B) after analysis considering TI noise (black line) and RI noise.

A SEDFIT tool in the Loading Options submenu allows one to split standard two-channel intensity data from the AUC into two sets of files with corresponding pseudo-absorbance data.

TI noise is also ubiquitous when using the interference optics, but of much larger magnitude and possibly with some low spatial frequency features or slopes, dependent on the alignment of camera and cylinder lens (Fig. 4.10). Because we can routinely measure less than a hundredth of a fringe shift with the red laser light, optical pathlength differences of only a few nanometers will contribute significantly to the signal! The high spatial frequency features are a reflection of the surface roughness and homogeneity of all optical elements in the light path. It is remarkable, and a sign of well-engineered mechanics and optics, that the optical pathlengths routinely stay constant to within ~0.001 fringes during a SV run. The magnitude of TI noise in interference optics typically exceeds the statistical noise by 10-fold or more, and at low sample concentrations it may even exceed the sedimentation signal of interest. However, when properly accounted for — which is absolutely essential for the analysis of most interference experiments — this is not a limiting factor for data analysis because of the excellent temporal stability of the interferometric TI noise [358].

It is very useful to occasionally inspect the TI profile obtained with the interference optical system in an empty rotor hole, particularly after any change to the

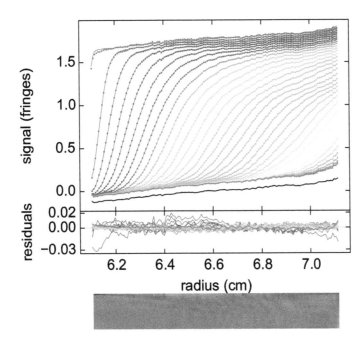

Figure 4.10 Rayleigh interference profiles from the same experiment shown in Fig. 4.8 and Fig. 4.9, with best-fit model (colored lines) considering TI noise (black line) and RI noise (see Fig. 4.15 below). For clarity, only every 10th data point is shown (dots). The rmsd is 0.0033 fringes; residuals are shown as overlay and bitmap.

optical alignment. The order of magnitude of the TI profile and its slope can then be compared qualitatively with that calculated from the sedimentation analysis. Although the details and to some limited extent the slope will vary from experiment to experiment, the absence of curvature and general stability are important features to note. Also, the analysis of a sequence of scans from an empty rotor hole can reveal the intrinsic statistical noise of the data acquisition system: Modeling of the TI and RI ('radial-invariant,' see below) noise contributions in such a case will produce the remaining rms noise that represents the intrinsic precision of a single data point.[23,24]

---

[23]There has been some confusion regarding the noise structure in interference optical data, whereby the magnitude of the TI baseline was considered as part of the noise of data acquisition. However, given that the TI baseline profile is constant and can be easily eliminated computationally, its magnitude is not of great importance; also improvements in the smoothness of the TI baseline are virtually irrelevant. What is highly relevant, in contrast, is the statistical noise of a single data point after the constant baseline has been accounted for.

[24]Related to this, TI noise analysis provides a convenient approach to average any set of time-invariant scans, since after modeling with TI (and possibly RI) noise the resulting TI profile $b(r)$ will capture the best local least-squares average radial-dependent signal, which can be saved in the format of an AUC data set. This can be a highly useful tool for the suppression of noise amplification in conjunction with the subtraction of buffer blanks in SE analysis of interferometric data (Section 5.2.4).

A SEDFIT function Save TI noise in file creates data files in AUC format from the calculated TI noise profile.

A problem occurs when switching rotor speeds, for example, as in multi-speed SE experiments. Whereas absorbance data at different rotor speeds only require consideration of the rotor stretch and translation of the TI noise profile, additional precautions must be met to ensure constancy of TI noise with interference optical detection. This is due to the higher sensitivity of the interference optics to any deformation and shifting of the cell components that may occur under different gravitational forces (where a single window may weigh 1 ton or more, dependent on rotor speed). For this reason, when a TI noise analysis is required for interference analysis across different speeds, the cell assemblies have to undergo a mechanical 'ageing' and stabilization procedure (see Section 5.2.4) [359]. With these precautions a stable TI baseline profile can be achieved for all rotor speeds [160].

The necessity of TI noise decomposition for the analysis of fluorescence optical data is unclear. In some data sets local offsets may occur, for example, due to contamination of window components with fluorescent molecules, but, at least at very low fluorophore concentrations (e.g., Fig. 3.1), it should not be necessary. Rather, in a strategy that should be avoided, it has occasionally been included in the model as a 'fudge parameter' to compensate for the lack of consideration of the particular structure of fluorescent data arising from optical design, and/or inadequacy of the model to describe the macromolecular concentration evolution.

4.1.4.2   Accounting for TI Noise

After an introduction to the nature and physical origin of TI noise, we turn to the practical treatment of TI noise in the data analysis. Even though details of the sedimentation analysis are not the focus of the present volume, the strategy taken to deal with this noise can considerably impact the experimental design.

TI noise correction can follow an experimental or computational path. Experimental approaches are aimed at the separate acquisition of data representing the TI noise component, whereas computational approaches extract TI noise information from the signal representing the redistribution of sample during the sedimentation experiment. This may be the change in SE profile across different rotor speeds, or the change in the sedimentation boundary shape and position with time in SV. These approaches are not mutually exclusive: for example, the subtraction of an experimentally acquired reference scan in SE does not preclude the determination of refined residual TI components in the joint analysis of SE at different rotor speeds [160].

The simplest, in some sense naïve, experimental approach to eliminate TI noise is through the subtraction of a separately acquired reference scan. This is the method of choice for SE analysis of interference data (especially for SE data acquired at a single rotor speed, where macromolecular signals and offsets cannot be distinguished a priori). It has two fundamental drawbacks: (1) It is not trivial to acquire such a reference scan. The presence of sample sedimentation will invariably contribute to the measured profile, and taking the reference at a very low rotor speed may not reproduce the correct baseline (at minimum from effects of rotor stretching). Water blanks taken prior to, or after, the sedimentation experiments may not reproduce the exact same TI noise pattern due to movement of the windows or other cell assembly components during acceleration and deceleration, and/or due to possible deposition of vacuum pump oil on the outside of the window. When taking this approach for the correction of SE data, it is prudent to acquire water blanks as TI baseline profiles both before and after the sedimentation experiment, to create an opportunity to verify the constancy of the TI noise. Furthermore, in conjunction with the interference optics in SE it is essential to subject the cell assembly to the ageing procedure mentioned above (and described in detail in Section 5.2.4) to mechanically stabilize it. (2) Subtraction of a particular reference scan will likely introduce a bias into the analysis, for example, a noise amplification by a factor $\sqrt{2}$, although this can be suppressed by averaging of baseline scans from multiple acquisitions.

Another simplistic and experimentally motivated approach is the pairwise subtraction of scans, which trivially eliminate the radially constant component. It is applicable where a change of sedimentation profile, usually with time but in principle equally with rotor speed, is generated in the course of the experiment, and it was originally conceived by Cohen and colleagues in the context of SV analysis [360]. However, this will transform the raw sedimentation data into a difference mode in which boundaries in SV and Boltzmann exponentials in SE are not easily recognized, and superposition of the difference curves obscures details of both the data and the fit. This is not desirable. Furthermore, it will lead to noise amplification. Most importantly, it will introduce more flexibility in the model than required, since the TI noise component is no longer constrained to be the same constant contribution for all scans, but rather only for pairs of scans. Thus, erroneous models implying a slow drift of TI profile may fit the data well and will not be excluded.[25]

When taking a differencing approach, after fitting the resulting data with a sedimentation model it is easily possible to back-calculate explicitly the TI noise profile for each scan pair *via* Eq. (4.5) [362]. This is very useful, since it allows assessing whether the TI offset implied by the model is realistic (see below) [362].

By far the most stringent and robust approach is the explicit computational consideration of TI noise, strictly as a set of unknown temporally constant offsets in the analysis, to be determined by least squares modeling of the raw data directly, as stated in Eq. (4.5). Like the differencing approach, it does require a change of the concentration profile of the sample either with time (in SV) or with rotor speed (in SE) that can be explicitly modeled. As we will see, the TI profile does not need to be separately determined and pre-defined, but its determination is

---

[25] If this approach is taken, the best statistical choice would be pairwise scan differencing for scans separated by a distance of 1/3 or 2/3 the total number of scans; the common choice of referencing with a distance of 1/2 the total number has a maximum of error for sedimentation velocity [361].

folded into the sedimentation model. Computationally the TI noise profile consists of a separate baseline for each radial point. Even though this amounts to a large number of unknowns, the TI noise profile can be very well determined if a large redistribution of sedimenting species is observed in the course of the experiment. Thus, experimental designs that create such shifts in signal patterns are to be preferred.

It is interesting also from an experimental point of view to follow the computational results. Eq. (4.5) can be efficiently solved by separating the linear and nonlinear variables [363]. The basic principle of algebraic noise decomposition, first described in [358], can be illustrated by considering n scans each with i experimental data points $a_{n,i}$ (short for $a(r_i, t_n)$, using the indices i and n to denote the dependence on discrete values of radius and time), and a model for the macromolecular redistribution $B_{n,i}(\{p\})$ dependent on a set of nonlinear parameters $\{p\}$ and subject to the radial-dependent TI offsets b_i (e.g., boundary in SV). The least-squares minimization then can be written as:

$$\underset{b_i}{Min} \sum_{n,i} \left(a_{n,i} - B_{n,i}(\{p\}) - b_i \right)^2 \qquad (4.7)$$

from which, after taking the partial derivative with respect to a specific radius j, it follows:

$$0 = \sum_{n,i} \left(a_{n,i} - B_{n,i}(\{p\}) - b_i \right) \delta_{i,j} \qquad (4.8)$$

(where $\delta_{i,j}$ is the Kronecker symbol) and the explicit form of the TI noise parameters arises as:

$$b_j = \frac{1}{N} \sum_n \left(a_{n,j} - B_{n,j}(\{p\}) \right) \equiv \bar{a}_j - \bar{B}_j(\{p\}) \qquad (4.9)$$

for any boundary model [358] (where N is the total number of scans, and the r.h.s. defines radial-point-averaged values across all scans of data and model). The combination with orthogonal radial-invariant noise (Section 4.2.5) is straightforward.[26] Since this computation is very efficient, it will be carried out at each evaluation of the boundary model for different nonlinear parameters as required in their optimization. The implicit (re)evaluation of TI noise as part of any boundary model is crucial to understanding the impact of TI noise on the statistics and optimization of AUC data. In fact, we may now re-write the remaining general AUC fitting

---

[26]Computationally more complex situations arise in the presence of multi-speed SE analyses, where the TI noise profile experiences a radial translation to higher radii at higher rotor speeds in a well-defined way due to rotor stretching [160], and in partial boundary modeling in SV, where only partially overlapping radial ranges are considered from each scan. In the latter approach, scans share a constant TI baseline [339], but the partial overlap creates a mild nonlinearity in the TI noise calculation. However, these cases follow the same principles as outlined here, and have no additional experimental consequences.

problem Eq. (4.7) as:

$$\underset{\{p\}}{Min} \sum_{n,i} \Big((a_{n,i} - \bar{a}_i) - \big(B_{n,i}(\{p\}) - \bar{B}_i(\{p\}) \big) \Big)^2 \tag{4.10}$$

namely, a form in which TI noise parameters do not appear anymore. The reader will note that this does actually correspond to fitting with a differencing scheme; the particular choice implied by Eq. (4.10) is the choice of the average scan (as opposed to any particular scan) as a reference to be subtracted from each scan. This has minimal noise amplification and eliminates excess flexibility in comparison to pairwise differencing, as shown in detail in ref [361]. Once any particular parameter set defining the model has been identified, Eq. (4.9) can be used to determine its inseparably linked TI noise profile, while eliminating the necessity of an extra measurement of a baseline profile.

It is very useful to explicitly calculate the TI noise pattern in the raw data space for further inspection.[27] For example, TI noise should usually be free of any long-range slope and low spatial frequency structure. If these occur, they are an indication that TI noise is capturing systematic deficiencies of the model, rather than local optical imperfections, in which case the TI noise model (along with the sedimentation model) should be rejected. In these cases it may be prudent to first carry out the data analysis without allowing for TI noise, and, if possible, eventually model the TI noise in the final refinement. This situation can arise, in particular, in analysis of multi-speed SE data.

After inspection of the calculated TI noise profile, especially in conjunction with the SV analysis of interference optical data where TI noise is usually large and extremely well defined, it is often useful to subtract the calculated TI noise from the raw data. This will allow for a better visual comparison of the experimental sedimentation boundaries, now without the time-invariant features, with the best-fit sedimentation model. This can inspire extension and improvements of the sedimentation model, and/or allow for the verification of detailed features such as boundaries of trace components, otherwise obscured by large quasi-random TI offsets. It is important to realize that TI noise subtraction does not change the information content of the data, as long as the degrees of freedom of TI noise are maintained in all subsequent data analyses. As has become clear from the derivation of Eq. (4.10), even though any particular instance of explicitly calculated TI noise is linked to a particular boundary model, the error surface to be optimized and its best-fit parameters are not dependent on any specific set of TI noise parameters, since the TI noise parameters are re-evaluated for any combination of boundary model parameters [361].

---

[27]There has been some confusion in the literature, caused by a lack of appreciation for the fact that even sedimentation models motivated by fitting scan differences imply underlying TI noise and offer the same or higher degrees of freedom, whether TI noise is explicitly calculated or not. It would be naïve to assume that simply neglecting to explicitly calculate the underlying TI noise, as advocated by Stafford [364], would in any way improve the model; rather this would just provide an opportunity for overlooking potential problems in the implications of the fit.

Menu functions and buttons on the SEDFIT and SEDPHAT graphical data displays allow for the subtraction of TI noise, as well as restoration of the original raw data, to both permit the inspection of TI noise and provide a clear view of the sedimentation boundaries.

4.1.5 Radial Gradients of Signal Magnification

A radial-dependent feature specific to the fluorescence optics with confocal design are gradients of signal magnification (Fig. 4.11 Panel A). A mismatch between the plane of rotation and the plane of scanning can result in a change of the focal depth with radius [98], which, in turn, can result in a different magnitude of the detected emission signal for equal fluorophore concentrations. This is often particularly pronounced at shallow focal depths, which are preferable due to the suppression of inner filter effects [100] and to minimize the zone of optical artifacts close to the end of the solution column (as discussed in Section 4.1.2 above). It can be easily accounted for with a linear correction for signal magnification:

$$s(r,t) = s_0(r,t) \times \left\{ 1 + \frac{d\varepsilon}{dr}(r - r_0) \right\} \tag{4.11}$$

with a signal increment gradient factor $d\varepsilon/dr$. An example of the effect of spatial signal magnification gradients and its analysis is shown in Fig. 4.11 Panel B and Panel C [100], where data acquisition results in signal changes of 28% along the solution column. Even though the shape of the data might at first sight suggest the presence of a broad distribution of large oligomers and aggregates (due to the 'sloping solution plateau' in the first several scans), there is in fact very little correlation with the occurrence of large species: if there were larger species, these would rapidly deplete and the boundaries would shrink at later times, rather than grow in size. Correspondingly, a size-distribution model cannot yield a satisfactory fit (Fig. 4.11 Panel C). Rather, the data shown in Fig. 4.11 is from a single species consistent with the size and shape of EGFP [100], and only a signal model of the form of Eq. (4.11) can fit the data well (Fig. 4.11 Panel B). As a consequence, the signal increment gradient is experimentally well-defined.

Small mismatches of the focal plane with the plane of rotation will not significantly impact the absorbance or interference optical detection, as in these optical systems light traverses through the entire cell with constant optical pathlength.

4.2 TEMPORAL DIMENSION

The temporal dimension is of crucial importance in SV, but practically irrelevant in SE. To enable a modern SV data analysis that results in high-precision parameter values, it is imperative that the sedimentation process is observed and data are recorded over the entire time-range of the experiment, and that temporal characteristics of experiment and data acquisition are accurately known and accounted for.

4.2.1   Rotor Acceleration and Effective Sedimentation Time

The rotor cannot be brought to the required speed ω_0 instantaneously, and with a maximal rotor acceleration of ~280 rpm/sec, attainment of 50,000 rpm requires approximately 3 minutes on the Optima or ProteomeLab XL-A/I. Neglecting this acceleration time would lead to significant errors in the analysis. However, in addition to the elapsed time t, the recorded scan files additionally store information on the integral $\int_0^t (\omega(t'))^2 dt'$. Therefore, a simple approach to account for rotor acceleration is to calculate an 'equivalent sedimentation time at full speed' as:

$$t^{(sed)} = \omega_0^{-2} \int_0^t \left(\omega(t')\right)^2 dt' \qquad (4.12)$$

Substitution of actual scan time with $t^{(sed)}$ provides correct sedimentation coefficients.[28] However, this introduces errors in the description of the diffusion process, which takes place throughout the entire rotor acceleration phase as soon as concentration gradients arise (Fig. 4.12).[29] The errors associated with incorrect diffusion times lead to an overestimate of the diffusion coefficient and underestimate of the buoyant molar mass [366]. Such errors are exacerbated with slower rotor acceleration, which may be sometimes desirable to avoid temperature changes caused by adiabatic stretching of the rotor on acceleration [366]. For example, with an isothermal rotor acceleration of 5.6 rpm/sec, the overestimate of diffusion in the 'effective sedimentation time' approach is 26% [366].

A more elegant and accurate approach is based on a reconstruction of the full rotor speed profile $\omega(t)$, and incorporating this as a time-dependent rotor speed directly into the numerical solutions of the Lamm equation Eq. (1.10) predicting

---

[28] In a time-varying field, the differential equation for particle motion Eq. (1.6) can be separated as

$$dr^{(p)}/r^{(p)} = s\omega(t)^2 dt$$

and integration leads to

$$r^{(p)}(t) = m \exp\left[s \int_0^t \omega(t')^2 dt'\right]$$

therefore, inserting $t^{(sed)}$ from Eq. (4.12) into the sedimentation equations with constant ω_0 produces the correct sedimentation terms in the standard form with

$$r^{(p)}(t^{(sed)}) = m \exp\left[s\omega_0^2 t^{(sed)}\right]$$

[29] Another error in the 'effective time' arises in the transition from the resting to the spinning state of the meniscus at the start of centrifugation. However, this will introduce a negligible time offset usually absorbed into the best-fit meniscus position [365].

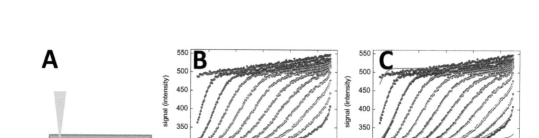

Figure 4.11 (*Panel A*) Principle of the changing focal depth as a function of radius caused by mismatch of the plane of scanning and the plane of rotation. (*B and C*) Fluorescence optical scans for the sedimentation of EGFP at 50,000 rpm, with a focal depth of ~1 mm [100]. (*Panel B*) A sedimentation model accounting for radial-dependent magnification gradient *via* Eq. (4.11). The remaining TI noise component (black line) is nearly constant throughout the cell. (*Panel C*) In contrast, the model not allowing for radial-dependent signal magnification changes, results in a bad fit with a compensatory artifactual slope in the calculated TI noise profile.

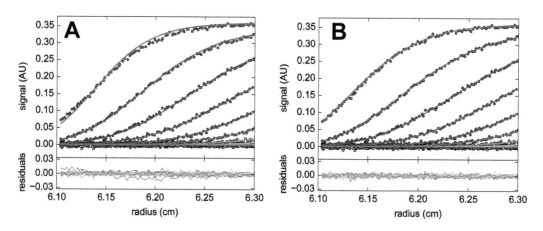

Figure 4.12 Sedimentation boundaries of BSA as acquired in an experiment with slow acceleration at 1,000 rpm/min (symbols), modeled with the $c(s)$ method, either accounting only for an effective sedimentation time $t^{(sed)}$ (see Eq. (4.12)) with instantaneous acceleration (*Panel A*), or accounting for a continuous acceleration based on true elapsed times (*Panel B*). Shown is the region near the meniscus, where the effect of boundary broadening is most noticeable. The 'effective sedimentation time' approach cannot account for the broadening of the boundary from initial diffusion, and yields an apparent molar mass that is 17% too low. This difference is exacerbated at lower acceleration rates. The correct value is obtained when the time-dependence of the rotor speed is explicitly accounted for in the model of the sedimentation process [366].

particle sedimentation and diffusion [151, 366] (Fig. 4.12 Panel B).

$$\omega(t) = \begin{cases} (d\omega/dt)_{acc} \times t & \text{for} \quad 0 < t < t_{acc} \\ \omega_0 & \text{else} \end{cases} \tag{4.13}$$

This is possible, knowing that the rotor acceleration is practically constant, from the difference between t and $t^{(sed)}$, i.e., using jointly the t and $\int \omega^2 dt$ entries of the files, for example those of the first scan (here denoted with index 1):

$$t_{acc} = \frac{3}{2}\left(t_1 - t_1^{(sed)}\right) \tag{4.14}$$

$$\text{and} \quad (d\omega/dt)_{acc} = \omega_0/t_{acc}$$

This approach requires ramping the rotor speed from rest to full speed in one continuous acceleration with constant rate, which applies to standard SV experiments (Section 5.6).

By default, SEDFIT and SEDPHAT will assume a standard SV experiment in which constant rotor acceleration from rest has taken place. Accordingly, the file header entries for $\int \omega^2 dt$ are used to calculate an effective sedimentation time following Eq. (4.12), and from comparison with the file header entry for the elapsed time since the start of sedimentation, the rotor acceleration time and the rotor acceleration rate are calculated following Eq. (4.14). Sedimentation models then use the constant acceleration model Eq. (4.13) in all Lamm equation solutions.

If the calculated acceleration rate is significantly below the expected rate of 280 rpm/sec, SEDFIT will flag this as improbably slow rotor acceleration, and suggest that the user either switches to the effective time model for further analysis, or specifies the history of rotor acceleration.

A more complex situation arises if an intermittent interruption of rotor acceleration is required, for example, when encountering unexpected difficulties in optical adjustment, or when, in an ill-advised approach rooted in AUC practice prior to detailed computerized SV data analysis, temperature equilibration is carried out at 3,000 rpm, rather than at rest (see Section 5.3.2). This makes it impossible to use the standard assumption of constant acceleration producing a well-defined relationship between t and $\int \omega^2 dt$.

There are three possible remedies: In a rigorous experimental strategy, the run should be stopped, the sample should be resuspended as described in detail in Section 5.2.4, and the cell assembly should be re-inserted into the rotor. After a new temperature equilibration phase, the rotor should be accelerated smoothly from 0 rpm to full speed.

Alternatively, if the experiment cannot be restarted, depending on the objective of the data analysis, one may accept errors in the diffusion parameters and carry out the analysis on the basis of effective sedimentation times. Sedimentation coefficients should be well determined, but diffusion related parameters (such as frictional ratio and molar mass) will be underestimated.

In SEDFIT, the Lamm equation options (Options ▷ Fitting Options ▷ Lamm equation parameters) allow the user to switch from the true elapsed time (default, termed ramp rotor speed) to effective sedimentation times (termed times from wwt). Lamm equation models can be switched in SEDPHAT in an analogous manner.

A more sophisticated computational consideration of non-uniform acceleration has recently become possible by a generalization of the SV analysis to sedimentation in a time-varying field [249].[30] The centrifugal rotor speed controls currently allow only the change of target rotor speed to be executed at a specific time, which causes a constant acceleration or deceleration to the new speed. Thus, in the case of an initial low-speed phase the temporal rotor speed profile consists of a constant initial acceleration to a speed ω_{low}, typically 3,000 rpm, initiated at time 0, and a second step initiated at time t_{low} to accelerate to the full target rotor speed ω_0. This adds two intermediate steps to the function $\omega(t)$ in Eq. (4.13).

When loading SV data, SEDFIT will automatically scan the data directory for a user-generated file speedsteps.txt. If it is present, it will be loaded and used to specify the temporal changes of rotor speed in the numerical solution of the Lamm equation. The file speedsteps.txt must be an ASCII text file containing one row of white-space delimited numbers for each rotor speed change. Columns are 1) the time (in seconds) when the speed change is initiated (corresponding, e.g., to the time when the 'Enter' button is pressed at the AUC console to execute a new rotor speed); 2) the new target rotor speed (in rpm); and 3) the rate of acceleration/deceleration (in rpm/sec) [249]. This file can be generated from an experimental .EQU file (assuming there was no delay in the execution of the equilibrium method file caused by scan time delays), or be created from scratch.

When analyzing data that have not been designed with this capability for data analysis in mind, the exact time t_{low} when acceleration to the full speed was initiated may not be known. However, in analogy to Eq. (4.14), it is possible to calculate t_{low} retrospectively from the difference of the true elapsed time t and the effective sedimentation time $t^{(sed)} = \omega_0^{-2} \int_0^t (\omega(t'))^2 dt'$ based on the integral $\int \omega^2 dt$ recorded

---

[30] This approach is possible in a general way due to the numerical solution of the Lamm equation Eq. (1.10). However, a special case analytical approximate solution for the Lamm equation in a time-varying field based on the Faxén approximation has already been developed by Nossal and Weiss [367].

in the file headers Eq. (4.12), which consists of:

$$\omega_0^2 t^{(sed)} = \int\limits_0^{\omega_0/\dot\omega} (\dot\omega t')^2 dt' + \omega_{low}^2 \left(t_{low} - \frac{\omega_{low}}{\dot\omega} \right) + \omega_0^2 \left(t - \frac{\omega_0 - \omega_{low}}{\dot\omega} - t_{low} \right)$$

with the rate of rotor acceleration $d\omega/dt \equiv \dot\omega$, leading to:

$$t_{low} = \left(t - t^{(sed)} \right) \frac{\omega_0^2}{(\omega_0^2 - \omega_{low}^2)} - \frac{2}{3\dot\omega} \frac{\omega_0^3 + 3\omega_0^2\omega_{low} + 3\omega_{low}^3}{(\omega_0^2 - \omega_{low}^2)} \tag{4.15}$$

In order to account for the extra steps, the rotor speed of the low-speed phase must be known, and the rotor acceleration rate must be assumed to be always the standard rate ($\sim$280 rpm/sec) [249].

> The SEDFIT function Options ▷ Loading Options ▷ Recreate Speedsteps.txt From SV Data will, more generally, extract the temporal changes of rotor speed across an entire variable-speed experiment, provided that at least a single scan is available to report on each step. If an impossibly low rate of rotor acceleration is flagged by SEDFIT when loading SV data in the default mode *via* Eq. (4.14), indicative of an unspecified initial low rotor speed phase, SEDFIT will suggest the creation of a suitable speedsteps.txt file *via* Eq. (4.15), or alternatively, the use of effective sedimentation times Eq. (4.12).

4.2.2 Scan Repetition Rate and Number of Scans Required in the Analysis

The scans for an SV analysis need to reflect the entire sedimentation process. This comprises the very early depletion at the meniscus, up to the very late stage where all particles of interest have seemingly cleared the optically accessible radial range. The significance of the early stages is that they provide access to very large sedimenting material, and better precision in the meniscus position (and/or the detection of initial convection) which translates into improved accuracy in the sedimentation coefficients and better resolution and diffusional deconvolution of the sedimentation coefficient distributions. The late stage allows for a resolution of the smallest sedimenting species such that they don't confound the interpretation of the main boundary of interest.[31] It is not critical that the scan times be evenly spaced, only that their scan times be known accurately (see below).

The acquisition of a fringe profile using the interference optics is essentially instantaneous, and the minimum time delay between successive scans is very small.

---

[31]The consideration of only a sequence of scans within a narrow time interval in which the boundary has sedimented midway, coined the 'rule of thumb', was dictated by the approximations required for the now outdated dc/dt data transform [368, 369]. Consideration of only these scans is poor practice, as it sacrifices valuable information that can be easily extracted given modern data analysis.

Converting a snapshot of the fringe image from the whole cell into radial fringe shift data and storing the radial region of interest takes only a few seconds, such that scan intervals of 10 sec or less can be routinely achieved. This provides a sufficient temporal resolution, even when scanning cells in all 8 rotor positions, for all but the largest sedimenting particles. In conjunction with a judicious choice of rotor speed, sedimentation coefficients as fast as 10^5 S can be measured [284]. With the interference optics, the number of available scans is so high that only a representative subset needs to be considered for the data analysis, at least in the preliminary stages. In this case, after an initial survey of the available data, a selection should be made of typically 50–100 scans that are equally spaced and representatively report on the entire sedimentation process of all components.[32]

> In SEDFIT and SEDPHAT, the default color scheme of scans is that of a color temperature scaled with scan number. With equal temporal spacing and typical boundary shapes, boundaries in the middle of the solution column are usually cyan or green, those near complete depletion are orange or red.

This situation is considerably different in the current absorbance scanner. At the fastest setting, when performing a single intensity measurement at each radial point, acquisition of the absorbance profile across a 10 mm solution column takes approximately a minute or longer. When using 7 cells, the time between absorbance scans is of the order of 7 minutes at 50,000 rpm. However, this is usually not limiting as in a properly set-up experiment, with typical signal-to-noise ratios of ∼100:1, as few as 10 scans can be sufficient for a good data analysis, provided they are reporting on the entire macromolecular sedimentation process across the cell.[33] Under these conditions, beyond 10–20 scans, any improvement will be of statistical nature and refine only the detailed aspects of the sedimentation process. For a typical experiment of a small protein at 50,000 rpm sedimenting at < 10 S, up to 100 scans are usually recorded.

The scan repetition rate will be lower when scanning the solution at multiple wavelengths, a technique allowing for the SV analysis of multi-component systems (see Section 4.4), because it proportionally increases the time interval between successive scans at the same wavelength [85, 90–92]. However, because the scans at every wavelength will report on the sedimentation process at different times, just from a changed viewpoint, the effect of scan rate on the information content regarding the temporal evolution of the concentration gradients is not very large if the data

---

[32]When making the scan selection, the majority of scans should report on the progress in sedimentation. Late scans that replicate the baseline should be avoided; these provide little information and diminish the role of the rmsd as a meaningful measure of the fit to the sedimentation process.

[33]In theory, even a single scan can be enough to define an apparent sedimentation coefficient distribution, knowing the sample to have uniform concentration at time $t = 0$. However, a minimum of 3–5 are desirable to distinguish sedimentation from diffusion.

are analyzed in a multi-signal analysis [85,90,91]. In many cases, multi-signal analyses can be conducted conveniently using interference optics and absorbance optics concurrently. Unfortunately, in such a case, interference data collected is limited by and obtained at the same low rate as the absorbance scans; noting that conversely, the interference scans do not appreciably slow absorbance data collection further.

The time required for a single fluorescence scan is similar to or slightly longer than that of the absorbance optics. However, since the fluorescence system is scanning all cells that have the same photomultiplier voltage and gain in each rotor revolution (parallel data collection) within the fixed radial range of 7.3–5.8 cm, it can yield a higher repetition rate than the absorbance system when using multiple cells.

For experiments with very low signal-to-noise (Fig. 3.1), different considerations apply [80, 99]. Here, as many scans as possible should be included in the data analysis, including scans reporting at times well after complete sedimentation, because these can help define the baseline and also report on the absence of slower-sedimenting material [99].

4.2.3   Speed of Data Collection

One question that arises with the relatively slow scanning speed in the absorbance and fluorescence detection systems is whether this can lead to a detectable distortion of the sedimentation boundaries, in particular for large particles that sediment rapidly and will have migrated during time required for scan acquisition. This will impact the measured boundary position, the boundary spread, and the shape of the plateau region, each to a different extent.

The most important effect is on the boundary position. Noting that the time-stamp in the absorbance data file corresponds to the start of the scan and that the absorbance scanner starts at the smallest radius, the finite time for data collection results in a time delay until the scanner reaches the experimental boundary position. During this time, the boundary will have moved further, by a magnitude that depends on its sedimentation velocity (Fig. 4.13), and yet it retains the original, start of scan time stamp. This can be approximated as follows: if sedimentation during a time interval Δt allows the particle to migrate the distance $\Delta r = s\omega^2 r \Delta t$, then the extra time before detection amounts to $\Delta\Delta t = \Delta r/v_{\text{scan}}$ (where v_{scan} is the linear velocity of the scanner), and it will appear to have moved by $\Delta r^* = \Delta r + s\omega^2 r \Delta\Delta t$, corresponding to an apparent sedimentation velocity $s^* = (\Delta r^*/\Delta t)(\omega^2 r)^{-1} = s(1 + s\omega^2 r/v_{\text{scan}})$ [296]. Applying these considerations to the movement across the whole solution column leads to [296, 339]:

$$\frac{s^*}{s} = \left(1 - \frac{s\omega^2(b-m)}{v_{\text{scan}} \log(b/m)}\right)^{-1} \tag{4.16}$$

resulting in an overestimate of the sedimentation coefficient, dependent on the ratio between the scan velocity and the sedimentation coefficient. Analogous considerations hold true for the fluorescence detection system, which starts out at the highest

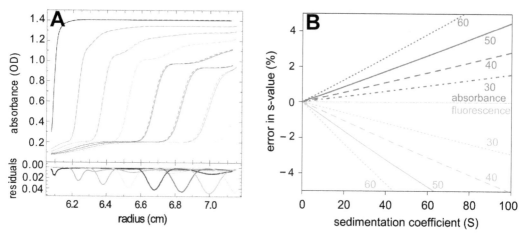

Figure 4.13 The effect of finite scan time on the apparent sedimentation coefficient. *Panel A:* Absorbance data for a SV experiment on thyroglobulin ($s = 19.1$ S) at 50,000 rpm. Shown are the experimental data (solid lines), the calculated absorbance profiles that would have been recorded with an instantaneous detection (dashed lines), and their difference in the panel below [339]. *Panel B:* Relative error of apparent sedimentation coefficients due to the finite speed of scanning as a function of s-value and rotor speed. For the absorbance system (magenta), standard data acquisition parameters (velocity mode, interval 0.003 cm, one repetition) are assumed, leading to a scan velocity of 2.53 cm/min from low to high radii. Resulting errors are shown for a rotor speed of 30,000 (dash-dotted line), 40,000 (dashed line), 50,000 (solid line), and 60,000 (dotted line) rpm. For the fluorescence detection (cyan), a scan speed of ~1.3 cm/min is observed, with scans proceeding in the opposite direction from high to low radii. The correct s-values are only obtained when accounting for the time-dependence of the scanning velocity directly in the evaluation of the numerical solutions to the Lamm equation (Eqs. (1.10) and (4.17)).

radii and scans in the opposite direction from high radii to small radii, and therefore produces underestimates of the sedimentation coefficients.

For small errors, one may predict the effect of a finite scan velocity using Eq. (4.16) and multiply the final result accordingly. Based on standard scan parameters for the absorbance system (see Section 5.7.3), with v_{scan} ~2.5 cm/min [339], the ~4.3 S BSA monomer will have an error in s of 0.18% at 50,000 rpm. Thyroglobulin, on the other hand, with a larger sedimentation coefficient of ~19.1 S will experience a larger error of 0.89% under the same run conditions. The scan rate for the fluorescence acquisition system is slower at ~1.3 cm/min at 50,000 rpm. Unless corrected for scan time, significant errors in s-values can therefore be expected with absorbance and fluorescence acquisition when studying very large particles, such as nanoparticles, viruses, or multi-protein complexes, which will migrate fast even at very slow rotor speeds.[34] In these cases, a more accurate description of scan time effects directly in the sedimentation model is prudent.

The impact of a finite scan velocity on the recorded boundary shape can be

---

[34]Such errors do not occur with the interference optical system, which can contribute to differences in s-values measured with this and absorbance or fluorescence detection [339]. The possibility to recognize such scan time errors is an additional advantage of using multiple detection systems.

qualitatively estimated by approximating the boundary shape as an error function with width σ. The time interval required to scan through the boundary is $dt = \sigma/v_{\text{scan}}$. During this time the boundary will have moved by $dr = s\omega^2 r dt$, leading to a relative error in the measured spread of the boundary by $dr/\sigma = s\omega^2 r/v_{\text{scan}}$. We would expect the apparent diffusion coefficient D^* characterizing the boundary spread to be overestimated by the square of this factor, and based on numerical simulations reported in [339], for standard SV absorbance settings the error would exceed 1% for particles sedimenting in excess of 21 S. However, this does not seem to be as significant as the error in s-values due to the inherently lower precision of molar mass measurements.

Finally, we should also expect scan-speed dependent artifacts in the shape of the solvent plateau, which during the scan will continuously decrease due to the ongoing sedimentation (Eq. 4.19 below). Quantitatively, however, even in cases of strong scan time effects this will barely exceed typical signal-to-noise levels [339]. Furthermore, this may be partially compensated for by the theoretical slight increase in the plateau from radial changes in solvent density due to pressure (see Section 2.3.1).

All of these effects can be naturally and accurately accounted for by including the scan velocity into the data analysis model when fitting the sedimentation boundary data. It is possible to explicitly add the finite scan speed by recognizing the experimental data $a^*(r,t)$ as:

$$a^*(r,t) = a(r, t_{\text{scan}} + (r - r_0)/v_{\text{scan}}) \tag{4.17}$$

where r_0 is the first radial position in the scan. As such, rather than evaluating the equation of motion of particles in the centrifugal field at a constant time for each scan, it will be evaluated at slightly different times for different radial points, mimicking the scanning of the solution column [339]. With numerical solutions of the Lamm equation, this is a trivial modification of the model, and it eliminates any concern about scan time effects for any experimental scan velocity.

An analytical approximation is available for the case of non-diffusing particles, where Eq. (1.7) will be replaced by [339]:

$$r^{(p)*}(t) = me^{s\omega^2 t} \times e^{s\omega^2(r^{(p)}-m)/v_{\text{scan}}} \tag{4.18}$$

and the plateau shape is the decaying exponential:

$$c^{(p)*}(r) = c_0 e^{-2s\omega^2 t} \times e^{-2s\omega^2(r^{(p)}-m)/v_{\text{scan}}} \tag{4.19}$$

replacing the radially constant plateau of an ideally sedimenting particle with ideal detection [339].

In SEDFIT and SEDPHAT, the Lamm equation options offer a switch to account for finite scanning velocity, dependent on the optical detection parameters used. There, the user can enter an experimentally determined scan velocity value.

4.2.4 Ensuring Temporal Accuracy

Obviously, the accuracy of the time intervals assigned to different scans will directly impact the s-value resulting from the sedimentation analysis. Unexpectedly, due to a bug in the AUC operating software, severe errors have been discovered in the recorded scan times and $\omega^2 t$-times stored in the header within the ASCII scan files of the Beckman Coulter absorbance and interference data. There is no known time error in data from the AVIV fluorescence optics. The magnitude of this error depends on the version of the GUI and firmware running on the AUC. Gross errors resulting in ~10% underreported elapsed times have occurred with the combination of the ProteomeLab GUI 6.0 and firmware 5.06 (Fig. 4.14), distributed by the manufacturer from 2011 to 2013 [370], and currently still installed on some instruments [297]. Subsequently, smaller but significant errors of 0.1%–2%, dependent on the rotor speed and scan method, were reported for prior software and firmware versions reaching back to the early 1990s and still present in more recent versions [296]. Unfortunately, this means that some of the earlier published data was subject to this error, and that the elapsed time entries from the scan files header still cannot be indiscriminately trusted.

Fortunately, the data acquisition software of the commercial instrument is designed such that there is only a small and nearly constant delay of ~10 sec between the end of the scan and the creation of the ASCII scan data file. This has been

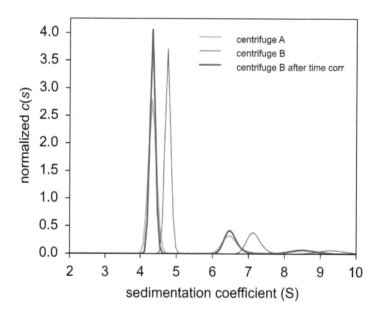

Figure 4.14 Overlay of the sedimentation coefficient distributions $c(s)$ for the same BSA sample from two analytical ultracentrifuges having different versions of the data acquisition GUI and firmware, before and after scan time corrections obtained by replacing the reported scan time by the data acquisition software with the file creation time stamps from the operating system [370].

verified in multiple ways [296, 370].[35] Thus, the computer time-stamp located in the file attributes can provide an independent measure of the elapsed time between scans [296, 370], which can be used for more accurate sedimentation analysis. Even though computer clocks are by no means high precision time measurement devices, their accuracy of measuring time intervals is typically better than 0.1% over the time-scale of hours, and therefore sufficient for SV analysis and superior to the centrifuge times. It is advantageous, therefore, to read the scan time intervals from the scan file creation times.

> A flag can be set in the SEDFIT Options to automatically screen sets of loaded data for discrepancies between file header time increments and operating system time stamps, and to create and reload time-corrected scan files. These files contain a flag indicating they have been time corrected with a dilation factor placed in the comment line, which can be recognized by SEDFIT upon reloading. By default, this option is switched on. In order to retain the original file attributes from the computer's operating system used for data collection, data files need to be transferred *in toto* within the folders used to create them.

One problem that arises with an accurate knowledge of only the time interval between scans is the unknown start time of centrifugation: by extrapolating the time dilation between scan header times and operating system times back to the start of centrifugation, potential errors of the order of seconds may occur as an offset in the overall scan time assignment (resulting in part from slight variations in the delay between the end of the scan and scan file creation). These uncertainties are small and can be easily absorbed into the degree of freedom of the effective meniscus position. This is consistent with a focus on the analysis of artifact-free scans and their accurate examination of progression with time, rather than constraining the analysis with imperfectly known spatial and temporal starting points.

4.2.5 Radial-Invariant (RI) Noise

The RI noise is a radially constant baseline that can change with time from scan to scan. It will take the form

$$a^*(r, t) = a(r, t) + \beta(t, \omega) \tag{4.20}$$

Depending on the type of experiment conducted, the change from scan to scan may involve the same rotor speed at different points in time (as in SV), or different rotor speeds (as in SE and variable-field SV). It can have multiple origins, dependent on the optical system used, the sample under investigation, and the type of AUC experiment considered. RI noise is orthogonal to TI noise (Section 4.1.4) and can be calculated efficiently and implicitly as part of the AUC data analysis. Since

---

[35]This was tested under conditions where no significantly time-consuming processes, other than AUC data acquisition were running on the computer.

time-dependent baselines of the RI noise form are very significant, accounting for them in the data analysis will often significantly improve the fit of the AUC data and increase confidence in the derived parameters.

> In SEDFIT and SEDPHAT, RI noise calculation can be switched on by checking the respective checkbox in the parameter menus. This will render the fixed baseline offset inactive.

4.2.5.1 Physical Origin

RI noise is invariably present when using the interference optics, and this was the application where $\beta(t, \omega)$ terms were first introduced in the SV boundary modeling [358]. It arises *a priori* from the lack of an absolute zero for the fringe shift, the interference pattern being periodic and only constrained by a neighboring relationship in data from consecutive columns of camera pixels. This requires the consideration of an arbitrary integer as an offset in the data, and this offset can and will arbitrarily change during the Rayleigh interference data acquisition (or rather in the processing of the image captured by the camera). This offset can be readily discerned, for example, in Fig. 1.12. In principle this integral fringe shift displacement can be removed 'manually' in the data analysis, given knowledge, for example, of a hinge region of the concentration evolution.

However, this is not necessary since a second source of RI offsets produces displacements that are typically smaller, yet highly significant, and not constrained to integer values. This is referred to as 'jitter,' physically originating from periodic temperature fluctuations and mechanical vibrations in parts of the optical system. As a consequence, the time-dependent RI offsets must be considered in the data analysis and included as degrees of freedom in the fit of interference SV data. Computationally this is very straightforward, when using modern boundary modeling (see below), and orthogonal to the computation of TI noise (Section 4.1.4). For example, the best-fit RI noise component from the analysis of the BSA data in Fig. 4.10 is shown in Fig. 4.15.

> A SEDFIT Loading Options function can align fringe scan data to coincide over a certain radial region (adding either integral or arbitrary offsets such that the mean signal in the specified region is identical). With judicious choice of the radial region (for example, in the solvent free region above the solution column), this can remove integral fringe shift and diminish jitter components. Use of this function is not required for normal data analysis, as this is accounted for, usually much more accurately, by the computation of RI noise.

This type of noise is not usually very large with the absorbance optics, and occurs infrequently in absorbance SE and virtually not at all in absorbance SV.

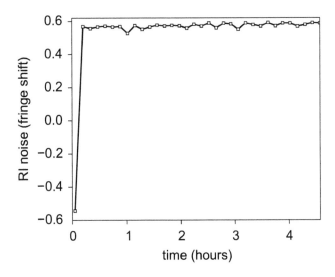

Figure 4.15 The best-fit RI noise from the Rayleigh interference profiles shown in Fig. 4.10. The first scan differs by an integral fringe shift, whereas the remaining scans are subject to only low-level RI noise (jitter).

However, it can be identified clearly, for example, when concentration distributions in SE show a higher absorbance close to the meniscus at higher rotor speeds. Possible causes might be small wavelength shifts, or drifts in the delay time that determines the angular position of the rotor when the lamp flashes upon a change of rotor speed. Buffer components that show increasing absorbance with time, such as DTT or β-mercaptoethanol, also cause a shift in baseline that can be described and properly accounted for as RI noise. Consideration of RI noise is usually also necessary when working with pseudo-absorbance data.

Fluorescence optical data do not usually show RI noise, although it could be conceived that a non-sedimenting fluorophore may provide a baseline signal and be subject to photophysical processes leading to a change in the baseline from scan to scan. None of the published data seem to exhibit such a feature.

4.2.5.2 Accounting for RI Noise

Accounting for RI noise poses the equivalent analysis problem as for the TI calculation discussed in Section 4.1.4. Again, the optimal strategy is to account for this explicitly in the analysis of the raw data and, accordingly, use a model structured as in Eq. (4.20).[36] Calculation of the RI offsets $\beta(t)$ proceeds completely analo-

---

[36]Instead, empirical alignment of scans in the air-to-air region is a pre-requisite of some analysis strategies in the absence of an explicit boundary model. However, this *ad hoc* approach increases errors [358]. Further, fringe shift offsets may not be identical at either side of the meniscus. Alternatively, empirical alignment in the solvent plateau requires exclusion of scans and assumptions on the absence of small sedimenting species. Therefore, in the context of explicit direct modeling of the raw data the computational introduction of RI offsets is more accurate and more straightforward.

gously to TI noise calculation (see Section 4.1.4): Assuming N experimental scans identified with index n, each consisting of a total of R_n radial data points $a_{n,i}$, radially numbered with index i, and a model for the boundary $B_{n,i}(\{p\})$ dependent on a set of nonlinear parameters $\{p\}$ and subject to time-dependent offsets β_n, the least-squares minimization then requires:

$$\underset{\beta_n}{Min} \sum_{n,i} \left(a_{n,i} - B_{n,i}(\{p\}) - \beta_n \right)^2 \tag{4.21}$$

from which follows, after taking the partial derivative with respect to a particular time n':

$$0 = \sum_{n,i} \left(a_{n,i} - B_{n,i}(\{p\}) - \beta_n \right) \delta_{n,n'} \tag{4.22}$$

and the explicit form of the RI noise parameter arises as:

$$\beta_n = \frac{1}{R_i} \sum_{i} \left(a_{n,i} - B_{n,i}(\{p\}) \right) \equiv \bar{a}_n^* - \bar{B}_n^* \tag{4.23}$$

for any boundary model (and with the r.h.s. defining scan-averaged values across all radii of data and model) [358]. The analysis with combined TI and RI noise proceeds analogously, with the additional condition that $\sum b_i = 0$, after a first solution with regard to RI noise is inserted as the boundary model in the determination of TI noise [358]. As with TI noise, a more complex scenario arises in partial-boundary modeling in SV, where only partially overlapping radial ranges are considered from each scan [339]. With the help of Eq. (4.23) we can also rephrase the general AUC model problem to:

$$\underset{\{p\}}{Min} \sum_{n,i} \left([a_{n,i} - \bar{a}_n^*] - [B_{n,i}(\{p\}) - \bar{B}_n^*(\{p\})] \right)^2 \tag{4.24}$$

which does not have an explicit RI noise parameter. After optimization of the parameters $\{p\}$, the particular values of the RI noise parameters that are inextricably linked to each parameter set can be calculated *via* Eq. (4.23) allowing for a construction of the full boundary model.

Finally, it is highly useful to subtract the calculated RI components (along with TI components) from the raw data for inspection of the macromolecular signals and assess in detail how well they are modeled. It must be recognized that the particular estimates of RI and TI noise parameters are estimates associated with the particular boundary model; as long as any further analysis will keep the same degrees of freedom of unknown RI offsets, this subtraction from the raw data will leave their information content invariant, and the change in display is of no consequence (other than clarifying the data display).

Menu functions and buttons on the SEDFIT and SEDPHAT graphical data display allow for the subtraction of RI noise, as well as restoration of the original raw data, to permit inspection of both TI noise and provide a clear view of the sedimentation boundaries. Also, any particular RI noise parameter estimates can be copied to the clipboard for separate inspection, if desired.

4.2.6 Time-Dependent Signal Intensity

In nearly all of the AUC literature, systems that were studied provide an invariant signal increment which is constant over time. Trivial exceptions are samples with buffer components such as DTT or β-mercaptoethanol, which exhibit an increase of UV absorbance with time, but are small molecules and show little sedimentation, so that their absorbance can usually be modeled as radially constant RI noise contributions. In contrast, time-varying signals have to be considered when working with fluorescence detection. In fact, as shown by Zhao et al. [100], time-dependent overall signal intensity changes are a general characteristics of fluorescence-detected AUC data, and their consideration in the data analysis is key to achieving a good fit that allows detailed quantitative interpretation of the data.

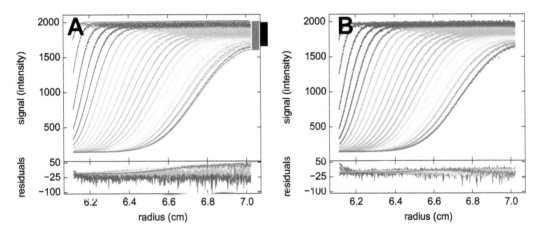

Figure 4.16 Fluorescence signal obtained during the sedimentation of EGFP at 50,000 rpm [100]. Conditions are such that no significant photobleaching of the fluorophore occurs [340]. *Panel A*: The best-fit naïve boundary model that does not account for time-dependent signal changes predicts, based on the boundary migration, a drop in the plateau signal as indicated by the length of the red bar (calculated from Eq. (4.26)). The actual drop in the fluorescence signal due to sedimentation alone is indicated by the black bar. This difference causes a significant systematic misfit of the data. *Panel B*: After allowing for a drift of the signal with time Eq. (4.27), in this case caused by a drift in the laser power increasing at a constant rate of 0.0127/hour, a good fit is achieved with correct description of the plateau region.

In general, FDS data can be influenced by a combination of both the laser drift and photophysical processes of the fluorophores. Fig. 4.16 highlights the effect of laser intensity drifts, in this case leading to a small increase in the signal intensity

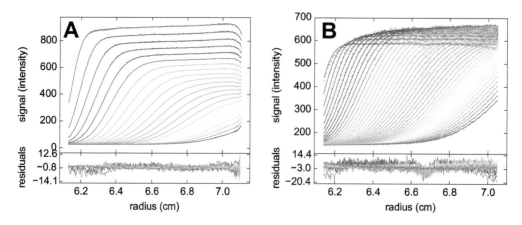

Figure 4.17 Fluorescence signal obtained during the sedimentation of photoswitchable fluorescent proteins exhibiting different photophysical behavior from the illumination at 488 nm with 50 mW during the sedimentation experiment [340]. *Panel A*: Dronpa [371], bleaching, at 25%/hour with a limiting value of 17%; *Panel B*: Padron [371], quickly increasing at a rate of 142%/hour to 150% of the initial loading signal. Fits are based on Eq. (4.28).

with time. It can be recognized clearly from an apparent diminution of the drop in plateau signals (Fig. 4.16 Panel A black bar) relative to the expected effect of radial dilution (Fig. 4.16 Panel A red bar).

With regard to photophysical effects, even for the highest available laser power settings in the current FDS system (50 mW), they will usually be negligible for GFP and commonly used FITC derivatives in standard buffers, mainly due to the transient illumination in the spinning rotor which exposes the sample to less than 1% of the laser power of a stationary system. By contrast, photophysical effects may be significant, however, for fluorophores designed to be photoswitchable, such as those recently introduced and now commonly used for super-resolution imaging [340]. An example for the fluorescence signal profile for a photo-switchable fluorescent protein designed to be easily bleached is shown in Fig. 4.17 Panel A. Likewise, molecules have been designed that are photoactivatable, i.e., increase their fluorescence during illumination at the excitation wavelength, and an example of increasing signal boundaries with time from FDS AUC studies with such molecules is shown in Fig. 4.17 Panel B. In any event, an excellent description of the sedimentation boundaries can be achieved with Eq. (4.27) and Eq. (4.28) (below), resulting in well-determined parameters for the sedimenting molecules [340].

Formally, the structure of the data $I(r,t)$ assumes the form:

$$I(r,t) = c(r,t) \times \varepsilon(t) \tag{4.25}$$

where $c(r,t)$ describes the evolution of the molecular concentration distribution, and $\varepsilon(t)$ is the time-dependent signal change. For SE experiments, this signal drift essentially hinders the application of mass conservation principles. Fortunately, SV experiments have an internal reference for the signal intensity: We have seen that the concentration in the plateau region is directly linked to the boundary position by the square dilution law Eq. (1.9), which predicts a mildly exponential decay of the plateau concentration with time, strictly linked to sedimentation. Therefore, any deviation from the predicted change in the plateau concentration with time (compare Fig. 4.16 Panel A black bar and red bar) must directly reflect the changes in the signal increment $\varepsilon(t)$.[37] Conceptually, with reference to the square dilution law Eq. (1.9), we may express this as:

$$\varepsilon(t) = \varepsilon_0 \frac{I_{xp}(t)}{I_{\text{squaredil}}(t)} = \varepsilon_0 \frac{I_{xp}(t) r_{\text{mid}}(t)^2}{I_0 m^2} \tag{4.26}$$

where I_{xp} is the observed plateau signal, considered relative to the geometrically expected $I_{\text{squaredil}}$, as a function of the initial signal I_0, and the ratio of the boundary midpoint r_{mid} and the meniscus m. However, while this highlights the source of information, Eq. (4.26) is not advantageous for actual analysis.

Computationally, as always in AUC analysis, it is preferable that the unknown time-dependence be explicitly included in the model, based on a physically meaningful parameterization, and that the unknown parameters be refined in the direct least-squares fit of the raw data. In this case, $\varepsilon(t)$ can be folded into the computation of the the Lamm equation solutions as a time-dependent multiplicative amplitude factor. Parameterizations that have been useful are [100, 340]:

$$\varepsilon(t) = \varepsilon_0 \left(1 + \alpha t\right) \tag{4.27}$$

describing a linear change, such as encountered in first order approximation of the laser intensity drift, and a decaying exponential,

$$\varepsilon(t) = \varepsilon_0 \left(1 + \beta e^{-\kappa \phi(r,s) t}\right) \quad \text{with} \quad \beta > -1 \tag{4.28}$$

such as expected in photophysical processes that reach some steady-state. In the latter, the exponential will usually be decreasing, with β at -1 for complete photobleaching (or conversion to a dark state, respectively), or between -1 and 0 for partial bleaching (or conversion). In principle, more complex temporal signal changes can be conceived. The factor $\phi(r, s)$ in Eq. (4.28) accounts for a small correction due to the radial dependence of the incident photon flux across the sample which is moving on an arc through a beam with radially invariant width (Fig. 4.18) (unpublished): Due to the rotational motion, volume elements closer to the center of rotation remain in the beam for a longer time than those further away from the center of rotation, and therefore experience a different dose of photons at each rotation. Additional small corrections can be applied for the exposure history arising

---

[37]In a Gedankenexperiment, we can imagine sedimentation in a rectangular cell, where radial dilution is absent and the solution plateau concentration is constant with time and always equal to the loading concentration. It is very clear that time-dependent changes in the signal in the plateau region, if they were to occur in such a configuration, would trivially report on $\varepsilon(t)$. In the sector-shaped cell the concentration in the plateau region is not constant, but no less an internal reference due to its predictable decrease that is geometrically linked with the boundary movement.

from the radial migration of the fluorophores accumulating different total photon counts depending on their sedimentation coefficient. This leads to the form:

$$\phi(r, s) = \phi_0 \frac{2\pi\delta_{\text{beam}}}{m\omega^2 s} \left(1 - \frac{m}{r}\right) \tag{4.29}$$

with ϕ_0 the photon flux emanating from the FDS onto the spinning rotor in a beam of width δ_{beam}) (unpublished). As a result, for example, signal profiles for bleaching fluorophores that at higher radii are illuminated less will exhibit slight positive slopes, weakly time-dependent in magnitude (compare Fig. 4.17 Panel A). The slopes appear similar to those caused by radial gradients of signal magnification (Section 4.1.5) but differ due to their dependence on the nature of the fluorophore.

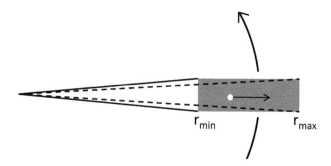

Figure 4.18 Illustration of different exposure times for samples volume elements moving on an arc across a beam with constant width (gray rectangle). The range of rotation angles at which a molecule at radius r_{min} is in the light path (indicated by the solid black lines) is larger than the angular range for molecules at radius r_{max} (indicated by the dashed lines). Since the all molecules have the same angular velocity, the ratio of angular exposure range equals the ratio of exposure times, leading the slightly lower photon flux, and smaller effect of photo-activation or photo-bleaching at higher radii. A fluorophore (white circle) migrates along the trajectory Eq. (1.7) to different radii, which allows back-calculation of the exposure history for a molecule found at a given radius at a given time, leading to the s-value and radius-dependence of the photon exposure Eq. (4.29).

4.3   SIGNAL DIMENSION

4.3.1   What Are We Detecting? The Choice of Optical System

Choosing the best optical detection system(s) for a particular application requires an anticipation of the signal contributions from all solution components (including the particles of interest, potential impurities, solvent and co-solutes) for the different optical systems, in addition to considerations of signal linearity, the concentration range of interest, and the anticipated noise structure, selectivity and sensitivity of the optical systems. It can be very informative to simulate sedimentation data with different hypothesized parameters to judge whether the question under study could possibly be answered with the planned experiment, based on the expected signal and signal-to-noise ratio. The present section recapitulates some of the necessary practical considerations.

The interference optical system offers the greatest precision in radius, time, and signal-to-noise ratio. It has the benefit that virtually all solution constituents contribute to the signal without requiring a specific label. At first sight, this would appear to make the interference optical system the default choice in most situations. However, this indiscrimination can be a drawback, as this may introduce unwanted signals from species such as buffer components. These confounding signals can be eliminated by adding these co-solutes to the buffer in the reference sector with precisely the same concentration and solution volume. If the buffer match is not completely successful, it is possible to accommodate the differential buffer signals through appropriate terms in the SV model [237]. A precise buffer match is particularly difficult to achieve with high co-solute concentrations (e.g., salts at 1 M concentrations; solutions containing sucrose or glycerol), making it sometimes more advantageous to leave them out of the reference solution and explicitly account for their signals by adding appropriate terms in the data analysis [237] (Section 4.3.4). The interference signal will also include contributions from detergents and carbohydrates when present; this can be advantageous, as we have seen, when determining particle composition in the context of a multi-signal interference and absorbance detection. IF detection will be helpful in the presence of absorbing solution components, such as nucleotides, since they have a small refractive index signal contribution. As noted above, a great advantage of the IF system is its speed of data acquisition, which makes it possible to acquire these data without incurring the time penalties of other systems (see below).

Although the absorbance system performs worse in terms of radial and temporal resolution than the interference system, this is not crucial for all but the largest particles, and it excels with regard to selectivity in that it only detects macromolecular solutes that have a specific chromophore, such as aromatic amino acids found in proteins. Signal offsets from sedimenting buffer components are rarer and generally smaller than in the interference optics, but can still occur. Unlike the interference optics, these cannot be simply compensated for in a matching reference buffer, because their absorbance will still diminish the intensity of transmitted light and consequently increase the noise of the measured signal and limit the signal linearity. Nevertheless, as discussed in more detail in Section 3.2, buffer components that absorb in the far UV, such as HEPES, TRIS, DTT, and nucleotides, can be tolerated at moderate concentration. The presence of nucleic acids or oligonucleotides at high concentrations can essentially block UV absorbance. When using the absorbance system, one should be aware that apparent absorbance signals are generated by scattering (turbidity), as well as optical artifacts from high concentration gradients, such as Wiener skewing mentioned above (Fig. 3.2).

The fluorescence detection is, of course, the system of choice for the study of samples at low concentrations, and naturally suited for molecules that have a fluorophore that can be excited at 488 nm. Extrinsic labeling is often possible but requires careful considerations and control experiments [102]. Having a confocal design, it is also best suited to study tracer sedimentation in highly concentrated solutions. As described in more detail in Section 3.2, when working at low nM or sub-nM concentrations, it is important to add carrier proteins that prevent surface adsorption of the protein of interest, requiring additional control experiments defining the absence of any interactions between the two. The carrier protein may also contribute significantly to the signal, as is the case with BSA, and these signals need to be accounted for in the analysis [99]. Similarly, other proteins present at high concentrations that do not carry the fluorophore of interest may contribute signals through inelastic scattering and/or contaminations (see Sections 3.1.1 and 5.2.4).

The choice of optical system can also be influenced by factors other than the optical properties of the sample and the solvent. This includes the question of whether SE or SV experiments are conducted. In the absence of other configurations, due to the higher temporal and spatial resolution, interference detection is usually the first choice for SV. For SE, in contrast, due to the greater complications arising from RI and TI noise with the interference optics, frequently the absorbance system is chosen.

The most common choice in our laboratories is the acquisition of data with both interference and absorbance, in particular for the study of multi-component systems allowing for a subsequent multi-signal analysis. For example, many proteins have significant differences in their interference signal increment and molar extinction coefficient at 280 nm, due to different fractions of aromatic amino acids. Simultaneous absorbance and interference analysis can frequently allow for an unraveling of their individual concentration profiles. This can be exploited both in SE [160, 166] and SV [85, 90–92] (Section 4.4). Another example where the use of multiple detection systems is extremely useful are SV experiments of solutions with detergent-solubilized membrane proteins, where the availability of signals from the protein as well as refractive index signal contributions from the co-sedimenting detergent can permit the determination of the composition of the protein/detergent micelle (see above) [97, 372].

Finally, sometimes the number of samples that need to be studied can be a deciding factor: The absorbance system requires a counterbalance, which leaves only seven rotor positions for samples in an 8-hole rotor (3 in a 4-hole rotor).[38] As described in Section 4.1.4 above, using a pseudo-absorbance data collection strategy it is possible to double the number of samples for SV experiments [356, 357]. However, this comes at the price of low-spatial frequency TI and additional RI noise components (see above), which may be detrimental when working with small particles exhibiting shallow boundaries or at very low signal levels. This detection mode is not supported by the manufacturer, and places constraints on the sample placed

---

[38]In some instruments it is possible, though not recommended, to run the absorbance data acquisition without a counterbalance, if no delay calibration is chosen. This requires a determination of the delay calibration in a separate run with the same rotor, just prior to the experiment of interest.

in the reference sector.[39] Use of the interference optics does not require a counterbalance, and can therefore be operated with samples in all rotor positions. The fluorescence system does require a specialized counterbalance, but no reference solutions, so that fourteen samples can be studied in a single run with an 8-hole rotor.

4.3.2 Signal Increments

Signal increments are important when designing experiments to make sure that the loaded sample concentration is adequate for the optical detection in the experiment. Conversely, when using the absorbance or interference optics in the proper signal range, it is possible to measure concentrations directly if a signal increment is known. Knowledge of the concentration is crucial, for example, when determining binding constants, or for the quantitation of relative subpopulations in polydisperse samples. Signal increments may be derived independently from knowledge about the sample (e.g., the amino acid composition of proteins), or they may be measured in multi-signal AUC experiments.

4.3.2.1 Absorbance/Pseudo-Absorbance

In the absorbance system, the signal from sedimenting particles at a certain concentration is determined by their extinction coefficient ε, following the Beer–Lambert law $A = \varepsilon \times c \times l$ (Eq. 1.13) (with pathlengths of $l = 3$ mm or 12 mm in standard AUC centerpieces). For proteins not carrying any absorbing prosthetic groups or absorbing co-factors, it can be convenient to base the interpretation of the signal on the extinction coefficient predicted from amino acid composition. Using the method of Pace et al. [82], the average deviation between predicted and measured extinction coefficients is ~4%, though some proteins deviate from predictions by as much as ~20%. With some exceptions, the theoretically predicted value will be satisfactory to establish a relationship between absorbance and concentration.

> The Options menu of SEDFIT has a function that calculates protein extinction coefficients on the basis of amino acid composition, requiring a 1-letter code sequence to be pasted in a text edit window.

However, the exact extinction spectrum will depend on the protein secondary and tertiary structure [373]. Therefore, in the context of a study of interacting systems with multi-wavelength or multi-signal acquisition, where data with signal-

---

[39]It can conflict with the adjustment of the photomultiplier gain which is carried out automatically at 6.5 cm in the reference sectors, if a change in absorbance by more than ~0.2 OD occurs during the run. This also requires the reference sector to have a smaller absorbance than that of the corresponding sample sector. In addition, the absorbance of the reference sector should not exceed 0.5 OD, as this may result in very high photomultiplier voltages.

to-noise ratio of 100–1000 need to be modeled, the precision of the theoretical estimate of extinction coefficient ratios is not adequate for relating different signals to each other.

Nucleic acids absorb ultraviolet light with a maximum extinction at 260 nm and average extinction coefficients for double-stranded DNA, single-stranded DNA and RNA are 0.02 $(\mu g/mL)^{-1}cm^{-1}$, 0.027 $(\mu g/mL)^{-1}cm^{-1}$ and 0.025 $(\mu g/mL)^{-1}cm^{-1}$, respectively. Extinction coefficients for single-stranded DNA oligomers can be calculated based on their sequence [374–376],[40] provided they do not partially fold into double-stranded structures. Sequence based predictions can also be carried out for wholly double-stranded DNA, after accounting for hypochromicity [377].[41] In the case of RNA, which can adopt various secondary structures, the extinction coefficient can be readily determined by comparing the absorbance of the intact molecule with that of a limit hydrolysate.[42] As in the case of protein extinction coefficients, the predicted value will be sufficient to establish a relationship between absorbance and concentration. However, the predicted coefficients for nucleic acids are only accurate in the pH range of 6.5 to 8.5 and may depend on the actual solution conditions; specifically the presence of cations that bind to the phosphate backbone may result in a slight decrease of the extinction coefficient.

In multi-wavelength experiments, it is usually sufficient to assume absolute knowledge on the extinction coefficient at a single wavelength (see Section 6.1.4 for wavelength calibration), and treat the extinction coefficients at the other wavelengths as floating parameters to be determined from the experimental data. For interacting systems of multiple protein components, this should be done for each component in separate experiments, such that the precise extinction ratio (as it appears based on the absorbance optics of the AUC) can be fixed for the subsequent analysis of the interacting mixtures.

In a similar manner, the absorbance extinction coefficient can be experimentally determined from combined absorbance and interference detection multi-signal experiments, where the interference signal increment is used as a reference. This is based on the observation that the refractive index increment can usually be predicted with greater precision (see below). An experiment designed to measure the extinction coefficient is best carried out as a SV run applying conditions such that

---

[40]Websites such as http://biophysics.idtdna.com/ may be used to calculate extinction coefficients for single-stranded and double-stranded DNA oligomers.

[41]Extinction coefficients for double-stranded DNA may be calculated directly, after accounting for hypochromicity [377]. Alternatively, the spectrum of the double-stranded DNA may be compared to one obtained for the identical sample heated to 95 °C, resulting in the formation of two single-strands. Care needs to be taken that no solvent losses occur due to evaporation. DNA may also be digested with nucleases and the spectra compared.

[42]Websites such as http://www.scripps.edu/california/research/dna-protein-research/forms/biopolymercalc2.html may be used to calculate extinction coefficients for RNA. A limiting hydrolysis with NaOH is required [378] to account for hypochromicity from possible secondary structure and double-stranded regions. A comparison of spectra for the RNA sample and hydrolysate will provide the extinction coefficient.

clearly defined boundaries can be generated. The sedimentation boundaries can then either be modeled globally using the multi-signal SV models in SEDPHAT to directly fit molar extinction coefficients. Alternatively, a separate fit with a $c(s)$ model may be carried out, followed by peak integration — in this case the ratio of the integrated peak signals will equal the ratios of the molar signal increments. The benefit of such a measurement of extinction coefficients over that performed in bench-top spectrophotometers is that absorbing impurities may be excluded from biasing the result, and that turbid aggregates are removed. The downside, however, is that use of the AUC for determining absolute extinction coefficients will require a calibration of the photometric accuracy of the absorbance system (Section 6.1.5).

For large particles, apparent absorbance can arise from their turbidity. This provides no further difficulty for the experimental detection, in principle. However, it can be far from trivial to predict the signal increments, which at sizes above $\lambda/20$ are dependent on particle size, shape and the acquisition wavelength [379–381].

For binding studies, it is usually assumed that the extinction coefficients of the components do not change upon complex formation, i.e., do not show hypo- or hyper-chromicity. This may not be the case, for example, in binding studies involving nucleic acids, or when studying proteins that undergo significant conformational changes. If undetected, a binding isotherm will misrepresent the complex concentration and the binding constant. A control for this is to verify whether the signal from a mixture (at concentrations populating the complex) is equal to the sum of the signals of the individual components when studied separately. If this effect is large, it becomes in itself a useful indicator for binding.

4.3.2.2 Interference

In interference optics, the measured fringe shift is proportional to the refractive index increment of the particles under study. We can rephrase the formula for the fringe shift Eq. (1.14) to define a molar fringe increment that is analogous to an extinction coefficient:

$$\Delta J = \left(\frac{dn}{dw} \right) \frac{l}{\lambda} w =: \varepsilon^{(IF)} c l \qquad (4.30)$$

which leads to the molar fringe increment:

$$\varepsilon^{(IF)} = \left(\frac{dn}{dw} \right) \frac{M}{\lambda} \qquad (4.31)$$

A compilation of reported refractive index increments of different compounds can be found in [382, 383], and a short list of representative molecules is tabulated in Appendix C. The refractive index increment of molecules with mixed composition can be calculated as the weight-average of the dn/dw values of its constituents, analogous to the partial specific volume Eq. (3.12).

With regard to amino acids, despite significant variations among the refractive index increments ranging from 0.165 mL/g for proline to 0.277 mL/g for tryptophan residues [384], these values correlate little with chemical properties of amino acids, and therefore usually average out to quite similar refractive index increments for proteins in water of (0.190 ± 0.003) mL/g (mean and 68% limit for human proteins) at 589.3 nm and 25 °C [95]. However, greater deviations can be expected for smaller proteins, and certain classes of proteins, such as fatty acid hyroxylases, reflectins, and eye lens crystallins [385]; a calculation based on the amino acid composition is always recommended [95].[43] The value of the refractive index increment is dependent on wavelength, through a Cauchy relation, and Perlman and Longsworth [386] proposed the formula:

$$\left(\frac{dn}{dw}\right)_\lambda = \left(\frac{dn}{dw}\right)_{578} \times \left(0.940 + \frac{20,000 \text{ nm}^2}{\lambda^2}\right) \qquad (4.32)$$

(where λ is the wavelength in nm) for proteins, on the basis of experimental data using different albumins and lactoglobulin as model proteins. A different wavelength dependence should be expected, however, near absorbance bands (due to the intimate relationship between the extinction coefficient and refractive index, each contributing to the imaginary and real part of the complex refractive index, respectively). As a rule of thumb, at a laser wavelength of 665 nm, with Eq. (4.32) this works out to be:

$$\varepsilon^{(\text{IF})} \approx 2.81 \times \left(\frac{M}{\text{Da}}\right) \frac{\text{fringes}}{\text{molar} \times \text{cm}} \qquad (4.33)$$

or approximately 3.38 fringes for a 1 mg/mL solution of (unmodified) protein in a pathlength cell of 1.2 cm.[44]

> SEDFIT has a `Calculator` function to calculate the refractive index increment, as well as the effective molar fringe increment, on the basis of the amino acid composition.

It should be noted that the refractive index increment dn/dw will also depend on hydration and preferential solvation, in a way that is formally analogous to the density increment $d\rho/dw$ discussed above [127,141]. Just like the density increment, this will significantly impact the refractive index increment of polyelectrolytes: the refractive index increment of DNA was found to be 0.168 mL/g in 0.2 M NaCl, but 0.131 mL/g in 0.2 M CsCl [141]. However, the availability of specific refractive index increments in common co-solutes is more limited [141]. Also, like density increments, preferential hydration in dilute solutions will not generate significant additional contrast in the refractive index.

---

[43] This is based on the compositional determination of refraction per gram, analogous to the compositional determination of partial-specific volumes, followed by the calculation of the protein refractive index via the Lorentz-Lorenz formula, and the transformation to a refractive index increment using the Wiener equation for dilute solutions [95].

[44] Typical rmsd values of 0.005 fringes translate to a minimal concentration of a few μg/ml, resulting in a refractive index change of $< 10^{-6}$.

Analogous to the density increments, the refractive index increments will also depend on the solvent for contrast. For instance, polymers with a refractive index equal to that of water will have a vanishing refractive index increment in water [241]. In solutions with high salt concentrations these polymers will exhibit negative signals in the interference optics (in analogy to negative density increments having the potential to cause flotation in sedimentation, however, these polymers have a high density and sediment). Likewise, in solutions of high salt, protein refractive index increments will be slightly reduced.

4.3.2.3 Fluorescence

The fluorescence detector only reports relative concentrations and the signal increments depend on many factors, including the optical configuration, excitation power density and wavelength, the extinction coefficient of the fluorophore, fluorescence quantum yield, the wavelength range of detection, and the photomultiplier voltage and amplifier gain setting. The fluorescence quantum yield of a fluorophore also depends on many factors, including its protonation and conformational states (which may in turn depend on the incident light), as well as surrounding chemical and solvent environment (for example factors governing the dielectric properties, dipoles and quenchers). It is therefore impossible to accurately predict *a priori* signal levels for a given fluorophore concentration. Instead, it is common practice to determine sample concentrations *via* absorbance measurements (using known extinction coefficients) on the stock solution used for dilutions. Fortunately, the magnitude of the fluorescence signal can be adjusted over a very wide range, using different photomultiplier voltages and amplifier gains, to produce adequate signals. In addition, the focal depth can be varied to minimize inner filter effects. The signal increments then follow from an *a priori* knowledge of the sample concentrations calculated from the dilution factors of the stock solutions.

As the fluorescence quantum yield depends on the local environment of the fluorophore, it is possible that this may change with macromolecular association state. This is the basis for the use of fluorescence quenching in binding assays. Such an effect can be experimentally established by comparing the fluorescence signal from a constant concentration of fluorescent molecules in the presence and absence of a large excess of non-fluorescent binding partner. This will impact the data analysis when the chemical reaction is folded directly into the model of sedimentation (i.e., direct modeling with a system of coupled Lamm equations [174, 175]). However, it will not affect the sedimentation analysis in the more common and robust multi-stage approach, where $c(s)$ sedimentation coefficient distributions are first determined based on the AUC data, leading to the determination of isotherms reflecting the concentration-dependence of the sedimentation boundary structure, which in a second stage are analyzed using mass action law and effective particle models [34, 169, 173]. In this approach, complex-linked quenching will lead to under-representation of the complex species, similar to absorbance measurements with hypochromicity, which in mass action law models can be trivially accounted for by appropriate changes in the binding constants. Specifically, the isotherm of s_w values in this situation will be formally analogous to a fluorescence anisotropy binding isotherm in which quenching can be easily accounted for.[45]

4.3.3 Signal Linearity

4.3.3.1 *Experimental Limits of Signal Linearity*

The main assumption made for most AUC data analyses is that the measured signal is proportional to sample concentration. Clearly, practical and technical limits exist for this assumption to hold.

In the Rayleigh interference optics these limits are far away from typical experimental conditions, even at concentrations of ~10 mg/mL, but their existence can be conceived in various ways when considering, for example, particle interactions that lead to changes in refractive index increments, refractive index mixing rules in concentrated solutions where simple additivity breaks down [387], nonlinear responses in the presence of high gradients [246], or other optical effects. The linearity of the absorbance signal is principally limited by stray light, bandwidth and wavelength polychromicity [64]. When acquiring data at the maximum of the extinction coefficient spectrum (280 nm for proteins, due to the presence of aromatic amino acids and 260 nm for nucleic acids), absorbance values should not exceed 1.3–1.5 OD. Lower limits of linearity are observed at wavelengths that are on a slope of the macromolecular extinction (see Section 4.4.1).[46]

When using fluorescence, many effects may lead to nonlinear detection, including inner filter effects and collisional quenching. For instance, limited linearity was reported for a prototype fluorescence detector [79]. However, when using FITC [388] and EGFP [100] as model fluorophores on the current commercial instrument, it was found that the signal can be sufficiently linear up to several hundred nM, provided the focal point was kept at short distances inside the solution, and the count-rate was far from saturating the detector (e.g., below 3000 counts). Signal count rates were found to be strictly proportional to the amplifier gain settings, however, these do not scale between different photomultiplier voltages or laser power settings.[47]

The question of focal depth and signal linearity warrants closer examination. As shown in Fig. 4.19 and Fig. 4.20, at larger focal depths and μM concentrations of EGFP the signal drops noticeably with focal depth. Thus is due to inner

---

[45] In principle, such quenching could present an additional source of information for quantifying the degree of binding, just as in steady-state fluorescence quenching, but to our knowledge no such application has been reported thus far in the literature in conjunction with SV.

[46] An apparent deviation from signal linearity can be observed under conditions when protein is adsorbed on the windows and centerpiece walls. In fluorescence detection, this can usually be rectified by addition of carrier protein (Section 3.2).

[47] The lack of predictable relationship between the photomultiplier voltage and signal is no drawback in practice: A large dynamic range spanning at least two orders of magnitude can be covered with a single photomultiplier setting, and cells can be scanned serially at multiple voltages, creating an empirical data set that can relate signal intensities from different settings.

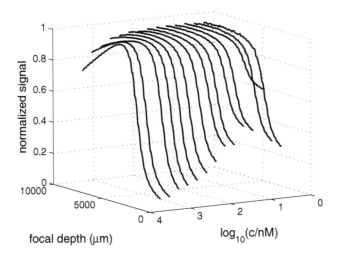

Figure 4.19 Focal scans in z-direction perpendicular to the plane of rotation into EGFP samples at various concentrations, taken radially in the middle of the solution column. The focal scans were normalized (division by the maximum signal in of each focal scan) to better compare their shapes at different concentrations [100].

filter effects, which are not observed at lower concentrations. It is therefore desirable to use shallow focal depths that minimize the optical pathlength. A perceived disadvantage of small focal depths arises from the angular mismatch between the optical scanning path and the plane of rotation which may create a slight change in focal depth with radius, resulting in a radial-dependent change of signal magnification [98, 388]. However, this can easily be accounted for in the data analysis (see Section 4.1.5), minimizing the potential for inner filter effects, and extending the range of signal linearity up to μM concentrations of EGFP (green and cyan lines in Fig. 4.20) [100].

The most sensitive approach to test for signal linearity at given experimental SV conditions is to use a rigorous $c(s)$ analysis, accounting for the characteristic FDS data structure, and determine the loading signal by a $c(s)$ integration over a well-defined peak. The resulting signal (corrected for amplifier gain) can then be divided by the molar loading concentration (as in Fig. 4.20), and the ratio tested for constancy at a given photomultiplier voltage.[48] Oversampling the same concentration at multiple photomultiplier voltages (each covering a different concentration range, but not approaching saturation) will permit the continuous extension of this test across an extremely large concentration range. This test will also highlight the dependence of signal on assembly state in the case of self-associating systems [102]. At very low concentrations, extreme care must be applied to eliminate contamination of cell components by surface-adsorbed fluorescent molecules remnant from prior experiments.

---

[48]Log-log plots of signal versus concentration will tend to obscure any nonlinear responses.

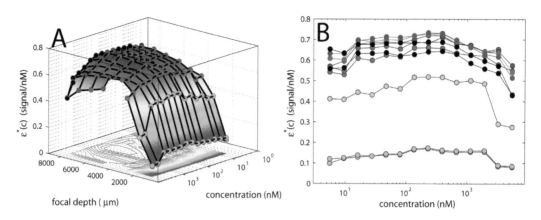

Figure 4.20 Normalized fluorescence signal as a function of EGFP concentration and focal depth [100]. The observed boundary amplitudes after a $c(s)$ integration were normalized to a gain setting of 1 and specific signal per nM concentration unit. A surface and contour plot (*Panel A*) and overlay (*Panel B*) are shown for the same data. For clarity, experimental series with different focal depths are highlighted with different color markers: 989 μm in green (two sets), 2055 μm in light blue, 3055 μm and 3095 μm in red, and 5055 μm in purple, 6055 μm in magenta, 7055 μm in dark blue, and 8000 μm in black (two sets).

4.3.3.2 Effect of Nonlinear Signals on Observed Sedimentation Parameters

It is worthwhile to consider, in theory, the impact of unaccounted-for signal nonlinearity on the analysis [100]. Qualitatively, the decreasing signal increments at higher concentrations will cause lower curvature in SE, and therefore an underestimate of the molar mass or binding constant. In SV this would impact the interpretation of the recorded sedimentation boundary if we consider the half-concentration point as the boundary midpoint, and compare this to the half-signal point (Fig. 4.21). Due to the compression of the signal from the lower signal increment at the higher concentrations, the boundary signal midpoint will be lagging the concentration midpoint, by a small amount at steep boundaries, but a larger amount at shallower boundaries. Since the boundary slope decreases with time due to diffusion, the mid-point lag will increase with time, causing an apparent reduction in the sedimentation velocity. Further, the slopes at the signal midpoints will be reduced relative to the concentration slopes, which will lead to an apparently lower molar mass in SV.

A more quantitative analysis has been provided in the Appendix of ref. [100]. Let us assume that experimental signals $f(r,t)$ can be empirically described over a suitable concentration range as a power law dependence on the local concentration $c(r,t)$:

$$f(r,t) = A\big(c(r,t)\big)^{\beta} \tag{4.34}$$

with a constant scaling parameter A and power law coefficient β, the latter capturing the nonlinearity in the signal response for values $\beta < 1.0$.[49] The power-law

---

[49]Strictly also for values $\beta > 1.0$, but this is not experimentally observed.

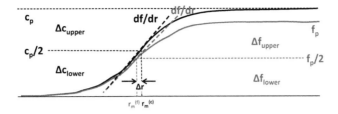

Figure 4.21 Schematic of a sedimentation boundary in concentration units (black) and nonlinear signal units (red). The boundary can be subdivided in a lower and upper half, in both concentration and signal units [100].

approximation of the signal lends itself particularly well to predict the potential effects of signal nonlinearity on the sedimentation analysis, and may be regarded as a first-order Taylor approximation of the logarithm of signal as a function of logarithm of concentration.

In SE, the apparent molar mass is determined by the slope $M \sim d\log(c)/dr^2$, from which it follows that the apparent molar mass derived from the signal units M^* is reduced by a factor that equals the power coefficient:

$$M^* = \beta M \tag{4.35}$$

When dealing with signal nonlinearity one may be tempted to truncate the signal subjected to analysis to a small concentration range where detection is deemed sufficiently linear. However, if the nonlinearity of the detection is described by the power-law Eq. (4.34), then the apparent molar mass would be the same, independent of the concentration range of the data set considered. For SV, the computational analysis leads to a first-order expression for the lag:

$$\Delta r \approx K_\beta \frac{c_p}{2} \left(\frac{dc}{dr} \right)^{-1} \tag{4.36}$$

with the abbreviation:

$$K_\beta = \left(1 - 2^{1-\beta} \right) / \left(\beta - \left(1 - 2^{1-\beta} \right) \right) \tag{4.37}$$

representing the fraction of the boundary half-width by which the signal midpoint is lagging the concentration midpoint. Using a Faxén approximation to the Lamm equation this can be rephrased as:

$$\Delta r \approx K_\beta e^{\omega^2 st} \sqrt{\pi Dt} \tag{4.38}$$

which highlights how the lag grows with time. In a first order approximation, this leads to an error in sedimentation coefficient of:

$$\Delta s^* \approx K_\beta \frac{1}{\omega^2 mt} \sqrt{\pi Dt} \tag{4.39}$$

or, in the mid-point of a typical 12 mm solution column extending from $m = 6.0$ cm to $b = 7.2$ cm, for proteins with $\bar{v} = 0.73$ mL/g spinning at 50,000 rpm:

$$\frac{\Delta s^*}{s} \approx K_\beta \times 55 \times \left(\frac{M}{Da}\right)^{-\frac{1}{2}} \tag{4.40}$$

For example, a relative error of 3% in s would be observed for a 50 kDa molecule with a β-value of 0.85 (i.e., conditions where the upper half of the boundary is compressed by ~80%, $K_\beta = -0.11$).

The error in the diffusion coefficient can be approximated by:

$$\frac{D^*}{D} \approx \left(\frac{1 + K_\beta}{\beta}\right)^2 \tag{4.41}$$

such that the molar mass from the boundary spread and migration appears as:

$$\frac{M^*}{M} \approx \frac{\beta^2}{(1 + K_\beta)^2} \left(1 + K_\beta \frac{55}{\sqrt{M/Da}}\right) \tag{4.42}$$

With the same β-value of 0.85, the error in the molar mass for the 50 kDa protein is ~10%.

4.3.3.3 Linearizing Data Transformation

The power coefficient β represents the slope in a plot of the logarithm of signal vs logarithm of concentration, and can be determined experimentally. If an estimate of the effects above seem significant, it is possible in principle to either directly fit for β with a nonlinear relationship Eq. (4.34), or back-transform the data onto a linear scale. Even though a direct fit of the original data with a nonlinear transformation of the model is preferred for fundamental statistical reasons, the difficulty is that linear parameters will now become nonlinear, eliminating optimization algorithms and rendering certain models unfeasible (particularly, distribution models). This motivates the use of a linearizing transform of the data, justified by the observation that error distortions will be minimal for small values of β. This may require an estimate of baseline signals (e.g., dark currents, etc.).

> SEDFIT offers linearizing transforms of SV data for carrying out computational steps of data analysis in a linear signal space, along with back-transforms into the original data space for inspection of the quality of fit.

4.3.4 Buffer Mismatch Signals

As we have seen above, buffer components can generate significant signal contributions, in particular, in the interference optics (Fig. 4.22). For example, the sedimentation of 400 μL 100 mM NaCl, which has a total signal amplitude of ~17 fringes in a standard 12 mm pathlength centerpiece, creates a signal difference of ~3 fringes from meniscus to bottom at 50,000 rpm [237], a signal magnitude similar to that of ~1 mg/mL of protein! (A more extreme example is shown in Fig. 4.27 below, where the sedimentation of 600 mM CsCl at 50,000 rpm results in signal difference of ~20 fringes from meniscus to bottom, i.e., equivalent in fringe shift amplitude to a boundary of ~7 mg/mL protein.)

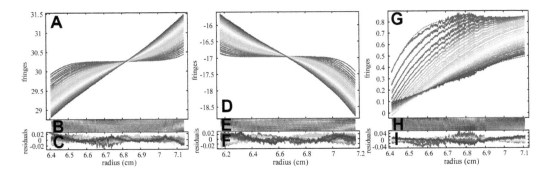

Figure 4.22 Signal profiles of 150 mM NaCl in SV experiments at 50,000 rpm, 20 °C, with best-fit single-species Lamm equation solution and residuals (lower panels) [237]. (*Panels A–C*) NaCl on the sample side and water as a reference; (*Panels D–F*) Signal from NaCl on the reference side, with water on the sample side; (*Panels G–I*) Difference signal of NaCl on both sides, in the presence of a volume mismatch with 300 μL in the sample sector and 400 μL in the reference sector. This mismatch exaggerates commonly encountered inadvertent volume mismatches in order to demonstrate the principle and better visualize the signal. However, even volume differences as small as 10 μL can create signal offsets impacting the data analysis. These effects are more pronounced at higher salt concentrations, or in the presence of added co-solutes such as urea or glycerol.

Since the buffer components are ordinarily small, their sedimentation profiles will always resemble a shallow tilt, starting at the ends of the solution column and migrating into the solution with time. Since the signal is positive from the molecules in the sample side (Fig. 4.22 Panel A), and negative from the molecules in the reference side (Fig. 4.22 Panel D), the standard strategy to eliminate such undesired signals is to match them, so that they cancel out. This is the origin of the requirement for a precise buffer (and meniscus) match for reference and sample, often using dialysis or chromatography, when using the interference optics.

Matching, however, requires the same concentration of buffer components, as well as the same volume of the solution columns in both sectors. If this condition is not fulfilled, net difference signals arise, which are smaller when compared to those observed when water is used as the reference, however, they have a more complicated structure. Fig. 4.22 Panel G shows the signals resulting from a volume mismatch of 150 mM NaCl, and Fig. 4.23 shows examples of net signals of an SV boundary of BSA sedimenting at 50,000 rpm in the presence of large mismatch in the concentration of buffer salts [237]. While the mismatch in concentrations is very large in these examples, chosen here in order to emphasize the signal amplitude

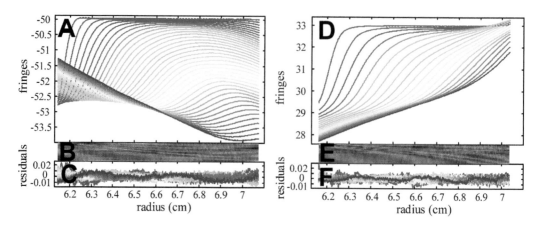

Figure 4.23 Signal profiles of BSA at 1 mg/mL sedimenting at 50,000 rpm, 20 °C, with mismatch of buffer salts. Shown are fringe shift data (*Panels A and D*), residual bitmaps (*Panels B and E*), and residuals overlay (*C and F*) of a fit with buffer mismatch model. (*Panels A–C*) twofold-concentrated PBS in the reference sector with PBS in the sample sector; (*Panels D–F*) twofold-concentrated PBS in the sample vs PBS in the reference sector [237].

and patterns, even much lower mismatches will usually bias the analysis of SV, if unrecognized.

Fortunately, it is possible to quantitatively model the profiles from small co-solute sedimentation very easily, as single species Lamm equation solutions (Eq. 1.10). This provides the basis for modeling of signal differences from the sample and reference sectors. The more significant the mismatch is, the more informative the data will be about the sedimentation and diffusion coefficients of the co-solutes, and the easier it will be to fit for these parameters. The residuals in Fig. 4.22 and Fig. 4.23 show the precision in modeling the signal from the buffer contributions and/or mismatch as an additional Lamm equation solution, or as a difference of Lamm equation solutions, superimposed to the model of the macromolecular sedimentation profile. Taking into account such difference in signals will require estimates of the difference between sample meniscus and reference meniscus. This may be treated as an adjustable parameter in the fit, although errors in the difference between the menisci only contribute second-order effects to the overall fit and may therefore be fixed (in contrast to the absolute position of the sample meniscus, see Section 4.1.2 above).

In some cases, e.g., with high glycerol or sucrose concentrations, or very high salt concentrations, it will be practically impossible to achieve a precise match. In this case, it can be easier to forgo any attempt at matching and simply treat the buffer signals, even if very large, as an independently sedimenting separate component (Fig. 4.27).

SEDFIT offers the opportunity to correct for buffer mismatch signal offsets, arising from mismatches in loading volume, concentration, and/or salt species. The switch to invoke these offsets and access to the relevant parameters can be found in the menu Options ▷ Fitting Options ▷ Corrections for Buffer Sedimentation, or with the keyboard shortcut control-B. The reference meniscus may be graphically initialized, after first switching on the meniscus mismatch mode, and then dragging the magenta line indicating the reference meniscus to the estimated meniscus of the reference solution column.

In SE, buffer mismatch signals will be much smaller due to the shallower sedimentation profiles of small molecules at the lower rotor speeds. Still, buffer matching is important in order to minimize systematic errors in the molar mass estimates and binding constants derived from the SE analysis.

Also, buffer signals are not confined to the use of the interference detection. Buffer signals can occur when using buffer components such as β-mercaptoethanol or DTT in the absorbance optics, even if a perfect concentration and volume match is achieved, due to the unavoidable variation in the degree of oxidation of these compounds, altering their UV absorbance. In these cases, the signal contribution due to buffer mismatch may change as a function of time, which can be modeled as an RI noise component. More on buffer signal contributions in the absorbance optics can be found in Section 3.2.

4.3.5 Noise

An understanding of the AUC data noise structure is indispensable for a meaningful analysis. Three main types of noise, or adventitious signal offsets, typically occur when recording concentration distributions in sedimentation experiments. The first is random noise arising from unavoidable statistical fluctuations whenever experimental data are acquired, from photon counts, electronic noise, and mechanical vibrations, among others. The second is a constant signal offset across the cell that varies from scan to scan (i.e., with time), termed RI ('radial-invariant') noise, originating from random assignment of reference phase shifts in interference optics and jitter, as described in Section 4.2.5 . The third is a signal offset characteristic for each radial position that is constant in time, termed TI ('time-invariant') noise, for example, due to surface roughness of optical elements, as described in Section 4.1.4. Thus, if the theoretical signal arising from the sedimenting particles is $a(r,t)$, the total actually measured signal $a^*(r,t)$ is given by:

$$a^*(r,t) = a(r,t) + b(r) + \beta(t) + \delta(r,t) \tag{4.43}$$

with $\delta(r,t)$ the statistical noise of each data point. As described previously, when RI and TI noise occur, their respective degrees of freedom can be built into the analysis by directly modeling the data with Eq. (4.43), and a quantitative least-squares estimate for their contributions can be obtained efficiently by using a separation of linear and nonlinear variables [358, 361, 363].

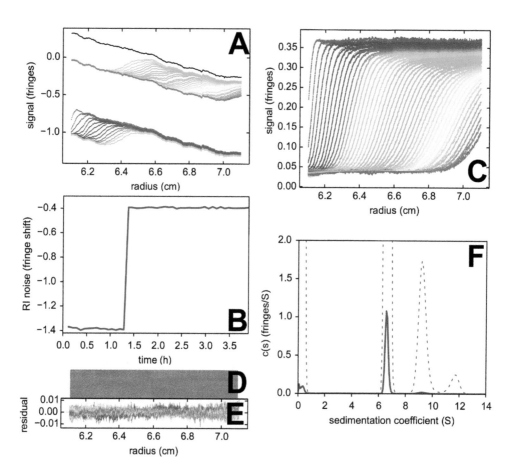

Figure 4.24 Illustration of the different noise contributions to experimental interference data collected for an IgG antibody sedimenting at 40,000 rpm in PBS [244]. The raw experimental traces (colored traces in *Panel A*, showing only every 5th point) are decomposed simultaneously into: TI noise (black line in *A*); RI components with integral fringe shifts (*Panel B*); macromolecular sedimentation signals (*Panel C*, solid lines showing best-fit $c(s)$ model; symbols showing TI and RI noise corrected data); and statistical noise (in bitmap presentation in *Panel D* and overlay in *Panel E*), as estimated from the best-fit $c(s)$ model. The resulting $c(s)$ distribution (*Panel F*, purple line, and in 100-fold magnification as blue dotted line) shows mostly monomer, but also significant contributions of dimer and higher oligomers at trace concentrations. The noise decomposition allows for a plotting of the raw data such that best-fit estimates of RI and TI components are subtracted, highlighting macromolecular migration.

A detailed example demonstrating how the raw data can be decomposed is shown in Fig. 4.24, which plots the raw interference SV data of an IgG antibody sample in Panel A. The integral fringe shifts (RI noise) and constant diagonal offset (TI noise) are dominant features, and their contributions are shown in Panel B, as estimated from the $c(s)$ boundary model of macromolecular migration in a direct fit of the structure of Eq. (4.43). The comprehensive and detailed model allows trace amounts of oligomers to be more clearly recognized (Panel C). Remaining residuals of the fit are plotted in an overlay representation in Panel D, and as bitmap in E. Ideally, these residuals should solely reflect statistical noise.

However, there are limits to the strict orthogonal distinction described in Eq. (4.43) as is sometimes apparent in interference optical data. For instance, when the RI fringe offsets contain a radial dependency from higher order vibrations of the optical elements, this will result in horizontal stripes in the residuals bitmap (faintly visible in Fig. 4.24 Panel D). Likewise, TI noise features may sometimes shift with time, due to instability and settling of the windows in the window cushions during the run, visible as localized vertical stripes in the residuals bitmap (Fig. 4.24 Panel D).

As mentioned above, it is highly instructive to subtract the estimated TI and RI noise signals from the raw data. This allows for a detailed comparison of the sedimentation data with the boundary model, and permits a better judgment of whether the remaining residuals impact the data analysis, or whether the data exhibit detailed features that are not yet captured by the model. This does not result in any change of information content, and does not introduce statistical bias as the data, for instrumental reasons, have the degree of freedom of superimposed arbitrary TI and RI noise.[50]

---

[50]Even though it will often appear as though noise subtraction has completely eliminated TI and RI noise contributions, it must be recognized that the subtracted quantities are estimated along with the sedimentation boundary model in a least-squares fit. Therefore, the same degree of freedom of TI and RI noise must be maintained in all following analyses, so as not to introduce a bias.

SEDFIT and SEDPHAT can perform TI and RI noise subtraction conveniently by pressing the keyboard shortcut control-N. Also, there are buttons above the data display in SEDFIT that allow for systematic noise subtraction, and restoration.

Once TI and RI noise are subtracted, it is possible to zoom in on particular features of the data, such as the leading and trailing edge of the boundaries, to examine the quality of the model with respect to species that sediment faster or slower than the main SV boundary. Zooming is accomplished in the data window by keeping the right mouse button pressed while dragging a rectangle describing the new plot limits. An enhanced comparison of model with data is possible after marking an integration range of $c(s)$ distributions (pressing the integration button and drawing a rectangle with the right mouse button in the $c(s)$ plot), which results in a colored highlight of the model in the sedimentation data space. After integration the boundaries are shaded gray to red with increasing red intensity reflecting increasing relative contribution of the particular marked species in the $c(s)$ distribution to the signal at each radius and time.

4.3.5.1 Statistical Noise

Statistical noise is frequently the limiting factor that determines the information content of the data. In order to have high confidence in the derived parameters and not destroy information, it is important that statistical noise is not amplified or distorted in any way. For this reason, modern analysis approaches attempt to fit the data, as much as possible, in the original data space.

The magnitude of statistical noise at zero concentration can be assessed accurately in the following way: Using an empty or water-filled cell assembly, bring the rotor to speed and take 50 scans. Load these scans into SEDFIT, and model the data with a discrete species model, switching off all parameters for macromolecular species, but switching on the RI and TI noise. After executing a RUN command, the rmsd will correspond solely to the statistical noise of data acquisition.

The statistical noise from the interference optics of an XL-I, should be well-described as a Gaussian distribution [389],[51] with a signal-independent magnitude of typically ∼0.005 fringes, though values between 0.002–0.010 fringe are common; higher values may be observed when analyzing boundaries with very high fringe count. When working with moderate maximal fringe shift magnitudes, assuming accurate adjustment of the camera exposure, noise higher than 0.01 fringes can indicate a loss of fringe contrast, frequently originating from the scattering of light. The latter can occur due to oil deposition on the lens or other element in the optical path in the rotor chamber. Also, scattering can arise within the windows from faults

---

[51]However, imperfections in the Gaussian distribution of residuals may be recognized in the residuals histogram (e.g., Fig. 4.30), which can be caused by unaccounted for systematic noise contributions, such as vibrations or higher-order interference patterns.

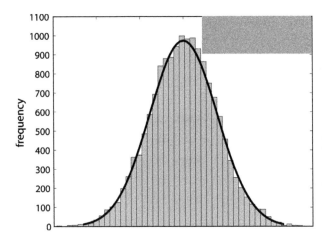

Figure 4.25 Histogram of statistical noise in the absorbance data acquired in a sequence of scans of a water-filled cell assembly, and best Gaussian fit (line) [389]. The gray rectangle in the top right corner shows the corresponding residual bitmap representation of this data set.

in the crystal lattice. Sapphire windows are preferred over quartz for their better mechanical stability.

Typically, the noise at 280 nm for absorbance optical data is of the order of 0.005 OD, although values between 0.002–0.020 OD may be observed depending on the lamp emission intensity and transmission of the solvent at the wavelength chosen for data acquisition. Here, quartz windows are somewhat preferred over sapphire for their slightly better UV transparence, especially in the far UV. The statistical properties of the noise in the absorbance data is more complicated. The absorbance data are composed of light intensity values measured with the photomultiplier, which has a Gaussian error distribution (counting light has Poisson statistics, approaching a Gaussian distribution at high count levels). However, what is reported as absorbance is the logarithm of the ratio of the reference and sample intensity values, and the resulting statistics are non-Gaussian, in theory. For this reason, Lewis and Dimitriadis have proposed directly fitting the intensity data, instead [355]. Nevertheless, in SV experiments with zero absorption (using water in both sample and reference sector) a histogram of the residuals of a fit to a large number of scans only accounting for TI noise (see below) suggested that the statistical noise can indeed be well approximated as a Gaussian distribution (Fig. 4.25) [389]. A possible origin of this apparent conflict may be the lack of consideration for the usually significant TI noise contributions in the former study, and/or the lack of absorbing macromolecules in the latter study. Statistical errors are certainly concentration dependent, increasing with higher absorbance, and individual noise estimates from the standard deviation of replicate measurements can be obtained for each data point in SE.

The statistical noise for the fluorescence detection system depends critically on the laser power, photomultiplier voltage, amplifier gain, and, to some extent, individual variations in the performance of the photomultiplier tube, making it impossible to compare absolute values. In principle, the noise should be Gaussian, but some experimental factors, such as errors in the rotor angle range for exposure, could potentially contribute to non-Gaussian behavior.

4.3.5.2 Residuals Bitmaps

In order to distinguish between the different sources of noise and diagnose the extent of their contributions to the misfit of a model following data analysis, we introduced the representation of noise as an image [151]. This novel tool provides an overview of the deviation between the data and the best fit, for many scans at once and makes use of the remarkable capability of the human eye to recognize patterns in the case of bad fits.

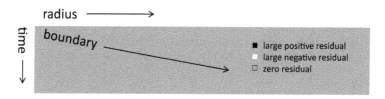

Figure 4.26 Schematics of the design of the residuals bitmap.

The image consists of a rectangular matrix of pixels, whereby each horizontal row corresponds to a scan in sequential temporal order from top to bottom, and each vertical column corresponds to a distance value at the given radial spacing of the neighboring data points (Fig. 4.26). This identifies each pixel as a data point from a particular scan at a particular radius. The color of the pixel is encoded by a gray scale, linearly mapping the magnitude of the local residual (i.e., the difference between data and fitted curves) from $-\delta_{max}$ to $+\delta_{max}$, assigned a value between 0 and 255. A residual value of zero will be neutral gray with a brightness of 128, positive residuals brighter and negative residuals darker.

SEDFIT uses different default scales δ_{max} for different optical detection systems. The default value corresponding to full black and white are 0.05 signal units in the absorbance and interference system. This value can be modified and set to different δ_{max}-values by toggling the switch for displaying the residuals bitmap in the Display ▷ Show Residuals Bitmap menu.(where a value of 0 will adjust the bitmap to a $\delta_{max} = $ rmsd). For fluorescence data, δ_{max} by default dynamically adapts to a value 3 times the rmsd.

In a data set with only statistical noise, a uniform gray pattern will appear with no discernible structure. We have already indicated the utility of the residuals bitmap in identifying contributions from different sources of noise. As shown in Fig. 4.24 Panel D, from horizontal lines it is possible to discern imperfections in the stability of the TI noise offset with time; whereas unaccounted for TI noise will show as vertical stripes, as shown most clearly in Fig. 4.8 Panel B, and unaccounted for

RI noise will show as horizontal stripes. A different form of non-random residuals is shown in Fig. 4.27, where higher-order interference patterns, likely from reflections or scattering of light in the optical path, become visible in the bitmap due to the change of solution refractive index with time, here caused by a high concentration of sedimenting co-solute. For FDS data, once the characteristic noise structure is accounted for, the residuals bitmap should look equally uniform as those from good fits of absorbance or interference data (Fig. 4.28).[52]

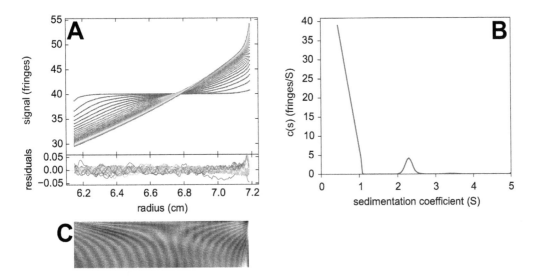

Figure 4.27   Sedimentation data of a 50 kDa protein at 50,000 rpm in the presence of 600 mM CsCl (*Panel A*). A 3 mm centerpiece was used, and no attempt to match the high salt concentration was made, leading to a strong signal offset from salt sedimentation. The data can be fit with a model of a discrete species (the salt) sedimenting at 0.445 S and diffusing at 159 F, at a total loading signal of 39 fringes, superimposed by the $c(s)$ distribution describing the sedimentation boundary of the protein at 2.3 S with a total loading signal of 1.0 fringe (*Panel B*). The residuals of the fit are 0.00992 fringes. The residuals bitmap (*Panel C*) shows oscillations indicative of a secondary interference pattern from scattering or refraction.

Importantly, the bitmap can display non-randomness of the residuals that co-migrate with the sedimentation boundary: typically these move to higher radii at later times, and therefore correspond to features in the diagonal (Fig. 4.26). Such features highlight systematic deviations of the model describing the data feature of interest, and such correlations in the residuals are particularly critical. An example is shown in Fig. 4.29, where a single-species Lamm equation model is applied on the left; this does not account for microheterogeneity in the glycosylated protein,

---

[52]One technical difficulty arising in the creation of the residual bitmaps is the necessity of a scaling algorithm: Absorbance and interference data have a roughly similar statistical noise which can be fixed to a constant default value against which the residuals grayscale is calculated. In contrast, the magnitude of noise in fluorescence data depends on photomultiplier voltage and signal gain, and is dynamically adjusted by default based on the rmsd and the minimum and maximum of the residuals, which can lead to a much more grainy and high contrast image.

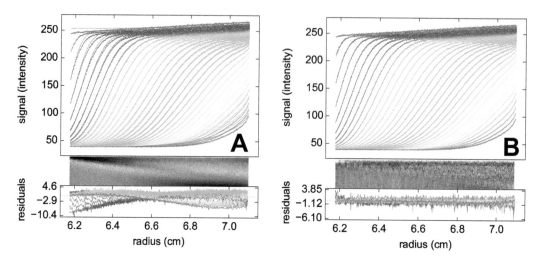

Figure 4.28 Example of the residuals bitmap prior to (A) and after (B) accounting for the specific structure of FDS data of 390 nM EGFP. (*Panel A*) Fit using the standard $c(s)$ model; (*Panel B*) Fit using a $c(s)$ model modified to account for radial signal magnification gradients Eq. (4.11), temporal signal intensity drift Eq. (4.27), shadow of the beam at the bottom of the centerpiece Eq. (4.1), and optical radial convolution Eq. (4.2) (the latter two factors only making very small contributions to the data shown, as the data set is truncated to exclude the shadow in the bottom region for better comparison).

resulting in a pronounced diagonal feature in the bitmap. In contrast, use of the appropriate model that allows for microheterogeneity as the origin of boundary spread, results in a much less pronounced diagonal feature. It is noteworthy, in this example, that the magnitude of the rmsd in the impostor fit might have seemed only slightly high. Therefore, one should apply caution in judging the fit based only on the rmsd, and instead examine correlations of misfits, such as in the residuals bitmap.

4.3.5.3   Other Measures for Randomness of the Residuals

Since it is hard to know *a priori* what the true noise in the data acquisition is in a given experiment, measures for the quality of fit that do not rely on the rmsd are very useful. More quantitative measures for randomness than the visual representation as a residuals bitmap have been developed [390]. A more quantitative approach to analyze the trends or systematicity in the residuals is a 'runs test,' which evaluates the frequency of consecutive residuals of the same sign. For normally distributed data, the number of runs R (i.e., the number of trains with consecutive residuals of the same sign) can be calculated from the total number of positive and negative n_p and n_n as [390]:

$$R = 1 + \frac{2n_p n_n}{n_p + n_n} \tag{4.44}$$

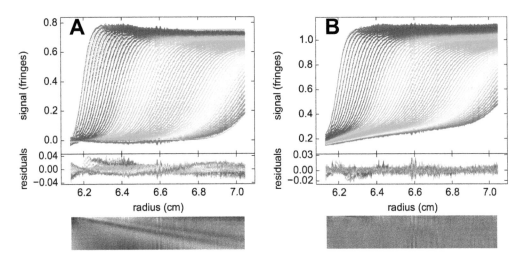

Figure 4.29 Alternative fits to SV data of a heavily glycosylated protein [244] that highlight the utility of the residuals bitmap. (Due to a different data selection, the numeric values vary slightly from those in [244].) *Panel A*: a single-species fit leads to an rmsd of 0.0090 fringes, which may be deemed almost acceptable from the overall level of noise. However, the residuals bitmap highlights an unacceptably large systematic misfit of the boundary in the pronounced diagonal feature. This correlation of errors with boundary position may go unnoticed when inspecting the overall fit of the raw data or the overlay of the residuals (and even more so if data were presented in a pair-wise difference mode). In this model, the modeled boundary spread corresponds to a mass of 61.1 kDa. (*Panel B*) A fit that properly accounts for the microheterogeneity in a $c(s)$ model, leading to an average molar mass of 92.2 kDa indicative of the dimer. The rmsd of the fit is here 0.0040 fringes, and the diagonal feature in the bitmap is substantially reduced.

with a variance

$$\sigma_R^2 = \frac{2n_p n_n (2n_p n_n - n_p - n_n)}{(n_p + n_n)^2 (n_p + n_n - 1)} \tag{4.45}$$

Based on the experimental data, the number of observed runs n_R can be compared with the expected number of runs with the quantity:

$$Z = \left| \frac{n_R - R + 0.5}{\sigma_R} \right| \tag{4.46}$$

which will be scaled as the number of standard deviations by which the observed number or runs differs from the expected number of runs for an ideally randomly distributed set of residuals [390].

SEDFIT calculates the number Z for any fit and displays this value as part of the results in the SEDFIT window.

Ideally, one would expect values of approximately 1.0 or below for an adequate fit of the data. In practice, however, this runs test is extremely sensitive to unavoidable imperfections in the data acquisition, to an extent that makes the quantitative interpretation of Z usually impossible for SV data. This is particularly the case for interference optical data, which are often limited not by stochastic noise, but rather by systematic imperfections in the stability of the data acquisition, such as vibrations causing oscillatory slopes in the RI noise (corresponding to horizontal lines in the residuals bitmap), or secondary interference patterns such as shown in Fig. 4.27. However, this quantity still offers a good relative scale to compare and rank the quality of fits to data acquired side-by-side, or consecutively with the same system on the same instrument. In the absence of strong systematic contributions, it also highlights the potential to further improve the fit.

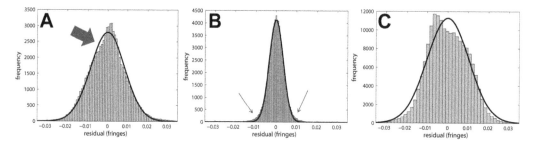

Figure 4.30 Histogram of residuals of interference optical data in comparison with the ideally expected Gaussian (solid lines). (*Panel A*) From the fits shown in Fig. 4.29 above, residuals of the impostor single discrete species fit. The arrow highlights a strong deviation from the expected Gaussian shape. (*Panel B*) From the same experimental data, the fit with the model appropriately accounting for sample microheterogeneity. The arrows indicate imperfections in the tails of the residuals statistics originating from the unaccounted for systematic errors (compare vertical stripes in the residual bitmaps of Fig. 4.29 Panel B). (*Panel C*) An extreme case of non-Gaussian residuals, from the data shown in Fig. 4.27.

Another practically useful criterion for the goodness of fit and the randomness of the residuals is the generation of a residuals histogram [391], such as presented in Fig. 4.25. In the case of poor data fits, the corresponding residuals may not follow a Gaussian distribution but instead display a more asymmetric or long-tailed distribution. The impostor and adequate fit shown in Fig. 4.29, for example, have histograms presented in Fig. 4.30. It may be discerned that the residuals from the impostor fit do not follow a Gaussian very well but are slightly asymmetric in the center. Again, the limitation of this measure for the residuals is the sensitivity to systematic imperfections in the data acquisition, which are most pronounced in the interference optical system, and results in broader tails in the residuals histogram, even in the adequate fit, than predicted by a Gaussian. On the other hand, this can also be an advantage and flag incomplete models, for example, lacking consideration of trace aggregates [391].

SEDFIT will optionally create, side-by-side with the residuals bitmap, histograms such as shown in Fig. 4.30 for the inspection and comparison of randomness in the residuals.

4.4   SPECTRAL DIMENSION

The spectral dimension can be tremendously information rich when studying heterogeneous interactions and multi-component solutions (Fig. 4.31). While there is currently no choice in the laser wavelengths for the interferometric detection system (which would only provide a small modulation of signal increments; see Eq. (4.32)), and the excitation wavelength in the commercial fluorescence detection (488 nm), an important component of the absorbance optics is the ability to freely choose any UV-VIS wavelength for absorbance measurements. As mentioned in the introductory Section 1.2.1, this is implemented through the use of xenon flashlamp that emits light across a broad range wavelengths from far UV to red, in combination with a diffraction grating located in the optical arm that serves as a monochromator and allows for wavelength selection between 200 and 800 nm, with a nominal optical bandpass of 2 nm.

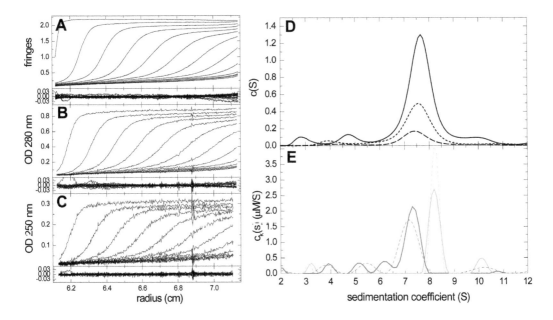

Figure 4.31 Exploiting different signals jointly in a global analysis can greatly increase the resolution of AUC experiments. Principle of multi-signal SV (MSSV) illustrated with mixtures of IgG and aldolase, both unlabeled [85]: (*Panels A–C*) Sedimentation data for a mixture of IgG and aldolase were acquired from the same solution concurrently by interference (*A*), absorbance at 280 nm (*B*) and absorbance at 250 nm (*C*). (*Panel D*) The standard $c(s)$ analyses when carried out separately on each signal, result in distributions are shown as solid (IF), short dashed (280 nm) and long dashed (250 nm) lines in the upper plot. (*Panel E*) By contrast, the joint MSSV analysis of the data using previously measured signal increments results in well-resolved peaks in $c_k(s)$ for IgG (red solid line) (and its Fab fragment) and aldolase (blue solid line), with peaks coinciding with $c(s)$ peaks measured in standard SV experiments of the separate proteins (dashed lines). Similarly, three-component mixtures can be spectrally deconvoluted.

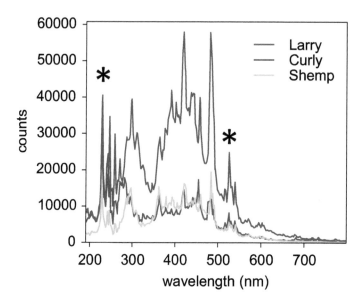

Figure 4.32   The spectrum of the xenon flashlamp measured in intensity mode for three different instruments. The absolute count rates differ from instrument to instrument. In the instruments Larry and Shemp, the relative intensity of the UV to visible light is lower than that observed for instrument Curly. The UV intensity degrades over time, leading to higher noise in the UV scans acquired. The rmsd of radial scans acquired at 280 nm was 0.0051 OD (Larry), 0.0052 OD (Curly) and 0.0026 OD (Shemp) highlighting the lack of correlation between the absolute count rate and data acquisition signal-to-noise (Fig. 4.34). Count rates of 20,000 (Larry), 15,300 (Curly), and 7,800 (Shemp) are measured at 280 nm. The stars denote the peaks at 229 nm and 527 nm that are used for wavelength calibration (Section 6.1.4).

The intensity of incident light across this range of wavelengths is shown for three different instruments in Fig. 4.32. The absolute and relative intensities are instrument dependent and, in most cases the intensity in the UV region is higher than that in the visible. Constant lamp features are the sharp intensity spikes at 229 nm and 527 nm used for wavelength calibration. To some extent, the UV intensity degrades with time and instrument use (compare instrument Shemp to Curly) and can be restored with lamp and optical path cleaning. However, while high count rates are preferable, this factor alone does not always translate into better signal-to-noise characteristics (Fig. 4.32). In a large benchmark study, only intensities below 800 counts were correlated with larger noise (as measured by the rmsd in a $c(s)$ analysis for a standard sample) [297]. It is generally true that the intensity in the visible above 550 nm is lower than that at shorter wavelengths, leading to higher noise in the yellow to red region. Similarly, spectral intensities in the far UV, below ~220 nm, are low. Therefore data collected at the maximum peptide bond extinction are usually very noisy, an effect often exacerbated by the absorption of buffer salts. However, there is a strong emission spike at 229 nm and this can be used very effectively to acquire absorbance data reporting on peptide bond extinction, with the sharpness of this 229 nm emission providing essentially monochromatic light.

In the current generation of analytical ultracentrifuges, the user interface allows for radial absorbance scans to be programmed to take place sequentially at three different wavelengths in the same run from the same solution column, in a cycle that can be repeated for a preset number of times with user-defined equidistant time-intervals. In addition, it is possible to acquire wavelength scans at a fixed radius or a series of equidistant radii.

4.4.1 Spectral Resolution

There are two main factors that limit the monochromaticity of the light in the absorbance system. First of all, higher order reflections from the diffraction grating monochromator will cause UV light to contaminate light in the visible. The diffraction grating is designed to provide high efficiency first-order reflections; consequently, higher-order reflections are much lower in intensity. Despite this, light at 280 nm will still be observed with the grating set to illuminate the sample at 560 nm. To suppress these higher-order UV contaminations in the visible region, a moveable cutoff filter on the monochromator housing can be placed into the light path when working at wavelengths greater than 400 nm (Fig. 4.33).

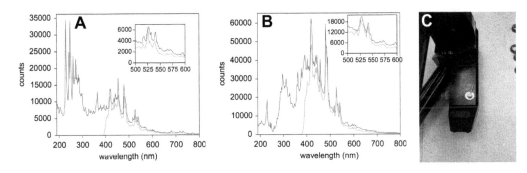

Figure 4.33 Lamp intensity spectra with (green) and without (purple) 400 nm cutoff filter for instruments Curly (*Panel A*) and Larry (*Panel B*). *Panel C*: location of the cutoff filter (shown in off position outside the light path).

The contribution of second order UV reflections to the visible may be discerned from comparison of the lamp intensity in the absence and presence of the filter (Fig. 4.33). An example of the effect of cross-contamination in the absence of the filter on the apparent signal increments in SV experiments is shown in Fig. 4.34, suggesting that at 560 nm approximately ∼7.6% of the incident light has a wavelength of 280 nm in instrument Larry, and 23.1% in instrument Curly. Clearly, this ratio depends on the spectral power output of the lamp and detailed optical properties of the monochromator, and will vary for different wavelengths and different instruments.

While use of the filter can eliminate this cross contamination, this can constrain the application of a specific type of multi-signal SV exploiting UV and visible

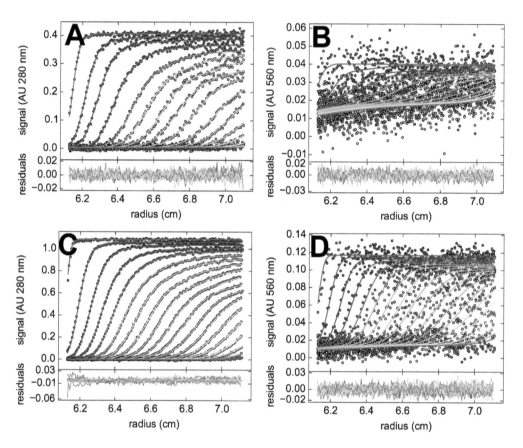

Figure 4.34 SV absorbance profiles for BSA at 280 nm (A, C) and 560 nm (B, D) collected using instruments Larry (A,B), and Curly (C,D), whose lamp intensity spectra are shown in Fig. 4.32. BSA is loaded at different concentrations (A, C), and no sedimenting boundaries are observed at 560 nm when using the UV cutoff filter, demonstrating that BSA does not absorb at this wavelength (not shown). SV data presented at 560 nm are collected without use of the UV cutoff filter, highlighting the extent of spectral cross-contamination. Based on the ratio of the total absorbance at 280 nm to the apparent absorbance at 560 nm, the contribution of 280 nm light intensity at the 560 nm monochromator setting can be calculated. Values of 7.6% and 23.1% are determined for Larry and Curly, respectively.

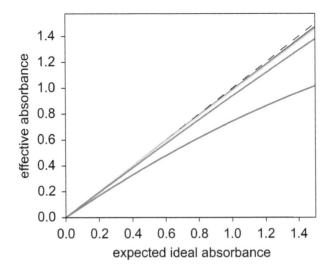

Figure 4.35 Effect of different sources and degrees of polychromicity on the linearity of absorbance detection. For reference, the black dashed line indicates an ideal detection. Blue and cyan illustrate the effect of polychromicity from a finite wavelength bandwidth: shown are the apparent absorbance values that would be expected for tryptophan when measured at its steepest slope at ~290 nm (blue) and close the low extinction tail at ~300 nm (cyan), with a finite bandwidth of 2 nm [65]. Red and magenta illustrate the effect of polychromicity from mixing of UV light in the visible absorbance detection: Red assumes 7% UV contamination (as seen in instrument Larry, Fig. 4.34) and a UV extinction that is 50% of the visible extinction. Magenta assumes 20% UV contamination (as seen in instrument Curly) and a UV extinction that is only 30% of the visible extinction. In either instrument, the effect would disappear completely if the extinction coefficients in the UV and VIS match (not shown). Use of the 400 nm UV cutoff filter will also lead to a disappearance of these effects.

absorbance simultaneously.[53] It is therefore useful to consider errors involved when not using the filter. Most chromophores absorbing at 560 nm will also absorb to some extent at 280 nm, and clearly if the macromolecular extinction at 280 nm and 560 nm is the same, then this spectral contamination will be without consequence, as long as the apparent extinction coefficient in the visible is measured experimentally in the same instrument as used for the following multi-signal SV experiment. If there is a difference in extinction coefficients at the detection wavelength λ and the contamination wavelength $\lambda/2$, then this will impact the linearity of the detection. The magnitude of this effect, for different extinction coefficient ratios, is illustrated in Fig. 4.35. At not too high absorbance values in the visible this effect is negligible. It can be minimized by using a chromophore (and degree of labeling) that results in a species having similar extinction coefficients in the UV and visible.

---

[53] This situation may arise, for example, in three component systems that rely jointly on a single chromophore in the visible, aromatic UV extinction, and interference detection for unambiguous discrimination of components. Binary systems relying only on a chromophore in the visible may usually be studied effectively with a combination of visible and IF detection (i.e., without UV signals, therefore allowing use of the filter).

The second limit of monochromaticity is the finite wavelength bandwidth $\Delta\lambda$ of the monochromator. As noted by Schachman and colleagues [64], polychromicity (from either source) limits the linearity of detection with concentration, similar to the effect of the finite radial resolution of absorbance scans Eq. (4.3): Considering the leading term of a Taylor series for the wavelength-dependence of the molar extinction coefficient,

$$\varepsilon(\lambda') \approx \varepsilon(\lambda) + (\lambda' - \lambda)\,(d\varepsilon/d\lambda)$$

we find the deviation between the measured and 'ideal' absorbance:

$$a^*(r) - a(r) = -\log_{10}\left\{\frac{1}{\Delta\lambda}\int_{\lambda-\Delta\lambda/2}^{\lambda+\Delta\lambda/2} 10^{-a\frac{(\lambda'-\lambda)}{\varepsilon}\times\frac{d\varepsilon}{d\lambda}}d\lambda'\right\} \qquad (4.47)$$

The deviations increase with the slope in the extinction profile and with increasing concentration. As illustrated in Fig. 4.35 the errors are also strongly dependent on the wavelength bandwidth, but should be small at the nominal bandwidth of 2 nm in the XL-A [65]. Therefore, it is best to measure at wavelengths where the slope vanishes, i.e., at a minimum or maximum of the extinction spectrum.[54] This will also minimize the effect of limited reproducibility of the wavelength control in the current commercial instruments (see below). For proteins, good choices usually are the maximum around 280 nm and the minimum at 240–250 nm. Furthermore, the narrow emission spike at 230 nm of the xenon flashlamp offers an opportunity to detect proteins *via* peptide bond extinction, as the narrow emission spike produces a more monochromatic illumination than would be predicted on the basis of the wavelength bandwidth.[55] Thus, absorbance at 230 nm is frequently used in absorbance optical detection of AUC experiments at low protein concentrations (Section 3.1.2), despite the strong extinction gradient observed for most proteins at this wavelength.

A proper wavelength calibration should be maintained for correct data acquisition (Section 6.1.4). Furthermore, especially when exploiting a relatively narrow absorption band, it is essential that the macromolecular absorbance spectra be measured with a wavelength scan in the actual AUC instrument to be used, and the scan wavelength will be determined by the peak absorbance wavelength from this scan.

---

[54]Conversely, poor results are often achieved when measuring on the flank of an absorption peak.

[55]If we account for the wavelength dependence of the lamp emission, $E(\lambda)$, Eq. (4.47) may be rewritten into

$$a^*(r) - a(r) = -\log_{10}\left\{\frac{1}{E^{tot}\Delta\lambda}\int_{\lambda-\Delta\lambda/2}^{\lambda+\Delta\lambda/2} E(\lambda')10^{-a\frac{(\lambda'-\lambda)}{\varepsilon}\times\frac{d\varepsilon}{d\lambda}}d\lambda'\right\}$$

which vanishes if the emission spectrum approaches a Dirac δ-function.

4.4.2   Wavelength Reproducibility of Radial Scans

Unfortunately, in the Optima and ProteomeLab XL-A/I analytical ultracentrifuges the motorized adjustment of the monochromator angle, and thereby the selected wavelength, is not always as well controlled as the precision of the wavelength measurement. This can lead to small discrepancies between the wavelength of data acquisition and the one entered in the user control software. The extent of this deviation, and whether this problem occurs in the first place, is strongly dependent on the particular instrument, and deviations are typically no larger than 2 nm. For example, if the desired wavelength is 280 nm, the scans may actually be acquired at 278 nm and, importantly, be reported as such in the scan file header.

These deviations are not reproducible, but strictly occur with wavelength adjustment. Therefore, if during an experiment absorbance data are acquired only at a single wavelength, this wavelength may not be exactly the one desired, but will remain constant throughout the experiment. If, however, the absorbance scan schedule requires the monochromator to be cycled through a sequence of alternating wavelengths, then in each cycle different wavelength errors may occur.

Usually, this is not a significant problem, considering that the optical bandwidth is about 2 nm, and, in order to achieve optimal linearity of the signal in consideration of this bandwidth, data are acquired at or close to minima or maxima of the extinction spectrum ($d\varepsilon/d\lambda = 0$), such that a small shift will have a negligible impact. Likewise, if the monochromator is properly calibrated, then a small shift in the few nm wide wavelength window around 230 nm will not significantly alter the transmission of the emission peak of the xenon lamp which is of course not changing in wavelength, even if the center of the transmission window is. Therefore, adjustment of the scan amplitudes based on an externally measured extinction profile $\varepsilon(\lambda)$ is not required and might increase data error (in particular for the 230 nm data).

In rare cases, wavelength variation may cause noticeable errors, which may be recognized from horizontal stripes in the residuals bitmap. In this case, a reliable approach to eliminate errors from wavelength variation in SV analyses is the selection of the subset of scans with identical reported wavelength. This is usually possible because the number of available scans typically far exceeds the minimal number required to define the sedimentation process well.

> SEDFIT can facilitate sorting the scan files *via* the function **Save Raw Data** and selection of the option to copy the data acquisition wavelength into the filename string.

4.4.3   Wavelength Scans

Wavelength scans $a(\lambda)$ are most frequently carried out in preliminary experiments or during the setup process of sedimentation experiments (see Section 5.6). This offers the opportunity for an accurate assessment of sample absorbance levels and

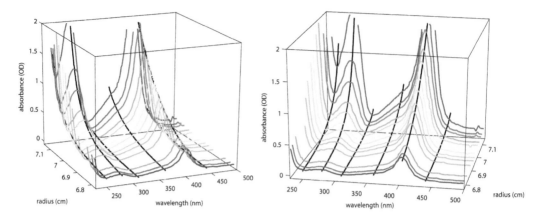

Figure 4.36 Two different perspectives of a sedimentation equilibrium absorbance data set of a non-interacting mixture of an immunoglobulin G and oxyhemoglobin in SE. The two-dimensional absorbance surface is composed of radial scans at multiple wavelengths (black lines) and wavelength scans at multiple radii (colored lines, color temperature increasing with higher radius). It may be discerned that the 280 nm peak has a steeper SE profile than the 405 nm peak, reflecting the higher molar mass of IgG, which does not contribute to the absorbance at 405 nm. Such a surface can be quantitatively decomposed into molar mass contributions and their associated extinction spectra [84].

for the selection of the optimal wavelengths for the radial scans of SE or SV experiments.

In combination with radial scans they can also be used to expand the analysis into a spectral dimension to produce an absorbance data surface $a(r, \lambda)$ in SE (Fig. 4.36), or a volume $a(r, \lambda, t)$ in SV. So far, however, incorporation of wavelength scans into the data analysis has been rather limited. In SE, such data has been used to facilitate the identification of components with different molar mass by simultaneously determining their extinction spectrum [84, 86]. Similarly, the radius- and time-dependence of wavelength scans $a(r, \lambda, t)$ acquired in a custom-built absorbance detector was exploited to help identifying large nanoparticles with different optical spectra sedimenting at different rates [392]. More quantitatively, it was shown how $a(r, \lambda)$ data acquired in the commercial AUC instrument in SE experiments can be exploited to enhance spectral discrimination of in a three-component system of interacting macromolecules, and to account for hyperchromicity on complex formation [84].

When acquiring wavelength scans, a high number of replicates and a small wavelength increment are recommended, especially if the data are part of further quantitative data analysis.

4.4.4 Multi-Wavelength and Multi-Signal Detection

The method of choice for incorporating spectral information into the sedimentation analysis is through the acquisition of radial scans at multiple wavelengths. As mentioned above, due to both potential wavelength errors and polychromicity limiting signal linearity, wavelengths should be chosen at a minimum or maximum of the macromolecular extinction spectrum, or at an emission spike of the xenon flashlamp.

In SE, data can be acquired liberally at many wavelengths, typically 230 nm, 250 nm, 280 nm for proteins, and signals in the visible spectral range if macromolecules carry chromophores that absorb there. This may be combined with interference optical detection, if the cell assemblies have been suitably prepared (Section 5.2). Data can be readily incorporated into the SE analysis [160] by globally modeling the equilibrium gradients acquired at all wavelengths, using either pre-determined extinction coefficient ratios, or treating extinction coefficient ratios as adjustable parameters, along with macromolecular concentrations. However, one of the extinction coefficients (or the loading concentration) must be known and fixed, in order to define the concentration scale.

In SV, due to the time-requirement for absorbance scans diminishing the temporal resolution, a more judicious choice of wavelengths is recommended. However, in most cases there is not complete spectral separation of the species of interest, such that the radial scans at all wavelengths jointly contribute information on the sedimentation progress of all species, and, to this extent, the temporal resolution is usually much better than the time-interval between scans of a single wavelength. In fact, it can be useful as an internal control to spectrally oversample the mixture, for example, to acquire three signals for two-component mixtures (a strategy applied, e.g., in [85, 90, 162]).

Interference optical data can be acquired quasi-simultaneously with absorbance data, without causing a significant temporal delay. For the purpose of data analysis and considering the information content of the data, the interference optical data may be considered equivalent to an absorbance signal to which all components contribute with specific signal increments (Section 4.3.2). Similarly, fluorescence optical data may be considered in the same data analysis framework as absorbance data, with specific signal increments that are non-vanishing only for the fluorescent species. Absent possible complex species that exhibit non-additive signal increments due to fluorescence quenching, which may be established independently (see Section 4.3.2) — and may similarly occur through hyper- or hypochromicity effects in absorbance — no additional considerations are necessary. Although the optical components of the FDS can be installed side-by-side with the absorbance and interference system, it is not yet clear at this time whether data can be collected *quasi*-simultaneously, or at least with cycles of scans with alternating optical systems.

The computational analysis of multi-signal SV data of heterogeneous proteins can proceed in multiple ways, dependent on the question under study. With the goal to determine binding constants, for example, for systems fulfilling very stringent purity requirements, data may be directly fit with a sedimentation/diffusion/reaction model while applying the respective signal increments to the temporal and spatial evolution of molar concentration distributions [174, 176], conceptually analogous to the SE data analysis of interacting systems. However, such an analysis in SV is much more sensitive to sample polydispersity than SE, due to the intrinsically much lower resolution of the latter. A more robust approach (with regard to contaminations and microheterogeneity) for SV is the determination of signal weighted-average s_w-values of the entire sedimenting system, and/or of boundary components, which can both be modeled simply by accounting for the respective species signal increments to the different wavelengths or signals acquired in the isotherm analysis [169].

Arguably the most powerful application of multi-signal strategies in SV is the multi-signal sedimentation velocity (MSSV) extension of the sedimentation coefficient distribution $c(s)$ to spectral component sedimentation coefficient distributions $c_k(s)$ [85, 90–92]. This approach can leverage spectral differences to strongly enhance the hydrodynamic resolution of SV [85]. Using separately measured signal increments, MSSV offers an opportunity to establish the number and stoichiometry of protein complexes (Fig. 4.31 and 4.37).[56] This is very important, as the determination of the interaction scheme is the first and often non-trivial step in the characterization of interacting systems. This technique has been key in the determination of mechanisms of protein assembly in multivalent three-component systems [393–395]. An example of a two-component system with mixed self- and hetero-association is shown in Fig. 4.37.

Due to the large number of data points in SV that provide an excellent statistical basis, and the inherent precision of the approach with signal-to-noise ratios often in excess of 100:1, relatively small spectral differences can be sufficient to spectrally resolve components in MSSV. A good quantitative predictor has been introduced in the form of D_{norm}:

$$D_{\mathrm{norm}} = \frac{\|\mathbf{det}\varepsilon_k^\lambda\|}{\prod\limits_k \|\vec{\varepsilon}_k\|} \tag{4.48}$$

which represents the relative volume of the parallelepiped formed by the extinction coefficient vectors of each component [91]. Dual-signal experiments with D_{norm} values > 0.065 usually result in spectrally distinguishable species in SV.[57] This often includes the discrimination of proteins that only differ in the number and ratio of aromatic amino acids, without extrinsic chromophores. One powerful combination is the detection by IF and absorbance at 280 nm, which can take advantage of refractive index contributions of post-translational protein modifications, as well as differences in the relative content of tryptophan residues.[58] The acquisition of absorbance data at both 250 nm and 280 nm allows one to exploit different ratios

---

[56] MSSV, with its combination of size and molar-ratio information, truly gives stoichiometry, whereas titration methods such as ITC only report the molar ratio (leaving ambiguous, for example, the presence of 1:1 versus 2:2 complexes).

[57] An alternative approach to predict feasibility of MSSV analysis implemented in SEDFIT is based on the condition number of a spectral sub-matrix for given s-values and scan times. This allows estimating the relative concentration errors of co-sedimenting components in MSSV given expected noise of data acquisition (unpublished).

[58] This combination also avoids issues of absorbance wavelength reproducibility by keeping the monochromator at a fixed wavelength throughout the experiment.

of tyrosine to tryptophan, the latter having a much larger $\varepsilon_{250}/\varepsilon_{280}$ ratio than the former. Obviously, the use of chromophoric labels absorbing in the visible can further enhance the spectral discrimination (but comes with the typical caveats concerning protein modifications, e.g., [102, 396]).

> SEDPHAT offers a calculator function to predict D_{norm} for given signal increments. Alternatively, a function for error propagation from data acquisition noise into relative component concentration errors in MSSV for species with given signal increments and s-value is available. These functions are useful to predict whether a MSSV experiment will be promising, or whether protein modification with extrinsic chromophores or expression with fluorescent fusion proteins should be pursued.

In cases where spectral discrimination is not sufficient to resolve different components in SV, it is possible to use mass conservation as a regularization constraint, in order to quantify the component contributions to the hydrodynamically distinguishable sedimentation boundaries [92]. Even though this does not lead directly to spectral discrimination, it was shown that signal conservation conditions lead to mathematical expressions equivalent to improved spectral decomposition. This strategy, termed MC-MSSV, is most effective in combination with constraints in the sedimentation coefficient range of uncomplexed species [92]. Thus, when carrying out an MSSV experiment it is advisable to design the individual sample mixtures such that mass conservation constraints are created, with well-defined loading volumes and loading concentrations.

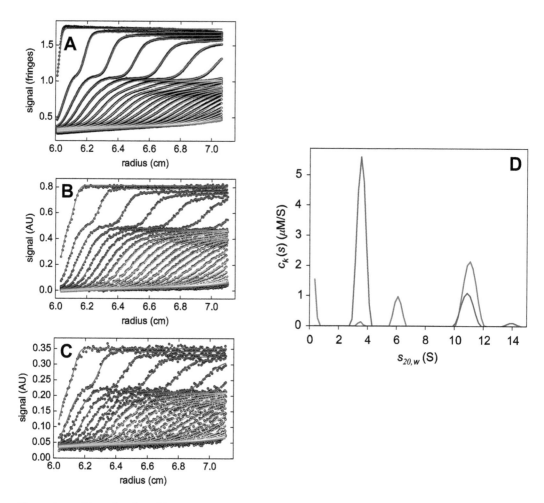

Figure 4.37 From [162], MSSV analysis of SV data acquired by interference optics (*Panel A*), and by absorbance at 280 nm (*Panel B*) and 250 nm (*Panel C*), producing component sedimentation coefficient distributions $c_k(s)$ (*Panel D*) for molecules HLA-A2 (blue) and U21 (red). In the mixture shown here, a complex at 11 S is formed with 2:1 composition and 4:2 stoichiometry HLA-A2:U21 [162].

CHAPTER 5

The Sedimentation Experiment

MAJOR considerations for the practical execution of an AUC experiment are discussed. Avoiding convection will be the one overriding principle, dictating the choice of centerpieces, their alignment, and requirements for strict temperature control. To avoid common pitfalls, the proper execution of a well-defined sedimentation experiment is considered in detail.

5.1 THE IMPORTANCE OF AVOIDING CONVECTION

In the course of an AUC experiment, it is critical that no macroscopic motion of any volume element of the solution occurs and that convection flows are absent [397]. These would add to macromolecular transport, substantially complicating data interpretation, particularly in SV experiments. Transient convection during the approach to equilibrium can usually be tolerated for conventional SE experiments.

Convective flows generally arise from density inversions, resulting from higher density regions at lower radii in the solution column. As these high density solution domains sediment into regions of lower density, solution stirring occurs. Therefore, to ensure gravitational stability, the solution density must monotonically increase with radius at all times [398]. In standard AUC experiments, no supporting density gradients such as those used in preparative ultracentrifugation are set up (e.g., sucrose gradients [399]). However, a small solvent density gradient will always occur due to the compressibility of the solvent. Co-solute concentration gradients, due to added salt and buffering agents included to stabilize pH and screen electrostatic charges, will reinforce this gradient.[1] This is sufficient and reliable in practice,

---

[1] A density increase of ∼0.008 g/mL would be expected across a 9 mm solution column at 50,000 rpm due to the compressibility of water. The presence of 0.15 M NaCl at sedimentation equilibrium would provide an additional density difference of just less than 0.002 g/mL.

however, convection can occur from different driving forces.[2]

Commonly, convective flows are generated by density inversions from non-uniform thermal expansion of the solvent due to imperfect temperature equilibration and control during the run.[3] In addition, if the walls of the centerpiece holding the solution are not sectorial and aligned radially with the axis of rotation, density inversions can be generated by local concentration gradients near the lateral boundaries of the solution column [397, 400]. In practice, these factors will limit the precision of SV [401]. Possible convection fed by mechanical vibrations of the drive [343] was problematic in the early days of AUC and towards the end of the Model E era, but is rarely observed in current instruments (XLA and XLI).[4] An additional source of gravitational instability and convection has been identified for macromolecular complexes which, due to partial volume changes upon macromolecular assembly, dissociate at higher pressures; experiments with such systems require stabilizing gradients from co-solutes [402–405]. Further, there is the potential for gravitational instability whenever boundaries are created by layering different solutions, such as in analytical zone centrifugation (Section 5.2.1), or other synthetic boundary techniques [93, 406, 407]. Conditions of high macromolecular volume occupancy can also drive convection in various ways [408].

Methods to prevent convection and ensure gravitational stability will be a common thread of discussion in the following sections.

5.2   ROTORS AND SAMPLE CELL ASSEMBLIES

5.2.1   Centerpieces

A variety of centerpiece designs have been developed for different purposes over the many decades of AUC. A comprehensive review is beyond the scope of this book, and only a few essential aspects will be discussed. Standard centerpieces have two sectors, accommodating one sample and one reference solution for absorbance and interference optical detection, or two sample solutions for fluorescence and pseudo-absorbance detection. Six-channel centerpieces can accommodate three times the number of samples, but they are only suitable for SE due to their non-sectorial geometry.

---

[2]Higher co-solute concentrations may be desired for other reasons, and thus generate stronger density gradients. However, complications may arise from the dynamically changing solution density and viscosity. These, together with contributions from preferential hydration and solvation, will complicate data analysis and interpretation.

[3]Convection driven by temperature gradients is theoretically possible in SE, but not observed in practice with current instruments. However, these factors were historically very significant in the early development of AUC [27, 343].

[4]However, some caution may be warranted when attempting SE at rotor speeds below 5,000 rpm.

5.2.1.1 Sector-Shaped Solution Columns for Unimpeded Sedimentation

While the sample compartments in most centerpieces are sector-shaped, some (not designed for use in SV) are rectangular, close to rectangular, or even circular. To avoid convection in SV experiments, it is imperative that radially-aligned sector-shaped centerpieces are used — the use of any other centerpiece would lead to local accumulation of material at the centerpiece walls and unbalanced local density increase, resulting in convection (Fig. 5.1).[5] The presence of convective flows may not be immediately obvious, and its driving forces may be reduced by diffusion, especially when studying small molecules. Despite this, if SV experiments are inadvertently carried out with non-sector shaped solution columns, the data must be dismissed, if only for the reason that all data analysis in analytical ultracentrifugation is based on radially symmetric flows, which is different from migration and diffusion in a rectangular geometry (such as in analytical electrophoresis [152]).[6]

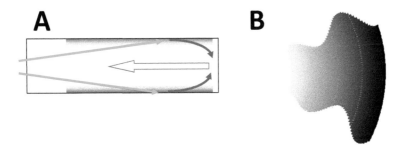

Figure 5.1 The role of centerpiece geometry in SV and SE. *Panel A*: If a rectangular solution column is used in SV, radially sedimenting molecules (green arrows) will concentrate at the side walls; the local increase in the solution density (red, wall-proximal color gradients) will result in convective flows (red arrows). *Panel B*: Any arbitrarily shaped solution column can be used for SE, as the radial increase in concentration will only depend on the distance from the center of rotation (lines of equal distance from center of rotation are drawn as thin dotted green lines), forming a barometric Boltzmann distribution (Section 1.1.3).

Misalignment of a sector-shaped cell will also lead to lateral concentration gradients and gravitational instability [234, 398]. In a careful study, Arthur and colleagues [412] estimated that inadvertent misalignment by as much as 0.5° may occur, accounting for the major source of cell-to-cell variability in the study of

---

[5]For example, the approach of using non-sector shaped solution columns for quantitative nanoparticle characterization [409] is fundamentally flawed in this respect. The instabilities described in [409], while not further examined, are likely of the same convective origin as observed by Svedberg and colleagues for non-sector shaped cells [397].

[6]Sedimentation models for rectangular geometry have been often used as a simplifying approximation to gain theoretical insights in the sedimentation process and boundary structure [34, 410, 411]. However, these models cannot account for convective flows that will in fact occur in rectangular solution columns.

Figure 5.2 Various centerpieces for AUC experiments. From left to right: standard double-sector centerpiece, six-channel centerpiece, band-forming centerpiece, and a synthetic boundary centerpiece.

trace aggregates.[7] Errington and Rowe attribute a 0.37% error in the repeatability of s-values to variation in cell alignment [295]. To achieve optimal alignment, Hersh and Schachman, in collaboration with Pickels, developed a tool that attaches to the rotor for a radial alignment of the side walls of the centerpiece [398]. This consists of a light source and a narrow slit that together create a diffraction pattern off the side walls of the centerpiece; this pattern is compared with precise reference lines on a screen. The rotor can be rotated to repeat this measurement on both sides.[8] This alignment is independent of the cell housing and scribe marks on the rotor. Unfortunately, current alignment procedures (including commercial tools) are indirect in that they rely on a visual alignment of reference marks inscribed on the cell housing and rotor [413], or on the precision of the grooves in the housing previously used for cell torquing. Both procedures are subject to error as a result of possible variations in the markings on the rotor, cell housing and/or centerpiece alignment within the housing.

For this same reason, it is critical in SV experiments that the centerpiece is manufactured with high precision and exceptional smoothness, with scratch free surfaces.[9] Scratches have been reported to be a source of convection [234]. Pekar and Sukumar have documented the effects of centerpiece imperfections on SV experiments designed to quantitate trace aggregates [414]. In a similar manner, Gabrielsen et al. have examined centerpiece quality and described higher precision centerpieces commercially available as of 2008 [401]. Any centerpiece found to be scratched on the interior walls should be designated solely for SE experiments.

In the case of SE, the shape of the centerpiece does not impact data quality, because the curvature of the gradient at equilibrium is independent of the shape of the solution column. However, this will affect the amplitude of the Boltzmann exponential for a given loading concentration, since the radial mass distribution

---

[7]Hearst and Vinograd estimated the precision of alignment by scribe marks to be ∼0.25°, and estimated the impact on the s-value to be 5% per degree of misalignment [234].

[8]A simpler device, without a light source, was commercially available from the Spinco Division of Beckman Instruments for the Model E.

[9]In order to avoid scratches during sample loading, we routinely use pipettors with polyethylene gel loading tips (Section 5.2.4). Unlike steel needles, these tips are soft and cannot damage the side walls of the centerpieces.

will depend on the solution column width. For instance, the parallel walls of six-channel centerpieces will lead to less material at higher radii when compared to sector shaped solution columns. The overall centerpiece geometry is therefore very important when implementing mass conservation models in SE analysis [160], but microscopic scratches or roughness on the side walls are irrelevant. Lateral concentration gradients may be an issue when approaching equilibrium, however these are usually smaller and on the long time-scale of the experiment they are effectively counteracted by diffusion. Consequently, even when using rectangular centerpieces for SE, the approach to equilibrium can be modeled quite well using Lamm equation solutions based on rectangular geometry [410].[10]

> For SE analysis in **SEDPHAT** it is important to enter the solution column shape in the local parameters, with a code of "2" for double sector centerpieces, and "6" for six-channel centerpieces. This is particularly critical when implementing soft mass conservation [160].

5.2.1.2 Synthetic Boundary Centerpieces

Synthetic boundary centerpieces have channels etched into one of the faces of the centerpiece.[11] Capillary forces prevent mixing when the cell is at rest, but allow liquid transfer from the reference to the sample sector at pressures encountered with rotor speeds of 5,000 – 10,000 rpm, depending on the depth of the groove [256,416]. This causes liquid transfer between sectors to essentially coincide with the start of centrifugation (Fig. 5.2) [415].[12]

Historically, synthetic boundary centerpieces had many uses, including overlaying solvent buffer (transferred from excess volume in the reference sector) over the macromolecular sample to provide initial solute-free regions suitable for various experimental and data analysis approaches [417–419]. Data from such a configuration are shown in Fig. 5.3 and 5.4. Even though this is not the standard AUC configuration, several useful applications for synthetic boundaries still exist [237,419–422].

Synthetic boundary experiments have been extremely useful for the determination of sedimentation and diffusion coefficients of very small molecules, such as buffer salts. The characterization of their sedimentation behavior may be impor-

---

[10]Fitting models with Lamm equations based on rectangular geometry are available in **SEDFIT**. These are usually implemented to examine theoretical relationships, but can be applied to the analysis of the approach to equilibrium in rectangular centerpieces.

[11]Synthetic boundary centerpieces were initially developed with a trapdoor design. The liquids were initially held in place by virtue of their surface tension, before entering the interior of the sample compartment at higher pressures [415].

[12]We successfully created such communication channels by scratching the surface of the centerpiece with a diamond knife (unpublished data).

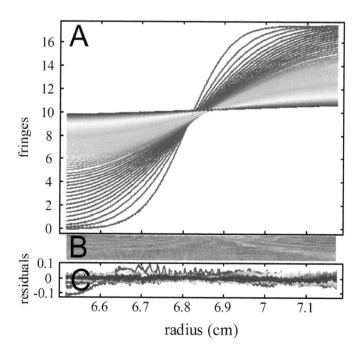

Figure 5.3 Synthetic boundary experiment with 100 mM NaCl in water at 30,000 rpm and 23 °C [237]. In contrast to a standard configuration (as in Fig. 4.22 Panel A), the synthetic boundary geometry can define the initial loading concentration very well from the amplitudes of the initial scans. The fit represented by the solid lines in *Panel A*, residuals bitmap and overlay in *Panel B* and *Panel C* is that of a single-species Lamm equation solution, with s of 0.139 S and an apparent molar mass (on the $\bar{v}$ scale of 0.73 mL/g) of 89 Da [237].

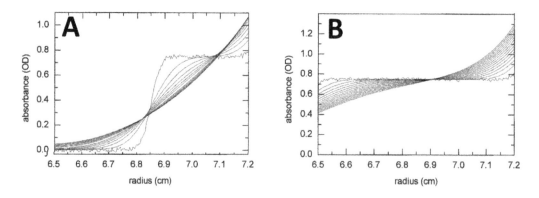

Figure 5.4 An experimental scan (noisy line) is taken as an initial absorbance distribution and propagated, using sedimentation and diffusion fluxes given by the Lamm equation, to later times. (*Panel A*) Synthetic boundary configuration; (*Panel B*) standard loading configuration [418].

tant when working with high concentrations of co-solutes, in order to determine whether dynamic density and viscosity gradients will occur from their sedimentation. The synthetic boundary configuration can accomplish this goal particularly well, by providing a clear view of the signal height for the initial boundary, which may be compared to the final concentration differences across the solution column. Knowledge of the loading signal avoids an otherwise persistent correlation between the concentration, sedimentation coefficient and diffusion coefficient resulting from the shallow, almost linear gradients in the standard configuration. If the relative concentration gradients are sufficiently large so as to cause significant density and viscosity gradients, then the s- and D-values determined from a synthetic boundary analysis can be inserted in the analysis of a conventional SV experiment. This is then used to describe the entire evolution of co-solute concentrations during the experiment in the presence of the macromolecule of interest [238, 418] (Fig. 5.4). In this manner, the dynamic density and viscosity co-solute gradients can be quantitatively accounted for in the analysis of macromolecular migration [238, 240].

The ability of the synthetic boundary centerpiece to keep reagents separate until the start of rotor acceleration offers significant experimental flexibility. Accordingly, synthetic boundary centrifugation is also of considerable importance in fields where macromolecular assemblies need to be analyzed after the rapid formation of an interface such as in supramolecular chemistry [420] and biomaterials [422]. Another application of the synthetic boundary configuration is the study of slow sedimenting boundaries in the presence of fast sedimenting macromolecules [398].

SEDFIT can analyze data from synthetic boundary experiments by switching to an initial condition that corresponds to a step function at the start of the run, with the radial position of the step corresponding to the location of the initial meniscus prior to any transfer. This model can be found in the menu Model ▷ Other Geometries ▷ Analytical Zone Centrifugation. The analytical zone centrifugation configuration (see below) is assumed as default, and the synthetic boundary configuration is indicated by the switch invert band (buffer lamella over sample) in the parameter box.

A drawback of synthetic boundary experiments is that the formation of the initial boundary is not always perfect [417]. The resulting mismatch between the perfect initial step function assumed in the data analysis and the true experimental initial concentration distribution creates a source of error that is absent in the standard SV configuration. As a result, synthetic boundary experiments are less reliable for quantitative analysis.

Such imperfections can be rigorously addressed by using the first scan collected once the rotor has reached the targeted speed as an initial condition in the Lamm equation. In this way, only macromolecular sedimentation and diffusion after the time of the first scan is modeled [418] (Fig. 5.4). The simulated diffusion quickly dissipates any noise in the experimental scan and thereby prevents statistical errors from propagating to later times. Obviously, the regions of optical artifacts close to the meniscus and bottom have to be excluded from the initial concentration distribution, and their values can be obtained by polynomial extrapolation [418].[13]

> SEDFIT provides the option to initialize the Lamm equation solutions with an experimental scan, which can be extended by polynomial extrapolation into the regions of optical artifacts towards the meniscus and cell bottom.

There are two limitations to the use of an experimental scan for initialization of the data analysis. First, it can only provide information on the total signal, which can be equated with the total concentration (in signal units) of a single species or single component. In the case of multi-component systems, additional assumptions are required to generate initial conditions for all species, such as the assumption that all species have the same initial concentration profile (the validity of which will depend on how soon after start of sedimentation the initial scan can be taken). Second, it will be difficult to distinguish between TI noise and initial signal, limiting the rigorous use of this method to absorbance and fluorescence data. In the case of interference data, an empirical estimate of TI noise will be required.

5.2.1.3 Band Centrifugation and Active Enzyme Centrifugation

The alternative strategy of overlaying a lamella of sample solution on top of a macromolecular-free solvent is also well-established in the form of analytical band centrifugation, alternatively referred to as analytical zone centrifugation [255, 256, 407]. An example of band centrifugation data is shown in Fig. 5.5. Obviously, all experimental designs of synthetic boundaries rely on establishing gravitationally stable configurations at any point in time. A critical feature of band centrifugation is that the leading edge of the lamella has a negative macromolecular concentration gradient, resulting in a negative contribution to the density gradient of the solution.[14] To ensure gravitational stability (i.e., a monotonously increasing solution density with increasing radius), this must be compensated for with an underlying positive density gradient of the solvent. This can be achieved through the use of a D_2O or $H_2{}^{18}O$ solvent in the macromolecular-free supporting column, and H_2O for the lamella of macromolecular sample. Gentler gradients based on salt concentration differences will work as well [255, 407], even with as little as 100 mM

---

[13]An initial condition determined experimentally can also be applied to SV data acquired using standard configurations as a remedy for convective processes observed in the beginning of the experiment. In contrast to the initial condition of uniform loading, which when violated will bias the entire set of scans at later times, an experimental initial taken after convective processes have subsided can provide a correct representation of the history of concentration distributions for the modeling of later scans.

[14]This limits the amount of material that can be loaded, since the negative gradient contribution increases with loading concentration.

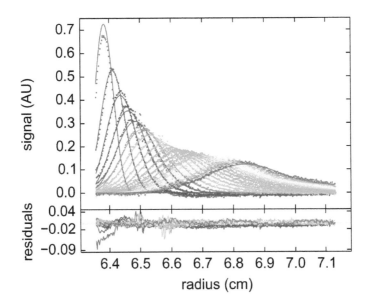

Figure 5.5 Band centrifugation of BSA at ∼1 mg/mL layered during acceleration as a lamella of 0.05 cm in H$_2$O/PBS over D$_2$O/PBS. Scans are recorded at 50,000 rpm and analyzed with a $c(s)$ distribution.

NaCl [256, 423], provided that macromolecular concentrations in the lamella are not too high. The solvent density gradient will initially be localized at the solvent interface, but with time broaden through diffusion and increasingly extend into the supporting solution column, such that the lamella is gravitationally stabilized throughout its entire migration process. Either strategy for the creation of a density gradient will cause signal offsets in the interference optics; therefore absorbance and fluorescence detection are the systems of choice for band centrifugation.

Band centrifugation, much like the synthetic boundary configuration, is not quite as quantitative and reliable as standard SV due to imperfections in the initial boundary. The initial state can be successfully modeled as a step function, or, to address potential imperfections in the overlaying process, initialized with an early experimental scan for single component samples.

SEDFIT can also accommodate the geometry of analytical zone centrifugation in the models of discrete species, as well as $c(s)$ distribution and the distribution of non-diffusing species ls-$g(s^*)$, which can be found in the Model ▷ Other Geometries ▷ Analytical Zone Centrifugation menu.

In addition to the low sample volume requirement, a key advantage of band centrifugation over standard SV is that the reagents can be kept separate until the start of the rotor acceleration. Due to the small volume of the sample lamella, diffusion also brings about a quick (though incomplete) buffer exchange. This may be exploited for the study of macromolecules in buffer conditions that induce rapid aggregation. Although this method was traditionally used for the study of nucleic acids [424], with some exceptions [256, 425], it led to the development of active enzyme centrifugation [416, 423, 426–428].

In active enzyme centrifugation, an enzyme is layered on top of a substrate-containing solution column. Enzymatic activity results in a product that can be distinctively detected by a change in absorbance, usually in the visible. This can be monitored using the absorbance optics and the leading boundary of the color change will reveal the sedimentation coefficient of the enzyme in its active form. Among the virtues of this approach is that the enzyme can be quite dilute. It does not need to be highly purified and may, in fact, be studied in the form of intact multi-protein complexes from crude extracts [426].

Active enzyme centrifugation can be analyzed in a semi-empirical fashion using difference curves to create traces resembling those of regular band centrifugation, the peak positions of which can be analyzed as a function of time. More rigorous data analysis is possible in the framework of a system of Lamm equations for species coupled by chemical reactions [428], with details on the practical implementation provided in [423].

5.2.1.4   Centerpiece Material and Design

Commercial centerpieces are made from a variety of materials exhibiting different chemical resistances to a variety of chemical solvents and aqueous co-solutes. Those most commonly used ones for aqueous applications are molded from charcoal-filled Epon[TM], a reinforced epoxy resin. Aluminum centerpieces and titanium[15] center-pieces are also commercially available, the latter requiring a special counterbalance due to their higher mass.[16] Both aluminum and titanium centerpieces require gaskets to seal the centerpiece against the window, and offer different chemical stability to organic solvents. Gaskets are not necessary for Epon centerpieces. According to the manufacturer, chemicals that soften Epon should not be used with these centerpieces. These include glacial acetic acid, >27% aqueous ammonia, chloroform, diethylene triamine, ethylene diamine, >40% aqueous formaldehyde, meta-cresol, n,n-dimethyl formamide, >10% nitric acid, >85% phosphoric acid, >70% sulfuric acid, and tetrahydrofuran [429], among others. A more detailed list of chemicals and their compatibility is supplied by the manufacturer [430]. Beyond solvent compatibility, a noteworthy distinction between metal and filled epoxy centerpieces is their permeability to gas, making aluminum centerpieces a better choice when requiring anaerobic conditions.

The two major physical parameters defining the centerpieces are the optical

---

[15]Titanium centerpieces can be obtained from Nanolytics, Potsdam, Germany.

[16]Kel-F centerpieces press-fitted into aluminum rings are no longer commercially available but may be found in long-established laboratories. These centerpieces should not be used at rotor speeds above 35,000 rpm [940].

pathlength and the number of samples that can be studied per cell.[17] Variation of the optical pathlength provides an approach to extend the dynamic concentration range suitable for the absorbance and interference detection systems. Standard centerpieces have a pathlength of 12 mm, and double-sector Epon centerpieces with a 3 mm pathlength are also available. The 3 mm ones need to be combined with spacer rings so as to position the centerpiece and windows in the middle of the housing barrel, allowing for proper imaging of the solution and for ease of external loading.[18] Titanium centerpieces are available also in 1.5 mm and 20 mm pathlengths. Ultrathin centerpieces have been created for the sedimentation equilibrium studies of hemoglobin and myoglobin at extremely high concentrations using the standard gasket of metal centerpieces [248].

Non-sector-shaped centerpieces can be used for SE. 6-channel centerpieces allow three samples of up to ~140 μL volume to be studied side-by-side in one rotor position; their installation in the cell assembly is described below (Section 5.2.4). Centerpieces with circular holes allowing the observation of 8 ultrashort (~0.8 mm) solution columns have been developed, but offer limited information content [346].[19]

5.2.2 Windows

Quartz and sapphire windows are two types of commercially available AUC windows. Sapphire windows are advantageous for most purposes because of their greater strength and scratch resistance. They are a single-crystal sapphire with the c-axis parallel to the face of the window. The axis, indicated by a scribe mark, should be aligned radially for best fringe contrast in interference optical detection [432]. Their higher strength is particularly important for interference optics [343], as it minimizes mechanical distortion in the high centrifugal field and fringe blurring observed with the quartz windows at higher fields. A disadvantage is their higher

---

[17] A design with a smaller sector angle (Spin Analytical, Inc., Berwick, ME) offers greater stability against breakage of the central septum arising from leaks during the experiment. However, the narrower solution columns can limit their use with the Beckman XLA/I absorbance optical system.

[18] The localization of centerpieces thinner than the standard 12 mm relative to the focal plane of the optics can be a potential concern, especially when studying very high concentrations, due to the different effects on optical aberrations [246]. There are different options dependent on the selection and position of spacer rings. Maintaining the mid-point of the cell assembly will create a 1.5 mm mismatch between the 2/3 plane of the centerpiece and the focal plane, if it is adjusted to the 2/3 plane of a standard 12 mm centerpiece. The potential effect of this mismatch was examined in a recent experimental study using thyroglobulin as a model system [339]: As expected, at low concentrations (0.5 mg/mL) no significant differences in the measured boundary spread and s-value were found, even if the mismatch was exacerbated to 3 mm. At high concentrations of 8 mg/mL, creating more significant refractive index gradients, significant effects were only observed in the boundary spread for the position with the larger mismatch, but not for the 1.5 mm mismatch.

[19] These centerpieces, developed by David Yphantis, have historically been very important [431], offering very rapid attainment of equilibrium due to the short solution column. Citation data show that their peak use was in the late 1960s, with no applications in the last decade.

cost and slightly lower transmittance in the far UV as compared to quartz windows [429].[20]

When using multi-signal detection exploiting both 230 nm UV absorbance and interference optical signals, a compromise may have to be made. However, good results have been achieved with both sapphire and quartz windows [85]. A good strategy may be to calculate the spectral discrimination utilizing D_{norm} based on absorbance detection at multiple wavelengths versus interference and absorbance excluding 230 nm. If component discrimination is better in the latter, sapphire windows appear to be the optimal choice, whereas quartz windows are optimal for the former. Data should still be acquired at all signals, even if one set may be noisier than usual.

As the quality of the windows has been noted to vary quite dramatically, it is good practice to examine the available sapphire windows for their UV transparence on a bench-top spectrophotometer. It is also prudent to monitor the quartz windows using the interference optical system and assess the stability of the fringe contrast and TI noise in the course of a test AUC experiment. This is particularly critical when using windows obtained from analytical ultracentrifugation laboratories that utilized the Model E and previous generation instruments.[21]

In practice, quartz and sapphire windows can be distinguished based on their mass, with the latter being heavier by approximately 2 grams. This difference in mass must be accounted for when balancing cell assemblies prior to an experiment.

Windows are held in place in aluminum holders, cushioned with gaskets and window liners. The window liners are important as they distribute the pressure on the windows in the centrifugal field, preventing breakage and mechanical distortions that would result in blurred fringes. They need to be inspected regularly and replaced if damaged. Standard window liners are made from Bakelite, but better results in terms of reproducibility of the TI noise pattern can be obtained with rigid polyvinyl chloride [359] or Teflon [434, 435] strips cut to the same dimensions as the Bakelite. This becomes important for multi-speed interference SE studies (see Section 5.2.4) [160].

5.2.3 Rotors

5.2.3.1 The Choice of Rotors

Two types of analytical rotors, made out of titanium, are available for the modern Beckman-Coulter AUC. The 4-hole rotor can be used at a maximum rotor speed of 60,000 rpm with room for three cells and a counterbalance at rotor position #4.

---

[20] Another difference between the quartz and sapphire windows is their refractive index (1.55 for quartz and 1.75–1.9 for sapphire), which leads to a different focal plane.

[21] In particular, wedge sapphire windows have to be identified and excluded from use in either absorbance or interference experiments. These were used to project images of different cells onto non-overlapping locations, allowing for data acquisition from several cells at once [433]. They may be used as a bottom window in cell assemblies when using the fluorescence detection system.

The 8-hole rotor, with a capacity of seven cells, is only rated for use up to 50,000 rpm. The choice of rotor requires a balance between the requisite sample capacity and rotor speed. Rotor speeds in excess of 50,000 rpm are often essential when studying macromolecules smaller than 10 kDa (see Section 5.5 below), but most studies on species with masses in excess of 30 kDa can be carried out using the eight-hole rotor. This is usually the default choice.

The difference in the number of samples that can be studied is exacerbated by the need for a counterbalance with radial and optical calibration markers in rotor position #8 for the absorbance system and in a user-defined position for the fluorescence optical system. The use of interference optics does not require a counterbalance.

An overspeed disk on the bottom of the rotor has a series of black and white sectors encoding for the maximum rotor speed. These sectors are read by an optical sensor near the rotor shaft, and problems often arise when the disk is damaged or scratched, resulting in a fault in the recognition of the allowable rotor speed.[22] In this case, an experiment designed to exploit the maximum rotor speed will be shut down prior to reaching the desired rotor speed. It is often possible to restart the run (after resuspending the sample and starting a new temperature equilibration phase) with a target rotor speed that is 1,000–2,000 rpm below the maximum. The rotor overspeed disk will need to be replaced prior to the next experiment.[23]

5.2.3.2   Rotor Stretching and Thermal Behavior

Even though the analytical rotors are made from titanium, they will stretch noticeably in the high centrifugal fields used [2, 436, 437]. As discussed in Section 4.1.1, a consequence of this is that the solution column will move to slightly higher radii at higher rotor speeds. The extent of rotor stretching can empirically be described well with a quadratic function $\Delta r = E\omega^2$, with the 'stretching modulus' E experimentally determined to be 8.6×10^{-12} cm/rpm^2 for the An-60 Ti four-hole rotor and 9.6×10^{-12} cm/rpm^2 for the An-50 Ti eight-hole rotor, amounting to maximal displacements of ~ 0.03 cm and ~ 0.024 cm, respectively. Although the magnitude of this displacement is small, it prevents a procedure for absolute radial calibration to be carried out at high rotor speeds.[24]

Shifts in the solution column from rotor speed-dependent rotor stretching need

---

[22]We pad the rotor stands with felt to protect the overspeed disk from scratches when handling the rotor. A newer generation of overspeed disks found on the preparative ultracentrifuges is less prone to malfunction due to scratches.

[23]When replacing the overspeed disk it is important to use the correct one for the rotor type. Overspeed disks for the analytical rotors have two magnets incorporated in them. One needs to make sure that the two magnets are present and aligned carefully with the scribe marks found on the bottom of the rotor.

[24]This does not represent a problem, since a given radial calibration of the detection system is in the resting laboratory reference frame and is valid independent of rotor speed.

to be accounted for in the global analysis of scans at multiple rotor speeds, specifically multi-speed SE [160] and variable-field SV analysis [249].[25]

Global multi-speed SE analysis offers the opportunity for a constraint in the total mass of all macromolecular components being constant across all experiments at different rotor speeds [160, 163, 165], where the total mass is determined by integration of all Boltzmann terms from meniscus to bottom. Without introducing unnecessary parameters for meniscus and bottom at each rotor speed, the translation of the solution column can be accounted for by using integration limits determined by:

$$\Delta r = E\omega^2 \quad , \quad b(\omega) = b_0 + \Delta r \quad \text{and} \quad m(\omega) = m_0 + \Delta r \tag{5.1}$$

with the zero-speed meniscus m_0 and bottom b_0 as underlying model parameters (the latter typically refined in the fit [160]). As we have seen in Section 4.1.4, another opportunity arising from the global analysis of SE profiles from the same cell at multiple rotor speeds is the determination of TI noise offsets. If these originate from imperfections in the windows, as is the case in the absorbance optical detection, Eq. (4.5) can be rewritten as:

$$a^*(r, \omega) = a(r, \omega) + b(r - \Delta r) \tag{5.2}$$

in order to account for the translation of the TI profile, for example, taking an unstretched position in the resting rotor as a reference. This leads to a slight variation in the numerical calculation of the noise profile, as described in [160].

In the direct analysis of SV carried out in time-varying fields, it is necessary to back-transform the radial dimension of scans acquired at different rotor speeds, in order to compensate for the rotor-speed and time-dependent translation of the entire solution column, thereby creating errors in the measured migration of the macromolecular sedimentation boundaries [249]. The back-transformation of the radial dimension from r to $r' = r - E\omega^2$ can be carried out on the basis of a known rotor stretching modulus, and will align the meniscus positions as well as TI noise contributions (for the absorbance system) of all scans. However, since the new radius scale would lead to an error in the calculated local centrifugal field $\omega^2 r'$ (amounting maximally to $\sim$0.5%), concomitant compensatory changes of the rotor speed value need to be carried out:

$$\omega' = \omega \sqrt{\frac{r}{r' - E\omega^2}} \tag{5.3}$$

to keep centrifugal fields invariant [249]. To arrive at a constant rotor speed for the entire scan, Eq. (5.3) may be evaluated, for example, in the middle of the solution column. This diminishes maximal errors from the radial translation to $\sim$0.05%, approximately a factor 10 below experimental errors in s-values [297].

---

[25] Neglected are additional rotor-speed dependent shifts in the radial position of the solution column at low rotor speeds from a translation of the rotor assuming a different axis of rotation [2, 437], which we believe to be small [2].

The stretching modulus for the 4-hole and 8-hole rotors is different due to their different shapes. The rotor type needs to be identified in the analysis of SE using SEDPHAT to quantitatively account for rotor stretching in the determination of TI noise.

In multi-speed SE experiments, rotor stretch modeling will account for the altered location of meniscus and bottom parameters, mass conservation, and (optionally) the translation of TI noise features. The opportunity to impose these detailed constraints in the analysis arise in multi-speed SE experiments, and is specified in the `Experimental Parameter Box`; but they are not available in the side-by-side analysis of single-speed SE experiments.

For variable speed SV analysis, the `SEDFIT` menu function `Options ▷ Loading Options ▷ Shift Scans to Undo Rotor Stretch` will back-transform ('unstretch') the loaded scans dependent on their rotor speed. This is required to achieve an analysis of macromolecular migration unbiased by translation of the entire sample solution column. In order to maintain the correct calculation of gravitational force, faithful to the force acting on the molecules in the experiment, a compensatory increase in the nominal rotor speed is applied in the scan file headers and in the associated 'speedsteps.txt' file.

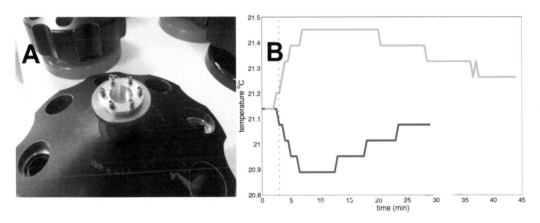

Figure 5.6 (*Panel A*) An 8-hole rotor handle modified to accommodate an iButton® temperature logger, and (*Panel B*) measured adiabatic heating (green) and cooling (blue) of an 8-hole rotor. Heating is observed when a temperature equilibrated rotor is decelerated (in this case from 50,000 rpm) to zero speed; cooling is observed when the temperature equilibrated rotor is accelerated from rest to high speed, in this case 50,000 rpm [366].

Any material with a positive coefficient of thermal expansion, such as titanium, will cool when undergoing adiabatic stretching [438, 439]. Upon acceleration to 50,000 rpm, a temperature-equilibrated 8-hole rotor will cool by ∼0.3 °C and require approximately 20 minutes to equilibrate back to the original temperature, once the external heating has compensated for the adiabatic cooling. The stored energy will be released in the form of adiabatic heating of the rotor on deceleration (Fig. 5.6). The magnitude of the adiabatic cooling depends strongly on the rotor speed since the stretching scales with ω^2 [438], and the acceleration schedule due to compensatory heat flows (Fig. 5.7) [366]. The console temperature reading will sometimes be obscured by artifacts of the radiometer [297]; however, adiabatic cooling of the rotor can be conveniently measured directly using iButton® miniature temperature loggers inserted in the rotor handle [366] (see Section 5.3 below). The current titanium rotors exhibit a ~3-fold higher stiffness than the aluminum rotors in use when this effect was originally discovered [440]. Although this has not entered standard practice, it is possible to carry out isothermal AUC experiments by slowing down the rotor acceleration such that the heat transfer matches the adiabatic cooling, resulting in a constant rotor temperature during acceleration [366] (Fig. 5.12).

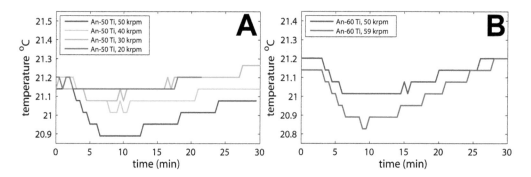

Figure 5.7 Dependence of the adiabatic cooling of the rotor on rotor type and rotor speed. (*Panel A*) 8-hole rotor; (*Panel B*) 4-hole rotor [366].

5.2.4 Cell Assembly

The AUC cell is assembled from a number of components (Fig. 5.8)— the centerpiece, which holds the sample, is sandwiched between windows, which are inserted into window holders with gaskets and liners. These components are aligned into an aluminum cell housing barrel and sealed together by torquing the cell housing screw ring. Details on how to fit these pieces together are provided by the manufacturer [429]; additionally, a detailed step-by-step protocol [441] and a video of the procedure [413] are available. It is of utmost importance that all components are in proper condition, dry and free of dust, lint or debris. Damaged or dirty window liners and gaskets will diminish baseline stability (through deformations of the windows), and increase the risk of window breakage and leaks.[26] Leaks will not only lead to sample and data loss, but can also damage the centerpiece through hydrostatic pressure differences on the centerpiece divider.

The seal between the window and the Epon centerpiece is solely based on the pressure between their flat surfaces; any dust or lint trapped on these surfaces can therefore result in leaks. Lens paper soaked with ethanol can be used to wipe these

---

[26]If the window gasket loses opacity it may need to be replaced. The liners should exhibit a certain degree of stiffness.

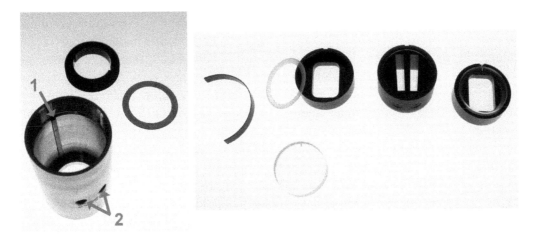

Figure 5.8 Components required for a complete cell assembly. The left panel shows the cell housing barrel with keyway (1) and holes for sample loading (2), which will be sealed with a plug gasket and cell housing plug (not shown). Also shown are the screw-ring washer and screw-ring used to hold the cell components firmly in place. The right panel shows two window assemblies and a 12 mm pathlength two-channel centerpiece. The disassembled window assembly on the left shows (from left to right on top) the window liner, window gasket and window holder. A quartz window is shown below.

surfaces. The quality of the seal can be visually inspected, once the assembly is in the barrel prior to and after cell torquing. In the case of a clean seal, the narrow air gap between the window and centerpiece divider should produce a uniform pattern of Newton rings visible after hand tightening the assembly (Fig. 5.9).[27] Once torqued, the interface should appear uniform and black, without the Newton rings.

Figure 5.9 Newton rings refracted from the sector divider of a charcoal-filled Epon centerpiece prior to cell torquing.

---

[27]Newton rings will not manifest on aluminum or aluminum-filled Epon centerpieces.

Different techniques may be used to successfully insert the window assemblies and centerpiece into the barrel. The key is to prevent any debris inside the barrel (for example, in the edges formed by the keyway slot in the thread) from falling onto the window or centerpiece surfaces, and to avoid touching these surfaces (or scratching them with a tool). The window assembly/centerpiece/window assembly sandwich may be pre-stacked on a table, aligned with the keyway so that the barrel can be carefully slid over the pre-positioned sandwich. Alternatively, the components may be dropped in one-by-one. If the aluminum cell housing barrels are slightly deformed and the centerpiece does not slide in easily (despite proper alignment of the keyway), a window assembly can be used, pre-positioned to its final orientation relative to the centerpiece, as a platform to push the window and centerpiece in place.

The assembly must be tightened and sealed by torquing within the recommended range of 120–140 inch-pounds (13.6–15.8 Nm).[28] Lower torque forces do not ensure a seal, and higher torques can deform the components resulting in leaks or cracked quartz windows. To have an accurate and reproducible torque, without producing rotational strain on the window holder [435], the screw ring washer and screw ring should be lightly lubricated.

By convention, when looking down on the cell from the screw ring side with the filling holes facing away, the right sector will be the sample sector, and the left one the reference. When utilizing the fluorescence optics, the left one will be designated 'A' and the right one 'B.' As mentioned previously, we recommend the use of pipettors with plastic gel-loading tips to load the samples and prevent scratches on the side walls of the centerpiece [413]. Precise volume matches between reference and sample side can be achieved with a proper pipetting technique, using the same pipette tip first for the reference solution (where necessary) and subsequently for the sample solution. It is essential, especially when working with buffers of low surface tension, to avoid bubbles or films that will wet the fill port. These will produce reflux of liquid out of the fill port. Air locks should also be avoided.[29] They can be formed by pushing the tip too far into the fill port, disallowing the escape of solution-displaced air. A useful technique is to slightly expel some of the remaining air in the pipette tip while retracting the tip from the filling port.[30]

One or two circular plug gaskets are used to seal each filling port after sample loading.[31] These gaskets are commercially available, however many types of flexible plastic sheeting can be used to punch out gaskets of the correct size and thickness. Plug screws should be carefully hand-tightened with a small screwdriver. Under-

---

[28] The torque stand and gauge need to operate properly for this purpose.

[29] There are centerpiece designs with two filling ports per sector. These help to avoid airlocks when loading, especially in the case of samples that contain detergents.

[30] It is virtually impossible to fix an air lock created by a wetted fill port, and to add any more solution into the centerpiece. In this case, the assembly should be taken apart, cleaned, and the whole procedure started over.

[31] This is user-dependent. A consistent technique should be developed.

tightened plug screws may result in a leak and loss of material, but over-tightened plug screws can deform the aluminum barrel, preventing the latter from sliding into the rotor hole.

A different protocol must be followed for 6-channel centerpieces, depending on their design. While 'external loading cells' have filling ports for each sector [359], standard 6-channel centerpieces do not. The current standard 6-channel centerpieces have a special cell housing that allows for the lower window assembly and centerpiece to be torqued into place. The filling port screws are used to further fix the centerpiece, and the cell is loaded in an open-face fashion. Once the centerpiece is full, the upper window assembly, screw ring gasket and screw ring are inserted and torqued.

This procedure can also be carried out using standard cell housing barrels — the entire sandwiched window assembly/centerpiece/window assembly should be inserted first into the barrel, followed by insertion of the screw ring and torquing it as specified above. The filling port screws in the standard barrels will have no matching filling port in the centerpiece, but can instead be used as 'set screws' to fix the centerpiece and lower window assembly in place while untorqueing the cell and removing the screw ring and upper window assembly [343].[32] In this state, the centerpiece is still kept in place tight enough to seal against the lower window and allow loading the sample into the 6-channel centerpieces. The upper window assembly can then be re-inserted, the screw ring screwed into the cell assembly, and the cell re-torqued after the filling port 'set screws' are taken out.

Before placing and aligning the cell assemblies into the rotor holes, their masses (which depends on the centerpiece, window and sample volume and density) need to be determined for the purpose of balancing the rotor. According to manufacturer's instructions, cells should be balanced in the rotor to better than 0.5 g. Within that tolerance, it is recommended that the counter balance be made slightly lighter than the opposite cell to anticipate possible sample loss from leakage.

The cell assemblies need to fit well into the rotor hole. Cells that are too loose, preventing a proper alignment, may be fixed by further tightening the cell housing screws. As mentioned above, care must be taken when tightening the screws to avoid damaging their slots. Sometimes an increase in the plug screw's torque can create a slight deformation of the barrel just sufficient to create a snug fit in the rotor. Alternatively, a single hair, thin fiber, or a single-ply tissue may be inserted between the cell assembly and the rotor — these items have the perfect thickness and stiffness to be used as a spacer to ensure a tighter fit.

As discussed in Section 5.2.1, especially for SV experiments it is extremely important to align the sectors so that the walls of the solution columns are perfectly radial.

---

[32]After the screw ring has been removed, this can be achieved by lifting off the upper window assembly with a clean hose attached to a vacuum pump or gently shaking the cell housing assembly.

Cell assemblies can be filled incorrectly or inserted improperly in the rotor, resulting in a variety of possible errors. A switch between the sample and reference sectors, equivalently achieved by inserting the cell assembly up-side-down into the rotor, will result in absorbance and interference data that will have a negative signal. Such data can still be analyzed when multiplied by -1, an operation that is available in the Loading Options menu of SEDFIT. When using fluorescence data acquisition, this error might be more difficult to spot, because it does not produce negative signals, but once samples are correctly identified, it is without consequence.

SV boundaries will also appear with negative signals in the absorbance and interference optics if the cell assembly is turned by $180°$ such that the sectors are divergent rather than convergent towards the center of rotation. These data cannot be analyzed and a detailed inspection of the data will reveal that sample radial concentration occurs, unlike the usual cases above where regular radial dilution occurs. In the case of SE, such data will still be fine after a change of sign; however no mass conservation models can be used.

Finally, it is possible to exchange reference and sample sector during loading, and then turn the cell $180°$. This will produce the usual positive signal, but can still be identified in SV by radial concentration rather than radial dilution. Again, such data will be unusable for SV, but fine for SE with the exception of mass conservation models.

5.2.4.1 Leaks

Leaks are a major source of anxiety, especially for beginners, as they result in sample loss and often damage of expensive components. However, with a consistently proper technique, the occurrence of leaks can be reduced to well below 1%. The key is to use intact and clean cell components and proper cell torquing, taking care not to over- or under-tighten any part of the cell assembly. Potential sources of leaks are window breakage or poor seals occurring at the cell-filling ports or at the interface of the window and centerpiece.

The risk of window breakage can be minimized by not over-torquing the cell assembly. Torquing beyond 140 inch-pound (16 Nm) will not generally help to further seal the window-to-centerpiece interface, especially in the presence of a scratch, lint or other debris wedging the window away from the centerpiece. Excessive torque can actually break quartz windows prior to application of any centrifugal field. Equally important in avoiding window breakage is an inspection of the window gaskets and liners, which cushion the window in its holder at $\sim$100,000 g, where they weigh the equivalent of hundreds of kg. As mentioned above, this cushion will also aid in reducing window distortions that diminish the fringe contrast with interference optical detection. Overall, in our laboratories our experience is such that quartz windows rarely break, in contrast to sapphire windows, which have never broken thus far.

Major factors responsible for leaks are the cleanliness and flatness of the window and centerpiece surfaces. As indicated above, visual inspection of this interface after assembly for Newton fringes and absence of any scattering indicating local elevations can effectively eliminate this source of leaks. It should be noted that with a proper cleaning technique and intact components, no grease of any kind is necessary to create a seal between the window and the centerpiece. Obviously, grease would be of concern as a source of potential sample contamination, degradation, or aggregation.

The seal at the filling port is also of concern and it is the experience of many colleagues that the use of two fill-port gaskets is good practice. The torque on the housing plug screw should be close to the comfortable limit of hand-tightening with a standard small screwdriver of the appropriate dimension with a blade width that matches the screw diameter. Over-torquing of the plug screw will be apparent in the inability to fit the then deformed aluminum barrel into the rotor hole.

Leaks may also occur across sectors if the centerpiece septum is scratched or deformed on its surface. Any indications of this may be tested with colorimetric indicator solutions to establish the presence and directionality of such leaks [419].

To minimize the risk for breakage and avoid sample losses, it is useful to identify leaks at the earliest time possible. For this reason, the pressure reading on the instrument console during the evacuation of the rotor chamber can be diagnostic, as leaks often reveal themselves as a sudden, transient increase in pressure of several hundred microns Hg before the chamber is fully evacuated.[33] If this is observed, it is worthwhile to interrupt the experiment, identify the leaking cell, disassemble, clean, and re-load.

On the rare occasions when leaks do occur, they can develop upon rotor acceleration when the pressure in the solution column increases. It is therefore prudent to monitor the samples during acceleration with the interference optics camera display using roughly adjusted laser angles and delays.[34] Even though interference data cannot be saved prior to reaching the target rotor speed, the location of the menisci can be clearly discerned, and one-sided leaks will reveal themselves by the formation of diagonal stripes in the camera picture (see Section 4.1.2). This inspection only requires a few seconds and in the case of a leak, the experiment should be stopped immediately (especially if the final rotor speed is in excess of 42,000 rpm).

It is possible, although not recommended as standard practice, to carry out an optical inspection of solution columns for leaks at 3,000 rpm. This eliminates the concurrent rotor acceleration and provides for more time, which may be helpful for beginners. However, in the case of an SV experiment, one will need to either 1) stop the low speed run, re-suspend the sample and temperature equilibrate anew prior to a restart, so as to have a well-defined and uniform loading condition for the

---

[33]If the instrument is equipped with an oil diffusion pump, a smaller increase of the order of a hundred microns Hg is commonly observed due to out-gassing of the diffusion pump oil. This occurs when the oil in the pump heats up; it is observed irrespective of whether cells are present or not in the rotor.

[34]Interference optical detection parameters need to be set up prior to rotor acceleration. The precise interference delay adjustments will not change much from run to run if the same rotor is used and can be roughly adjusted to monitor the solution columns at the earliest stages of rotor acceleration.

subsequent analysis, or 2) record the rotor speed and time of the low-speed phase and carry out a variable-field SV analysis (see Section 4.2.1).

Slow leaks that reveal themselves at full speed by depleting the solution column slightly with time result in a meniscus that travels to higher radii. Such a leak in the reference sector may be tolerable; in the data analysis one can exclude the affected radial range, apply TI signal corrections, and/or use partial-boundary modeling [339] that considers a sliding window of data unaffected by the artifact from the leak. However, the increasing pressure imbalance at rotor speeds greater than 42,000 rpm may lead to breakage or permanent distortion of the centerpiece. It may be tempting to consider a data analysis if there is a slow leak on the sample side; however, reliable results cannot be obtained in this case, as it is uncertain where the leak is and whether solution is drained from the solvent or solution plateau.

It is not general practice, but cell assemblies can, in principle, be tested for leaks prior to a sedimentation run by filling all the sectors with solvent and carrying out an experiment (no thermal equilibration required) at moderate rotor speeds. Solvent-solvent scans will indicate whether leaks are present through a meniscus mismatch or loss of solvent. Following the test run, the cells can be emptied by aspirating the solvent with a thin flexible vacuum tube bent to reach different corners of the intact cell assembly. Sample can then be inserted with minimal, for most conceivable analyses insignificant, concentration errors. This minimizes the risk for a leak.

5.2.4.2 Ageing and Baseline Blanks

AUC data analysis requires a distinction between the baseline and sedimentation signals. This is particularly important in interference optical detection, as the baseline signal may have significant low spatial frequency features that depend on the radius (Section 4.1.4). The significant feature is that these baseline profiles remain constant throughout the experiment. This is usually not a problem in SV, as the high centrifugal field will immediately compress the window cushions and cell assembly components into a new equilibrium position, which remains generally unchanged until the centrifugal force is released. In SE experiments, however, much smaller centrifugal forces are generated at the lowest rotor speed, which does not necessarily lead to a smooth and complete settling of the cell components, and instead allows for creep and adventitious changes. These pose a problem, as the calculation of TI noise in SE rests on the analysis of equilibrium data from different speeds [160], and even experimental approaches to measure baselines require stability across different speeds.

Ansevin and colleagues have therefore developed a protocol for stabilizing ('ageing') the cell components [359], to be used in conjunction with sapphire windows and reversibly compressible window cushions that replace the standard Bakelite cushions (see above). The ageing procedure consists of centrifuging at 50,000 rpm a cell assembly that is torqued to 140 inch pounds and loaded with water. After 30 min the run is stopped, and the cell assembly removed from the rotor. Usually, the components of the cell assembly will have moved somewhat, which can be diagnosed by a slight loosening of the screw ring. Therefore, the screw ring is re-torqued to 140 inch pounds. The centrifugation is then repeated. After a few cycles of high-centrifugal-field exposure and retightening, no change in torque is required after centrifugation, and the cell assemblies are stabilized [359]. In this state, water blank baselines can be measured at all the speeds required for the later experiment. This includes all speeds designed to be used in a SE experiment.

Even when using data analysis approaches to determine TI noise from multi-speed SE data, it is very useful (though not essential) to collect these water blanks and subtract them from the SE data. As a result, the calculated TI noise can be expected to only contain short-scale features, no overall drifts or steady increases or decreases towards the bottom of the cell, which would be indicative of deficiencies in the model.

The best statistical approach for baseline blank profiles is to acquire multiple scans (ten, for example). These are averaged in SEDFIT to reduce stochastic noise as described: Load all scans into SEDFIT and switch to a discrete species model in SV. Uncheck all sedimenting species and fit the data using just RI and TI noise. The function Data ▷ Save Systematic Noise ▷ Save Systematic Noise in File will save the TI profile, which in the absence of any sedimentation model will correspond to the average scan, in the same format as the original data scans. SEDFIT also allows the TI noise profile (i.e., the blank profile) to be subtracted directly from other scans, which are then to be resaved — under a different name — in the original format.

The cell assembly should provide the same reproducible baseline until disassembled (provided the laser scan settings are not changed). Water can be aspirated from the sample sector with flexible tubing connected to a vacuum source, and sample can be loaded.[35] Similarly, after the SE experiment, the sample can be aspirated with the vacuum tubing, the cell rinsed a few times with water, loaded with water, and a second set of water baseline blanks can be measured. To verify that no change has taken place in the baseline, it can be established that the pre- and post-experiment baselines coincide.[36]

---

[35] The Yphantis laboratory has developed special six-channel centerpieces ('external loading') that are kept permanently assembled [359]. They feature access to each solution compartment with double filling holes to allow cleaning by flow-through of large amounts of liquid. As SE experiments are not carried out frequently in our laboratories, such a setup is not worthwhile. Rather, if we require SE with interference detection, AUC cells will be assembled and aged specifically for the purpose.

[36] As an alternative to taking water blanks, Ang and Rowe estimate the baseline from the first scan once the sample-loaded cell is at speed, prior to significant sedimentation [442], whereas Horbett and Teller recommend resuspending the sample after the run and acquiring data at low speeds [435]. When these strategies are favored, it still may be very useful to collect, for comparison, additional water blanks, as described in [343].

The analysis of SE experiments in SEDPHAT allows for TI noise profiles that translate with rotor speed to account for rotor stretching. This selection can be made in the experimental parameter box. As discussed previously, this is used primarily for absorbance data.

Aged cell assemblies usually have baseline profiles that only depend slightly on rotor speed below 25,000 rpm (with rmsd deviations below 0.01 fringes) (Fig. 5.10), but larger changes may occur at higher rotor speed [160]. The baseline profiles in interference optics should not be expected always to shift to higher radii along with rotor stretching, as they do in absorbance optical detection, since a large part of the constant optical path length differences may not arise from the cell assembly, but rather from stationary lenses and mirrors [160].

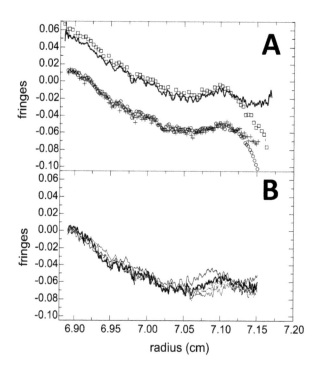

Figure 5.10 (*Panel A*) Comparison between the calculated rotor-speed-invariant noise (solid line) from a multi-speed SE analysis at five rotor speeds, as shown in Figure 4 in ref. [160], and the measured baseline profiles from water blanks before (+) and after (○) the SE experiment (only every third data point is shown) [160]. The rms error between all data shown after correction for the radial-invariant offset is 0.0031 fringes. For comparison, the rotor-speed-invariant noise is also shown (squares, every fifth data point) when the analysis was based only on the SE data at the three highest rotor speeds of Figure 4 in ref. [160]. (*Panel B*) Experimental baseline profiles from water blanks measured before the SE experiment at rotor speeds of 3,000 rpm (solid line), 5,000 rpm (dotted line), 8,000 rpm (bold solid line), 12,000 rpm (dash-dotted line), and 16,000 rpm (dashed line). The rms deviation among these profiles (except for a constant signal offset) is 0.0041 fringes

5.2.4.3 Resuspension, Disassembly, Cleaning, and Storage

Occasionally the samples need to be resuspended for another sedimentation experiment. The situation may arise after an interrupted sedimentation process (e.g., following a low-speed optical adjustment), or by design for a subsequent re-analysis of the identical sample (e.g., BSA reference material). Not all samples are suitable for analysis following resuspension, as irreversible pelleting and aggregation may occur at the high concentrations close to the bottom of the cell. This may be diagnosed from the signal amplitude of the resuspended material. However, a significant fraction of proteins studied in the authors' laboratories can be completely resuspended after a SV run with identical sedimentation behavior in subsequent runs. To resuspend the sample, the cell assemblies are taken out of the rotor and the solutions are gently agitated back and forth within the centerpiece, making sure that the air bubble over the meniscus dislodges and repeatedly moves to the bottom of the centerpiece compartments.

When all experiments are completed it is a good practice to recover any solution using a pipettor or vacuum tubing inserted into the filling holes, prior to disassembly of the cell components. This provides an opportunity to recover most of the samples and will prevent solutions from spilling into the housing and window holders when the cells are disassembled.

Disassembly is usually straightforward. First, the fill-port plugs are removed with a screwdriver. Next, the plug gaskets are removed with tweezers and discarded. At this stage the solutions are recovered or removed, and the cell disassembled. If the centerpiece and window holder assemblies do not easily slide out of the barrel, a cylindrical rod that fits into the barrel and touches the window holders — but not the windows — can be used to push the window assemblies and centerpiece sandwich out of the barrel.

In our laboratories, the cell components are cleaned and kept together as sets such that the history of the components can be traced. In this manner, cell components can be inspected if the data analysis raises questions about potential damage. There are a variety of possible methods for cleaning. We use one or more cycles of detergent solution and sonication followed in sequence with rinses with deionized water.[37] Windows can be cleaned efficiently with a final rinse with ethanol, and a single wipe with an ethanol soaked lint-free paper towel or lens paper across the area that light will be transmitted through during AUC. After cleaning and drying all components, we inspect for damage, re-assemble the cell,[38] hand-tighten or torque the screw ring, and store.[39]

---

[37]Common laboratory cleaning detergents have been used, such as Contrad 70 or Hellmanex, as well as solutions of sodium dodecyl sulfate. The physical inspection for pelleted material will aid with the cleaning process.

[38]In multi-user settings, it may be more practical to store the parts disassembled, so that each user can assemble the cells. In that case, it is advantageous to keep parts of each cell assembly together. This helps to recognize defects by tracing data anomalies in sequential runs.

[39]A good place for storage is a dessicator.

When working with the FDS system, due to the lower detectable concentrations and often 'sticky' fluorophores, it is necessary to be particularly meticulous with cell cleaning [102]. It is desirable to keep a dedicated set of cells exclusively for fluorescence detection at sub-nM concentrations. Sufficient cleanness can be tested prior to the run with FDS experiments using water or buffer-filled cell assemblies. After verifying the absence of signal, buffer can be aspirated and sample can be inserted.

5.3 TEMPERATURE

AUC experiments can be carried out at temperatures above 0 °C and below 40 °C. The particular temperature at which the run is conducted can be guided by the interests of the investigator, or dictated by the sample stability.

Low temperatures (usually 4 °C) are ordinarily used to ensure the stability of biological samples during long-column SE experiments, due to their extended time requirements. This is less of a problem for the rapid SV experiments. In fact, it can be cumbersome to carry out SV experiments at 4 °C due to the excessively long temperature equilibration time, which also requires overnight pre-cooling and temperature equilibration of the rotor and vacuum chamber. In such cases, temperatures of 8–10 °C may be more practical and easier to achieve for SV. The long equilibration times are not an issue for conventional SE experiments, since these can be started right after the cooling process has been initiated.[40] SV experiments, however, require a thorough temperature equilibration prior to the start of the run. Another possible complication arises from the fact that hydrodynamic standard temperature is 20 °C; SV experiments carried out at this temperature minimize potential errors arising from the density and, more importantly, viscosity corrections in the transformation of hydrodynamic parameters to standard conditions (Eq. 1.5). Higher temperatures > 30 °C can be chosen if dictated by the study, but constraints can exist for instruments without a turbomolecular pump, as the oil from the diffusion pump can deposit on the optical elements, degrading the detection. This behavior is instrument specific.

The experimentally set temperature does not always correspond to the true temperature [297], and this is often of little concern. However, the critical aspects of temperature control, especially for SV, are that (1) the temperature is uniform throughout the solution column and rotor; (2) it remains constant throughout the sedimentation experiment; and (3) that we accurately know the actual temperature. The latter is particularly relevant due to the strong temperature dependence of the solvent viscosity. In fact, the accuracy in the temperature calibration is one of the

---

[40]The one advantage that may be obtained by starting standard SE experiments at warmer temperatures is a slightly shorter time to reach equilibrium. However, the use of a toSE overspeeding protocol requires temperature pre-equilibration at least to within one degree in order to prevent convection destroying the established early concentration gradients [249].

principal factors that limits the accuracy of hydrodynamic parameters [295, 296] (see Chapter 6).

For experiments requiring the highest precision comparison in s-values, such as the study of protein conformational changes [295, 443] or density contrast SV [269], it is best to analyze the samples side-by-side in the same rotor in order to eliminate run-to-run variations in the temperature control.

5.3.1 Temperature Measurement and Control

The rotor chamber is internally lined by an aluminum can in thermal contact with series of underlying Peltier elements. These elements are, in turn, mounted on a heat sink with a large thermal mass. This arrangement heats and cools the rotor chamber directly, as well as the black anodized rotor through radiative heat transfer.

Temperature measurement is achieved through a thermistor in contact with the bottom of the can and a radiometer positioned below the surface of the spinning rotor [438]. The thermistor is interrogated at pressures above 100 μm Hg, whereas the radiometer reports on the rotor temperature below this threshold. The radiometer consists of a black metal surface in contact with a thermocouple, covered by a cellophane film transparent to IR. It cannot be used at higher pressures due to incorrect readings arising from the transfer of kinetic energy of gas molecules impacting the surface [438], and the absorptions of IR from moisture at higher pressures. The switch from thermistor to radiometer occurs automatically, and differences of a few °C may be observed during the transition. This is especially true during the initial stages of thermal equilibration, reflecting temperature differences between the can and the rotor surface. The radiometer readings are accurate to ± 0.5 °C, based on specifications provided by the manufacturer. However, in an extensive multi-laboratory study it was found that the calibration point may be off in individual instruments by as much as 1–2 °C [296, 297].[41]

Improved temperature accuracy can be achieved using iButton®[42] temperature loggers, which are miniaturized integrated circuits with an onboard memory [296]. They can be configured to collect temperature data at regular intervals to be retrieved after the experiment. Because of their widespread industrial use, they are inexpensive and can be purchased in calibrated form. Alternatively, they may be calibrated against a NIST traceable thermometer. Using the 11 bit high-resolution model DS1922L, temperature can be measured with a precision of 0.0625 °C and an accuracy of 0.1–0.2 °C [296, 366, 444], validated using orthogonal sets of reference

---

[41] This is in sharp contrast to an unsupported statement found in the review literature [26] that the temperature accuracy and stability in the Optima XLA/I is better than ±0.1 °C.

[42] Maxim Integrated Products.

experiments (see Chapter 6).[43]

When inserted into specialized holders designed for the standard AUC cell housing and placed into a rotor hole, they can withstand centrifugal forces at rotor speeds up to 3,000 rpm [296]. Optimal positioning is achieved using a modified rotor handle (Fig. 5.6), where they can withstand speeds of up to 60,000 rpm and thus log the temperature for the duration of an experiment [366].

The rotor temperature can also be measured by simply placing the iButton® on top of a counter-balance in the resting rotor [444]. As verified in a large-scale study [297], when the 8-hole rotor is temperature equilibrated in the ultracentrifuge chamber under high vacuum and in the absence of an optical arm (Fig. 5.11), the resting rotor temperature will then reflect the temperature of the spinning rotor to within $\pm 0.2\,^\circ$C.

After accounting for a set-point error of the radiometer calibration, which may undergo adventitious changes on the time-scale of months and after instrument service, the temperature control of the XLA/I becomes very effective. Based on our observations, there seem to be only minor heat flows during an SV experiment [444].

5.3.2   Temperature Equilibration

Irrespective of the set temperature and accuracy of calibration, SV experiments require a thorough temperature equilibration of the rotor, cell assembly, and sample and reference solutions within the rotor chamber before the start of the run. As discussed above, the absence of temperature gradients is important to avoid temperature-driven density changes leading to convection. Temperature equilibration is not trivial and the rotor chamber needs to be evacuated (by pressing the VACUUM button) for the temperature regulation to work properly. To measure the rotor temperature and thus interrogate the radiometer, fine temperature adjustment can only be done below 100 μm Hg. Therefore, the diffusion pump needs to be engaged, which can be achieved by pressing the START button of the ultracentrifuge even though the speed is set to zero rpm. The 'RUNNING' mode can be seen at the console. (In AUCs equipped with a turbomolecular pump, this is not necessary.)

Temperature equilibration has hysteresis and proceeds through decaying oscillations of over-heating and over-cooling. Consequently, the observation of a momentary coincidence between the temperature set-point and temperature reading

---

[43]Historically, other methods for measuring the temperature of the spinning rotor have been applied, including the observation of the melting points of solids [1, 439, 445], and calibrated thermocouples interrogated by radio telemetry [446] or direct electrical contact through a pin projecting from the rotor into a bath of mercury, as in the Model E [447]. The use of thermochromic indicator solutions [448] has not been validated and rarely applied in part due to the strong pressure and solvent composition dependence of the thermochromic shift [449], as well as large errors from the amplification of spectrophotometric calibration uncertainties discussed in [296, 297].

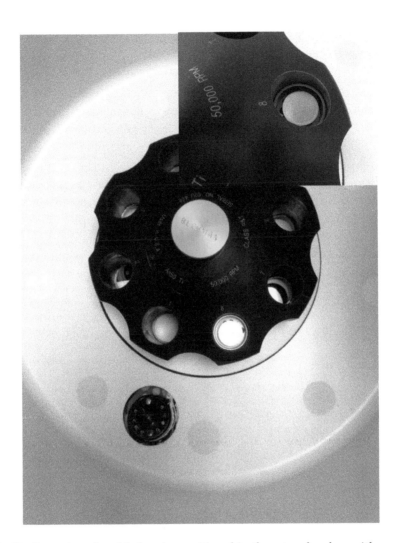

Figure 5.11 Configuration of an 8-hole rotor positioned in the rotor chamber, with counterbalance and iButton® temperature logger for determining the temperature of a resting rotor. Note the absence of the monochromator arm in the experiment. The iButton® is located on the counterbalance. The inset shows a close-up. Reproduced from [444].

cannot be taken as a sign that the system is equilibrated.[44] Once the oscillations have ceased and a constant rotor surface temperature is read by the radiometer for several minutes, an additional ~1–2 hours are required for equilibration of the rotor surface and interior, based on experience [450] and on measured differences in the time course of equilibration measured by the radiometer and the iButton® in the rotor handle [366].

Throughout the entire temperature equilibration for SV, the rotor should be at rest, to avoid the sedimentation of large material, and time-keeping errors in the number of revolutions and elapsed time since the start of the 'real' sedimentation run. Once rotor acceleration commences, the rotor will be transiently cooled by a few tenths of a degree due to adiabatic stretching (see Fig. 5.6 and Fig. 5.7 above). It is possible to minimize heat flows from adiabatic cooling by applying a slower acceleration (Fig. 5.12); this is achieved by applying a schedule of manual rotor speed changes [366], or automated speed changes in the 'equilibrium mode' of data acquisition. This approach will have implications for the data analysis due to the prolonged rotor acceleration phase (Section 4.2.1). Opposite effects can occur in solution from pressure-induced adiabatic heating of sample caused by the compressibility of the solution, as predicted by Mijnlieff et al. [451]. This also amounts to a few tenths of a degree and may counteract transient cooling of the solution from the adiabatic rotor cooling.

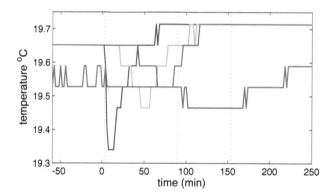

Figure 5.12 Demonstration of an isothermal SV experiment at 50,000 rpm (red trace) achieved by lowering the rotor acceleration to 5.6 rpm/sec [366]. The dotted vertical lines indicate the times at which 50,000 rpm was reached. For comparison, temperature traces at standard acceleration of 280 rpm/sec (blue) and at slower accelerations of 16.7 rpm/sec (cyan) and 11.1 rpm/sec (magenta) are shown. Temperature was measured with a calibrated iButton® located in the rotor handle.

---

[44]Pre-equilibrating the empty rotor and optical arm in the ultracentrifuge chamber can shorten this procedure. This can be done conveniently during sample preparation and assembly of the centrifuge cells. By setting the pre-equilibration temperature to ~2 °C below the run temperature, it is possible to partially counteract the unavoidable heating of the rotor during cell loading and alignment. The best-pre-equilibration temperature and magnitude of this effect will depend on room temperature, as well as the actual run temperature.

Unfortunately, on some instruments rotor acceleration occasionally initiates an unexpected increase in the radiometer readings by as much as 0.8 °C [297]. While there is no apparent explanation for this behavior, a comparison of the rotor handle iButton® temperature log with the radiometer readings indicates that these do not always reflect fluctuations in the rotor temperature. In some instances, we have correlated this with replacement of the diffusion pump oil, possibly a consequence of outgassing events that affect the radiometer. Other imperfections in the vacuum can be expected to impact the radiometer reading, as well. In any event, these temperature jumps will initiate heating and cooling cycles that can disturb rotor temperature equilibrium, and spawn convective events at the beginning of the run.

If temperature equilibration preceding an SV experiment is flawed, transient convective flows will occur at the beginning of the sedimentation process. Unfortunately, these will impact the entire SV data, as the evolution of the concentration profiles obviously is history-dependent.[45] One way to diagnose this is a misfit in the initial scans, and an apparent best-fit meniscus that is at smaller radii than expected [339] (Section 4.1.2).[46] It is important to recognize this, and repeat the experiment if best accuracy is required. Some limited rescue of information from the experiment may be possible by using Lamm equation solutions as fitting models that are based on initial conditions taken from the first experimental scan after convection has ceased (as in Fig. 5.4 in Section 5.2.1) [418].

5.3.3 Temperature-Dependent Experiments

The temperature dependence of macromolecular sedimentation and interaction parameters is important for some studies. Examples include studies on heat shock proteins and their activation [453, 454] and DNA-binding [455], the binding mechanisms of the Fc receptor FcRIII [327], temperature-induced conformational changes of plasminogen [456] and temperature-dependent oligomerization of a rotavirus non-structural protein [457], among others. Notably, a temperature dependence of the binding constants will only provide van't Hoff enthalpies of the reaction, unlike isothermal titration calorimetry, which directly measures binding enthalpies.

When carrying out temperature-dependent AUC studies, it is important to recognize the temperature dependence of the partial-specific volume (e.g., see Fig 3.5), as well as the solution density and viscosity. Likewise, the temperature dependence

---

[45]Importantly, a delay of the onset of scanning, advocated by some to eliminate the effects of initial convection [452], is not productive. The action of scanning has no influence on molecular fluxes, the evolution of the radial concentration profile and its history dependence. The action of not scanning while convection occurs hampers the recognition of the problem without solving it. This flawed concept appears as a corollary to the belief that merely not showing an implicit baseline profile will diminish the correlation of the baseline with sedimentation parameters [361].

[46]To reiterate from Section 4.1.2, the best possible estimate for a sedimentation coefficient under these circumstances will be obtained by keeping the meniscus as an adjustable parameter [339]. In this manner the s-value is governed by the migration of the recorded boundary positions, rather than any constraints in the visually apparent meniscus.

of the pKa value of the buffer needs to be considered as this may lead to a shift in pH if not compensated for. The pKa of TRIS, for example, will change by one approximately pH unit between $4\,°C$ and $37\,°C$, with a dpKa/dT of $-0.031/°C$; whereas MOPS has a fivefold lower temperature coefficient of $-0.006/°C$. In general, tables of dpKa/dT should be consulted, noting that a temperature-independent pH buffer system has been developed [458].

5.4   VACUUM

Vacuum ultracentrifuges were developed by Pickels and collaborators in the 1930s as a means to achieve higher rotor speeds at reduced friction with better thermal control [18, 447]. The vacuum chamber is now an integral part of the AUC and in the Optima XLA/I generation of ultracentrifuges a vacuum of a 5 μm Hg (micron) or better is required for operation [459]. Such a low pressure is also required for the operation of the FDS.[47] High vacuum in the AUC is achieved with two pumps in series — a mechanical roughing pump that achieves a vacuum of a few hundred μm Hg or below is coupled to an oil diffusion or turbomolecular pump for high vacuum. Turbomolecular pumps are preferred due to problems associated with oil deposition and outgassing of moisture in the diffusion pump.

Importantly, in the current generation of instruments, the diffusion pump will only be engaged when the centrifuge is in 'RUNNING' mode. As mentioned above, 'running' at zero rpm is crucial for establishing proper thermal equilibration, such that the radiometer is engaged at pressures below 100 μm Hg, and temperature controlled through feedback from the rotor surface rather than the thermistor-based reading of the rotor chamber lining (Section 5.3).

When setting up sedimentation experiments, it is usually worthwhile to observe the pressure reading as the rotor chamber is evacuated, which will be displayed once the pressure falls below 1,000 μm Hg. It should drop steadily until it reaches a few hundred μm Hg, at which point the diffusion pump (if engaged) will often outgas moisture resulting in a transient pressure increase of 100–200 μm Hg for a few minutes. The pressure should then drop rapidly towards zero or a few μm Hg. The switch in the temperature display should be noted at 100 μm Hg mark. If the transient pressure increase is larger than usual, or if there is any increase at all when using a turbomolecular pump, then this is a good indicator of a leak. In this case, it is advisable to stop the equilibration process or run, release the vacuum, and inspect the sample cells.

---

[47]Electric arcing from the FDS detection system through air is possible at intermediate pressures.

5.5 ROTOR SPEEDS

5.5.1 Rotor Speeds in Sedimentation Velocity

The choice of rotor speed depends strongly on the experimental goals, as well as the size and density of the particles under study. A few general considerations help set constraints and clarify the choice of rotor speed.

The first consideration is the s-range that can be observed for a given rotor speed. The vast majority of SV experiments on biological macromolecules are carried out at a rotor speed of 50,000 rpm. With a typical column length of 12 mm, and a first scan taken at least $\sim$3 min after the start of centrifugation, particles with an s-value of $\sim$350 S will have migrated close to the end of the radial data analysis window; only slower particles can be characterized and faster particles would not be observed. Based on a typical lower measurable limit of $\sim$0.1 S, this corresponds to a 3,500-fold range in s for a single run. Based on a minimal observation time of >1 hour for the species of interest, a more conservative upper limit for the s-value can be estimated for a given rotor speed in rpm:

$$s_{high} \approx \frac{4 \times 10^{10}}{\text{rpm}^2} \text{ S} \tag{5.4}$$

(based on the standard dimensions of the solution column outlined above). Using the arbitrary requirement that the boundary midpoint should have migrated 3 mm after about 6 hours for the smallest particles of interest, we arrive at an estimate for the lower limit at a given rotor speed in rpm:

$$s_{low} \approx \frac{2 \times 10^{9}}{\text{rpm}^2} \text{ S} \tag{5.5}$$

These limits are only meant as a guide and should not be interpreted very stringently, because the detection and analysis of faster or slower sedimenting particles is still possible. These limits may also be expressed in terms of a range of rotor speeds for a particle of given molar mass and density, using the approximate relationship $s_{20,w} \approx 0.012 M^{2/3} (1 - \bar{v}\rho) \bar{v}^{-1/3} (f/f_0)^{-1}$ (with s in units of S, M in Dalton, $\bar{v}$ in mL/g and ρ in g/mL) illustrated in Fig. 1.3:

$$4.1 \times 10^5 \times M^{-1/3} \sqrt{\frac{\bar{v}^3(f/f_0)}{(1 - \bar{v}\rho)}} < \text{rpm} < 1.8 \times 10^6 \times M^{-1/3} \sqrt{\frac{\bar{v}^3(f/f_0)}{(1 - \bar{v}\rho)}} \tag{5.6}$$

In the case of typical proteins with f/f_0 of 1.3 and $\bar{v}$ of 0.73 mL/g in a standard buffer, then:

$$\frac{5.6 \times 10^5}{M^{1/3}} < \text{rpm} < \frac{2.5 \times 10^6}{M^{1/3}} \tag{5.7}$$

(cyan and blue lines in Fig. 5.13).

The extent of boundary spread during the sedimentation process is a second consideration for rotor speed selection. A higher rotor speed results in shorter time

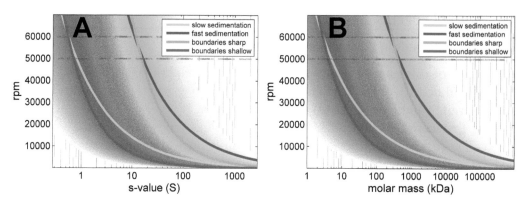

Figure 5.13 Visualization of estimated boundary migration and relative spread for typical proteins as a guide for the selection of rotor speeds in SV. The calculations assume an f/f_0 of 1.3 and $\bar{v}$ of 0.73 mL/g in water at 20 °C, with data shown as a function of s-value (*Panel A*) and molar mass (*Panel B*). The background color scales from shallow (magenta) to sharp (green) boundaries: the dark magenta line highlights a boundary half width equal to the midpoint migration in the center of the solution column, whereas the dark green line highlights a situation in which the spread is only one tenth of the boundary migration; the transition of white to lightly shaded region depicts 5-fold more extreme small or large boundary spread, respectively. Also highlighted is the line where boundary migration is only 3 mm into the cell in a 6 hour period (cyan line), and where the full length of a 12 mm solution column is traversed in 1 hour (blue line) following Eq. (5.7). Conventional SV experiments will be observed for rotor speeds bounded by the cyan and blue lines, and the magenta and green lines. Conditions corresponding to lightly shaded or white regions are of limited use. The gray horizontal lines mark the rotor-speed maxima for the An-50Ti (lower line) and An-60Ti (upper line) rotors.

for sedimentation to the bottom of the solution column, a concomitant lower diffusional spread during this process and higher resolution in s. On the other hand, a reasonable diffusional spread is necessary to determine the extent of diffusion for each species and thus the buoyant molar mass. Using simple calculations, we can develop *ad hoc* landmarks for the ratio of the diffusional boundary width (estimated as the FWHM of a point source) to sedimentation: comparably low information on diffusion is available if the width is only one tenth of the sedimentation distance when the boundary has traveled half the length of the solution column; conversely, high diffusion information and correspondingly low information on s is available when, at the same half-way point, the boundary width equals the sedimentation distance. Based on the Svedberg equation Eq. (1.8) these reference lines are easy to predict for a given system.

As illustrated in Fig. 5.13, these limits in boundary spread and migration describe a wide range of possible rotor speeds for medium-sized molecules sedimenting between 1 S and 20 S. In the case of smaller proteins (< 10 kDa) it is useful to carry out experiments in the 4-hole rotor at 60,000 rpm to diminish the boundary spread as much as possible. The choice of 50,000 rpm is for most purposes reasonable when studying species of intermediate size (20 kDa to 1 MDa and < 20 S). Therefore, we carry out most sedimentation velocity experiments at this rotor speed, optimizing resolution while still providing boundary spread information.

If particles larger than 20–30 S are the major focus of study, it is useful to reduce the rotor speed, more so for the higher the s-values. The largest particles of at least 1,000 S need to be studied at the lowest useful rotor speed of 3,000 rpm. An example is provided by an ensemble of unicellular organisms (*Spiroplasma melliferum*) with a modal s-value of 10,000 S and the largest detected species of 100,000 S at a rotor speed of 3,000 rpm using interference optical detection [284]. No diffusion information is available for species at this size, but the size resolution is superb.

These recommendations should only be used as a guide in the absence of other considerations. Sometimes the choice of an optimal rotor speed is more complex,[48] and in the case of uncertainty or an educated guess, the best course of action is to simulate data (with added typical instrument noise) at different rotor speeds with a model for the expected sedimentation parameters, and to compare the resulting confidence intervals from the analysis of the parameters of interest.

> SEDFIT and SEDPHAT can simulate both SE and SV data using the Generate function. In addition, boundary profiles resulting from any analysis model can be saved in the format of experimental data with optionally added random noise, to serve as simulated data.

5.5.2 Rotor Speeds and Solution Column Lengths in Sedimentation Equilibrium

The optimal rotor speed for SE depends largely on the desired shape of the SE profile. The curvature will essentially depend on the buoyant molar mass, rotor speed, and solution column length; the latter will strongly influence the time required to attain equilibrium and needs to be carefully considered when determining the feasibility of an experiment.

A convenient way to express the curvature of the Boltzmann exponentials in SE is often through the ratio of signal at the bottom to that at the meniscus, $c(b)/c(m)$, or better the ratio of signals that are closest to these points and reliably measurable such that they exceed the noise of data acquisition. Useful values start as low as 3.0 and will often exceed 100, exploiting the full dynamic range of the detection. With low curvature, the molar masses (or relative amplitudes of fixed exponents) are ill-conditioned and will correlate with the baseline signal parameter. Too high curvature in the SE profile, on the other hand, will push most of the material into optically inaccessible space close to the bottom of the solution column, and compress

---

[48]This is illustrated in a theoretical study on the effect of rotor speed on the quantitation of trace dimers in a predominantly monomeric IgG-sized protein [389]. In such a situation, the dimer sediments in the leading edge of the diffusional spread of the monomer boundary. A drawback becomes apparent at the highest rotor speed of 60,000 rpm due to the lower number of scans. The higher resolution, however, can partially compensate for the combination of better diffusional characterization and larger data set at the lowest rotor speed considered, 40,000 rpm.

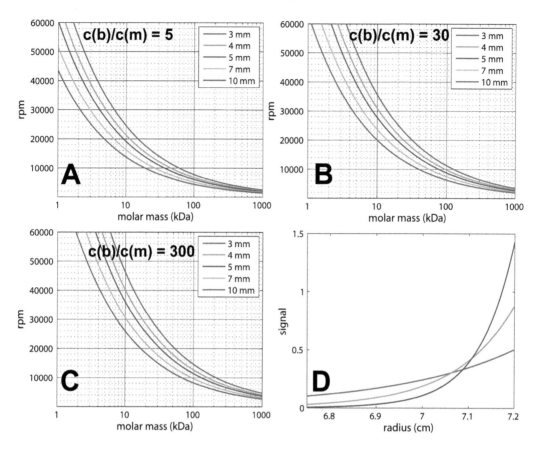

Figure 5.14 (*Panels A–C*) Rotor speeds required to generate SE profiles with $c(b)/c(m)$ ratios of 5 (*A*), 30 (*B*), and 300 (*C*) for different solution column lengths as a function of molar mass, assuming a partial-specific volume of 0.73 mL/g and a buffer density of 1.005 g/mL (Eq. 5.8). (*Panel D*) SE profiles for a 4.5 mm solution column of a 100 kDa protein at 4 °C and rotor speeds of 6,400 rpm (red), 9,200 rpm (green), and 12,000 rpm (blue), generating $c(b)/c(m)$ ratios of 5, 30, and 300. This illustrates the shapes that will be generated by the rotor speeds calculated in panels (*A*) through (*C*), at a constant loading concentration of 0.25 signal units in sector-shaped centerpieces.

the useful molar mass information into a few data points with steep gradients that are intrinsically more error-prone.

The buoyant molar mass of the particle of interest plays a key role in the selection of the rotor speed as the curvature of the Boltzmann distribution is governed by the exponent $\sim M(1 - \bar{v}\rho)\omega^2$ (Eq. 1.11). Thus a fourfold higher buoyant molar mass can be compensated by a twofold lower rotor speed without affecting the shape of the equilibrium profile.[49]

---

[49]The determination of the SE rotor speed can, in principle, be condensed to a 'reduced molar mass' σ, with $\sigma = \omega^2 M(1 - \bar{v}\rho)/RT$ [343]. Unfortunately, this nomenclature tends to be unnecessarily cryptic for the inexperienced users, and is therefore not adopted in the current work. It is also not a constant parameter, as different column lengths require different values of σ to obtain reasonable SE profiles for species of the same molar mass.

We can rearrange Eq. (1.11) to calculate a rotor speed that will result in a target $c(b)/c(m)$ ratio for a given buoyant molar mass:

$$\omega = \sqrt{\frac{2RT}{M(1-\bar{v}\rho)}} \times \sqrt{\log\left(\frac{c(b)}{c(m)}\right)} \times \sqrt{\frac{1}{b^2 - m^2}} \tag{5.8}$$

This relationship is depicted in Fig. 5.14 as a function of $c(b)/c(m)$ and the solution column height. It is the last term in Eq. (5.8) that relates the shape of the SE profile to the solution column length.

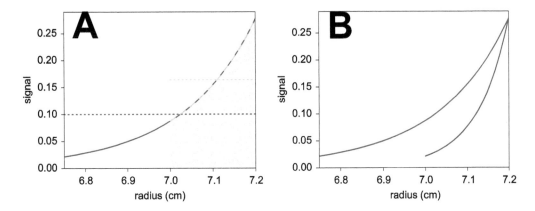

Figure 5.15 (*Panel A*) SE distribution for a 100 kDa protein, $\bar{v} = 0.73$ mL/g in PBS, at 8,000 rpm (purple solid line). The solution column is 6.75 cm to 7.2 cm and the initial loading concentration is 0.1 (purple dotted line). Under these conditions, $c(b)/c(m)$ is 12.7. The same profile, constrained to a radial range of 7.0 cm to 7.2 cm (cyan region) can be achieved with a loading concentration of 0.164 (cyan dotted line) at the same rotor speed, but with a $c(b)/c(m)$ of only 3.2. (*Panel B*) In order for the shorter solution column to achieve the same concentration range and $c(b)/c(m)$ ratio as the long column, 8,000 rpm data set (purple), the rotor speed needs to be increased to 11,894 rpm (blue) resulting in steeper gradients.

The relationship between rotor speed, the $c(b)/c(m)$ ratio, and column length can be easily understood in the following Gedankenexperiment: At equilibrium, any subsection of an SE profile across any radial interval $[r_1, r_2]$ within a longer solution column can be an SE profile for a correspondingly shorter solution column having $m = r_1$ and $b = r_2$ (Fig. 5.15 Panel A). However, at the same rotor speed, the shorter solution column would exhibit lower total curvature (as measured by the change in slope across the solution column). In order to obtain the same total curvature, a higher rotor speed is required (Fig. 5.15 Panel B). Even though shorter columns require a shorter time to reach equilibrium, their use is not recommended because the constant radial density of data points results in the equilibrium profile being sampled at fewer radial points, and because the required higher signal gradients have an inherent higher propensity for error.[50]

Even though the geometry of the SE profiles and optical detection point to the advantages of very long solution columns, a constraint is the time required to reach a state sufficiently close to sedimentation equilibrium (Section 5.7.2) for analysis.[51] Thus, the stability of the sample factors in, as well as the diffusion coefficient. Only for very small molecules does it make sense to exploit the full solution column available, with ~ 400 μL samples. For a different reason — namely the limit in accessible rotor speeds — for small molecules it is actually essential to use long solution columns at the highest available rotor speeds to achieve even a moderate curvature and reasonable $c(b)/c(m)$ ratios (Fig. 5.14).

For molecules larger than 5 kDa a compromise between long solution columns providing superior data sets and time to reach equilibrium must be found. In practice, when not using overspeeding techniques (Section 5.7.2), a column length of 4 to 5 mm is ideal for medium-sized proteins, with slightly longer columns for smaller molecules. Shorter solution columns of 3 mm to 4 mm work well for proteins with molar masses above 100 kDa. Solution columns shorter than 3 mm can produce only limited information, and are rarely a good choice unless proteins with very limited stability are being studied (e.g., [257]), in which case SV approaches may prove advantageous.[52]

Historically, two different schools of thought have promoted a $c(b)/c(m)$ ratio corresponding to either relatively low-speed [93] or high-speed conditions [343], respectively.[53] While low-speed runs will provide the best assessment of the highest molar mass species, the high-speed condition has the virtue of providing informa-

---

[50]For extremely short solution columns no curvature can be measured. Only the slope at the center is available to provide information on the average buoyant molar mass, assuming an average concentration and baseline. This results in much higher errors than standard SE [346]. This applies, for example, to solution columns of 2 mm or less in the absorbance optics, and slightly shorter columns in the interference optics (the latter being able to accommodate somewhat higher gradients due to the better radial resolution). Problems with short columns are further exacerbated by the optically inaccessible radial range close to the bottom of the cell, excluding a further > 0.5 mm from data analysis.

[51]When working with very dilute samples which show long term stability, longer solution columns than usual can be advantageous. Increased rotor speeds promote a higher build-up of material at the bottom of the cell, effectively concentrating the sample above the lower limit for detection.

[52]Van Holde and Baldwin have proposed ultra short-column SE utilizing 1 mm column lengths, which have the virtue of reaching equilibrium extremely fast [345]. The method was further developed by Yphantis [346] with specialized centerpieces, but this approach did not find much general use. The accuracy is poor, with errors of $\sim$10%, and surface adsorption limits the useful lower concentration. The method is superseded by SV which is just as fast but now provides a wealth of information, including the sample molar mass and polydispersity, from the shape of the boundary.

[53]These approaches coincided with a preferred choice of optical detection system. The absorbance system can measure absolute concentrations and does not require meniscus depletion conditions, and benefits from lower gradients due to the more limited range of signal linearity and radial resolution. This made it the preferred choice for low-speed SE. The interference system, on the other hand, benefits from meniscus depletion to provide a zero concentration reference, and can accommodate higher gradients, which made it the detection of choice for high-speed SE.

tion on the smallest species. High-speed conditions also allow a reliable measure of the baseline with 'meniscus depletion,' where all macromolecular solutes have essentially cleared the meniscus. Since modern methods for SE analysis are based on a global fitting of multiple data sets, it is advantageous to combine low- and high-speed approaches and carry out SE sequentially at a range of rotor speeds. As shown in [160], this can generate additional information because the redistribution of solutes between equilibrium conditions at different rotor speeds can implicitly define the total amount of dissolved material (along with the bottom of the solution column). Such mass conservation constraints greatly improve the reliability of the numerical analysis, and the best approach is thus a combination of both low- and high-speed SE.

As illustrated in Fig. 5.14, the combination of $c(b)/c(m)$ values of 5, 30, and 300 will usually produce excellent data sets. Eq. (5.8) can be used to determine the required rotor speeds; as this depends on the solution column length, a table for the solution volumes required to produce certain lengths of solution columns can be found in Appendix D. When using 4.5 mm solution columns ($\sim$160 μL in standard double sector centerpieces), one may simply scale the ω values of Fig. 5.14 Panel D to different molar masses, resulting in:

$$\omega_{4.5 \text{ mm}} = \sqrt{\frac{27 \text{ kDa}}{M\left(1 - \bar{v}\rho\right)}} \times \begin{cases} 6,400 \text{ rpm} \\ 9,200 \text{ rpm} \\ 12,000 \text{ rpm} \end{cases} \tag{5.9}$$

It is important that, in the course of the experiment, rotor speeds are adjusted from the lowest to the highest, making sure equilibrium is attained each time (see below). Unfortunately Eq. (5.9) has limited application as it is valid only for single species, not for multiple species of different mass in reversible association. Data simulations are the best approach for the design of such experiments.

The SE rotor speed calculator in SEDFIT calculates three optimal rotor speeds for a given solution column length with upper and lower molar mass values of interest. It will display for all rotor speeds the radial distributions of both species to allow adjustments toward the desired shape of the profile. In SEDPHAT, SE data of interacting systems can be simulated and re-analyzed to assess the information content of a putative experiment through robust error analysis.

5.6   STARTING THE RUN

Once the samples are loaded and the AUC cells are inserted and aligned in the rotor, it is time to start the run, with knowledge of the optimal data acquisition parameters and required rotor speeds. In the case of SE, this is quite straightforward: once a vacuum of $< 10\,\mu$m Hg (to allow for rotor acceleration) has been established, the first equilibrium rotor speed is entered on the AUC console and the experiment is started with rotor acceleration to the desired speed. No temperature equilibration is generally required prior to the rotor acceleration, and — because equilibrium is history independent — any acceleration schedule will work (unless overspeeding in time-optimized variable fields is conducted, as described below, in which case temperature equilibration is required).

The situation is more complex for SV experiments, as the run needs to start in a well-controlled manner. This is critical, because any imperfections during the start of the run will have repercussions for the rest of the experiment, since the boundary position and shape at any point in time are dependent on their entire history. Before starting the run, the operator should verify that the temperature has been equilibrated with the rotor at rest and the centrifuge in a 'RUNNING' mode at zero rpm (so as to engage the diffusion pump). The vacuum must show a reading of a few micron (μm Hg) or less, and equilibration requires that the temperature reading on the AUC console matches the set point for approximately two hours (Section 5.3.2).

An initial low-speed run for the adjustment of the data collection parameters and temperature equilibration is discouraged, as this will impact the time-stamps of the scan file (Section 4.2.1), and lead to sedimentation and diffusion of the largest species, resulting in a non-uniform, non-equilibrium state. Nearly all adjustments can be carried out during the acceleration phase or at full speed, and checks for leaks can be carried out in different ways (Section 5.2.4). The custom to include such a low speed initial phase has historic origins relating to the low-precision pre-computer analyses. In cases where the operator feels that such a low-speed pre-run is essential in an SV experiment, as described in Section 4.2.1, the rotor should either be stopped, sample be resuspended and equilibration started once more, or the low-speed pre-run needs to be computationally accounted for as an explicit rotor speed step during the Lamm equation solution. Unfortunately, the computational approach is currently not available in conjunction with all sedimentation models. In standard SE experiments, preliminary scans can be carried out conveniently at the target rotor speeds without penalty.

Before accelerating the rotor, the operator should also make sure that the scan settings file is prepared, so that data acquisition can commence as soon as the rotor has reached the desired speed. When using the interference optics, it is worthwhile to enter previous laser settings for the given rotor. The availability of previous scan settings will also allow for a visualization of the solution columns through the camera of the interference optics (if available) during acceleration.[54]

To start the SV run, the rotor acceleration setting on the centrifuge console

---

[54]Unless the overspeed disk of the rotor has been replaced between runs, each individual rotor/instrument pair will have essentially the same delay angle setting for interference data acquisition. Therefore, restoring the delay settings for each rotor position to the previous values for the given rotor will usually allow immediate visualization.

should be on maximum.[55] Entering the desired SV rotor speed on the console will initiate rotor acceleration. As described in Section 4.2.1, acceleration from 0 rpm to 50,000 rpm will take approximately 3 min. The combination of the $\int_0^t (\omega(t'))^2 dt'$ and the t entries of the scan file headers will allow for modeling of this acceleration phase in the numerical analysis.

While the rotor is accelerating, the operator should observe the temperature reading and take note of any fluctuations. As described in Section 5.2.3, one should expect transient cooling from adiabatic stretching of the rotor, which should stabilize within 20 min.[56] Likewise, the vacuum reading should be observed, as any sudden increase in pressure is usually indicative of a leak. For the same reason, the fringe patterns should be visually inspected while the rotor is accelerating. Diagonally striped fringe images indicate a leak (or loading volume mismatch), whereas horizontal fringes with a meniscus feature (single or double) at the expected radius indicates stable and accurately loaded solution columns. At this stage one can visually discern macromolecular depletion near the meniscus and accumulation of material near the bottom of the solution column, which can confirm the proper progression of the SV experiment.

5.7 COMMENCING DATA ACQUISITION AND RUN DURATION

5.7.1 Scan Parameters for Sedimentation Equilibrium

Scan parameters for SE are guided by the fact that at equilibrium there is ample time to exhaustively scan the solution column. In order to make up for the relative small amount of data that can be obtained from a single concentration gradient in SE relative to the multiple time-varying concentration gradients in SV, it is important to use the highest possible density of radial points when using the absorbance optics, currently 0.001 cm in a step mode. To improve the signal-to-noise ratio, 10 to 20 replicate data points are collected at each radial point, noting that no significant improvement is achieved with more replicates. It can be very useful to take absorbance scans at multiple wavelengths, even for studies of a single macromolecular component, in order to expand the useful concentration range. Specifically, studying protein samples at their local extinction minimum of 250 nm, in addition to 280 nm, extends available information to higher concentrations closer to the bottom of the solution column. Similarly, scans at 230 nm are routinely added in protein studies, for greater sensitivity and improved signal-to-noise at the low concentrations close to the meniscus.

It is also important to establish that equilibrium has indeed been attained for SE data analysis. Therefore, data need to be acquired at regular time intervals through-

---

[55] Any other setting will only temporarily delay acceleration at full rate; these alternative acceleration profiles were designed for use in preparative ultracentrifugation.

[56] As mentioned in Sections 4.2.1 and 5.3.2, it is possible to start the experiment with a slower acceleration (Fig. 5.12). By manual and step-wise adjustment of the rotor at regular time intervals, an isothermal acceleration can be achieved [366].

out the experiment, to properly assess the attainment of equilibrium. Strictly, the experiment will never reach equilibrium, but it will asymptotically approach that state, and become indistinguishable within the noise in the data. In this regard, it is important to note that concentration profiles that deviate significantly from equilibrium can often be well modeled by sums of Boltzmann distributions. Consequently, a deviation from the equilibrium state is usually not apparent from an SE data analysis which, in this case, would result in erroneous parameters. In practice, it is essential to monitor the progression toward equilibrium, as part of the experiment with scans taken at intervals of 3–6 hours.[57]

The best way to judge whether equilibrium has been attained is to plot the difference between earlier scans and the last scan as a function of time. An asymptotically decaying function should be observed, leveling to the value of the noise of data acquisition. This is the time at which equilibrium is reached and at this time, the residuals from the subtraction of scans should not show any trend. It is important to exclude regions of optical artifacts from this assessment.

> SEDFIT can test for the approach to equilibrium using a set of N scans taken during an SE experiment. The function `Options ▷ Equilibrium Tools ▷ Test Approach To Equilibrium` will display the rms deviation (not including baseline shifts) of scans 1 through N-1 relative to the last scan N. In addition, the radial residuals from the difference between the last two scans are shown.

5.7.2 Overspeeding and Expected Time to Equilibrium

An approximate time-scale for the attainment of equilibrium is useful to know *a priori*. Theoretical predictions differ for 'low-speed' and 'meniscus-depletion' type SE designs. In the case of 'low-speed' SE, a theoretical expression based on approximate solutions of the Lamm equation was given by van Holde and Baldwin [345]. The critical insight from this prediction is that the equilibration time is proportional to the square of the solution column length, $t \sim (b-m)^2$. Approximate predictions for 'meniscus-depletion' high-speed SE conditions were derived by Yphantis [343] and found to be proportional to the first power of the solution column height, $t \sim (b-m)$, indicating that the penalty of using longer solution columns is not as strong for high-speed SE as compared to low-speed SE. A more modern approach to this problem takes advantage of the efficiency of numerical Lamm equation solutions, and simply solves the Lamm equation numerically until the Boltzmann exponentials of SE are identical within a pre-defined precision over the observable radial range, given s, D, and the solution geometry [249].

---

[57]The strategy of using pair-wise differences of scans as an indicator for attainment of equilibrium will be misleading if shorter scan time-intervals are used in conjunction with standard height solution column. There may be significant changes that will not be observed on a shorter time scale using this method.

SEDFIT can calculate the minimal time to equilibrium using the van Holde and Baldwin low-speed approximation of [140], based on a pre-set precision in the molar mass (Options ▷ Equilibrium Tools ▷ Minimum Time To Equilibrium). Likewise, simulations based on Lamm equation solutions are carried out and reported.

Unfortunately, these calculations require prior knowledge of s- and D-values (or f/f_0) which are not always available. Furthermore, it is clear that significant concentrations of co-solvents that increase the solvent viscosity will extend the time to equilibrium, as will large frictional ratios. Reacting systems of interconverting states may be limited by the chemical relaxation constant, or the longest lifetime of complexes, k_{off}. All these extend the time required to reach equilibrium. On the other hand, initially stronger migration when the rotor is still at room temperature prior to reaching the target low temperature will shorten the time to equilibrium, though potentially counteracted by temperature gradient-driven convection, if the run is not started close to the target temperature. Following the 'low-speed' paradigm and using 4–5 mm solution columns, at least 24–36 hours should be allowed for SE at the first rotor speed.

However, it is often possible to significantly decrease the time required to reach equilibrium, by two- to ten-fold, using a carefully timed initial overspeeding [93, 163, 347, 460]. The idea is to start the SE experiment at a higher rotor speed ω' until the concentration profile resembles that desired for the lower equilibrium rotor speed ω, and then to decrease the rotor speed to ω. This involves accelerated depletion close to the meniscus, and may include concentration of the sample beyond the equilibrium concentration at the final target rotor speed. Steep gradients at the bottom produced by the latter process will rapidly decay, driven by strong diffusion fluxes once the final target rotor speed is reached; diffusional relaxation close to the bottom may be further accelerated by an under-speeding phase. For low-speed SE, Richards and co-workers presented an estimate for the time t':

$$t' = \frac{RT}{\omega'^2 s} \times \frac{c(b)/c_0 - 1 - \log(c(b)/c_0)}{M(1 - \bar{v}\rho)\omega^2 b^2 - \log(c(b)/c_0)} \tag{5.10}$$

where c_0 is the initial loading concentration, and $c(b)$ the final concentration at the bottom. In this approach, they recommended that the rotor be held for 1.1-fold t' at 1.4 times the equilibrium speed, followed by a brief underspeeding period of 0.05-fold t' at 0.7 times the equilibrium speed [93].

The overspeeding approach is not commonly used: A danger of initial overspeeding is that if too much material accumulates close to the bottom of the solution column, the time for attaining SE can be accidentally prolonged rather than shortened [163, 347]. Similarly, the very high concentration generated transiently very close to the bottom of the solution column may cause precipitation and surface film formation. This is particularly true if higher than expected molar mass species are present in the sample or when very high overspeed ratios ω'/ω are used. It is therefore prudent to monitor the overspeeding phase, for example, with an SV

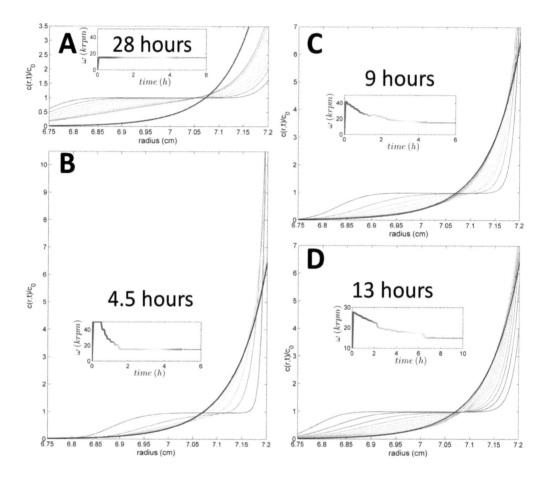

Figure 5.16 Calculated concentration profiles during the approach to equilibrium for different time-optimized varying-field overspeeding conditions, with corresponding temporal rotor speed profile shown in the insets, using the same color temperature map to indicate time [249]. All calculations are for a 65 kDa protein with 15% dimer, with frictional ratios of 1.4, in a 4.5 mm solution column corresponding to 150 μL sample in a regular double sector 12 mm pathlength centerpiece, with final equilibrium at 15,000 rpm at 20 °C. (*Panel A*) Conventional SE experiment with constant rotor speed; (*Panel B*) unconstrained optimization leading to transient 5-fold overconcentration at the base of the cell; (*Panel C*) optimization with overconcentration below a factor 1.6; (*Panel D*) optimization with overconcentration by less than a factor 1.07. Equilibration times, assessed by the criterion that the rmsd across the artifact-free portion of absorbance scans are within the noise of data acquisition identical to the theoretical Boltzmann distribution, were 28 h (*A*), 4.5 h (*B*), 9 h (*C*), and 13 h (*D*), respectively.

analysis of the approach to equilibrium. A conservative approach would ensure that the concentration $c(b)$ does not exceed the anticipated equilibrium concentration, although this would sacrifice a good portion of the potential savings in equilibration times. Alternatively, a more complex rotor speed schedule involving underspeeding or rotor slowing may be used [93, 249, 461].

The most powerful approach is the numerical optimization of the time-varying rotor speed profile, using Lamm equation simulation of the approach to equilibrium in time-varying fields ('toSE') [249]. Such optimization may be carried out with the constraint that the bottom concentration should never exceed a certain threshold to safeguard against excessive concentration build-up and possible aggregation or pelleting (Fig. 5.16). The general drawback of overspeeding requiring detailed prior knowledge of the molar mass and sedimentation coefficient can be eliminated by feedback from experimental scans that can be acquired during the overspeeding phase. As a result of the overspeeding process, sedimentation boundaries (albeit broad ones) are formed that separate from the meniscus and allow for an initial estimate of the molar mass or molar mass distribution. Analysis of the experimental data from this early time can spawn a new optimization of the remaining overspeeding phase, and, in this fashion, an iterative refinement of the rotor speed schedule can take place during the first few hours [249].

The function `Options ▷ Equilibrium Tools ▷ Minimum Time To Equilibrium` in SEDFIT may be used in this manner to optimize the rotor speed profiles for overspeeding. SEDFIT assumes an exponential decay of the rotor speed, modulated by segmented amplitude factors, which may assume positive or negative values to allow the description of both overspeeding and underspeeding phases, ultimately converging to the desired SE rotor speed.

Whether overspeeding is used or not to accelerate attainment of SE at the first rotor speed when carrying out a multi-speed SE experiment, the times required to attain the equilibrium at the next higher rotor speed will be significantly shorter than the time to reach the first equilibrium. Gains in the equilibration times from transient overspeeding may also be realized in the transition to higher rotor speeds [249].

5.7.3 Sedimentation Velocity

As a general guide, scan parameters for SV should be set to produce as many scans as possible. Standard parameters for the absorbance system use radial intervals of 0.003 cm, with a single acquisition in 'continuous mode.' The scan rate resulting from these settings is much faster when compared to that used for SE and is therefore suited for observing fast-moving boundaries. Related considerations regarding the speed of data acquisition and time-intervals of scans are described in Sections 4.2.2 and 4.2.3 and illustrated in Fig. 4.13.

The run length in SV is governed by the smallest species: if dominated by back-diffusion, these should approach an equilibrium condition; if they are sufficiently large to form a boundary then it should have migrated completely (i.e., even with the trailing edge) outside the radial window of observation.

5.8 STOPPING THE RUN

Before stopping an SV run, if it cannot be visually recognized that the sedimentation is complete, it is advisable to carry out a preliminary $c(s)$ data analysis in order to verify that the smallest sedimenting species has migrated sufficiently. These analyses can be performed in just a few minutes. If this verifies that sedimentation is complete, the run can be stopped without further considerations. If another run is to be started immediately, it is wise to enter a low temperature set point, e.g., 10 °C, to prevent the heater from engaging while the vacuum is released. As described in Section 3.1.4, some samples are sufficiently stable that they can be resuspended and re-run if necessary (Section 5.2.4). Otherwise, samples can be aspirated out of the cell assemblies and all components cleaned and stored (Section 5.2.4). Sample recovery is generally useful for later reference, whether to test for degradation by mass spectrometry or SDS-PAGE, or to determine sample concentrations.

Before stopping an SE experiment, it should be verified that equilibrium was attained (Section 5.7.1). At the end of SE experiments with interference optics, it is useful to acquire another water blank from the intact cell assemblies (Section 5.2.4). The sample should be aspirated through the filling ports and the cell assembly should be rinsed a few times thoroughly with buffer and water, and then filled with water. After re-insertion into the rotor and alignment, the same scan setting parameters of the interference optics must be used as in the SE experiment.

CHAPTER 6

Control And Calibration Experiments

T HE interpretation of AUC data is solidly based on thermodynamic first principles and for this reason it has fulfilled a critical role as a physical method throughout almost a century. Even though AUC does not rely on comparative reference standards, quantitative work requires an instrument that is properly calibrated with respect to all data dimensions. Most applications of AUC require quantitatively accurate results, such as the meaningful interpretation of hydrodynamic friction coefficients in the context of theoretical models of macromolecular shapes and conformation. This is also true for the accurate determination of buoyant molar masses, crucial for the deduction of particle molar masses, densities and composition, as well as protein oligomeric states and often binding constants.

We present the definitions of precision and accuracy of measurement in the context of AUC experiments and note that precision refers to the reproducibility (in different laboratories) or repeatability (in the same laboratory) of a measurement, whereas accuracy refers to how close the measured value is to the *true* value. The latter requires measurements using devices ultimately calibrated against absolute standards of metrology as defined by the General Conference of Weights and Measures, maintained by the International Bureau of Weights and Measures and national institutions such as the European Institute for Reference Materials and Measurement and the U.S. National Institute of Standards and Technology (NIST). In the case of spectrophotometric properties, additional protocols and reference material from pharmacopeias may be useful.

Historically, errors in sedimentation coefficients of less than 1% are expected from the AUC [295, 445, 462], and this can be achieved with the current instrumentation and analytical protocols [167, 297]. However, due to possible fluctuations and adventitious shifts in the performance of internal calibrations [296], systematic errors in the operating software [370], unnoticed malfunctions and possible operator errors, the accuracy of results from individual instruments can be quite poor [296, 297, 370], unless periodic calibration experiments are carried out. In fact,

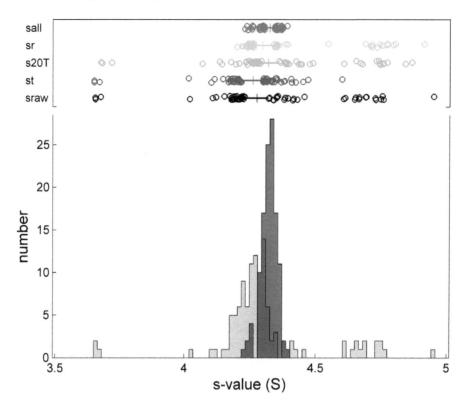

Figure 6.1 Histogram and box-and-whisker plot of s-values for the BSA monomer after corrections for calibration accuracy based on a multi-laboratory study [297]: Raw experimental s-values (black, with gray histogram), scan time corrected s_t-values (blue), rotor temperature corrected s_{20T}-values (green), radial magnification corrected s_r-values (cyan), and fully corrected $s_{20T,t,r,v}$-values (red with red histogram) are shown. The box-and-whisker plots indicate the central 50% of the data as solid line and draw the smaller and larger 25% percentiles as individual circles. The median for each group is displayed as a vertical line.

scan time, temperature, and radial calibration errors each translating to s-value errors of the order of 5-10%, or in some cases even more, have been observed (Fig. 6.1) [296, 297].[1]

Therefore, to ensure the accuracy of all data dimensions, calibration and control experiments carried out using external standards are essential. In the case of the sedimentation coefficient, any source of measurement error contributing 0.1% or more can be considered significant. Since the internal instrument calibrations are subject to slow drift or sudden change, often for poorly understood reasons, it is important that instrument validation using the external calibration methods be carried out on a regular basis.

---

[1]Such errors have historic precedent in the $\sim$10% discrepancies of s-values obtained from Svedberg's oil-turbine ultracentrifuge and the Spinco Model E. These differences were later traced to thermocouple calibration errors in the oil-turbine instrument [431, 445, 462–464].

6.1 DATA DIMENSIONS

In addition to the obvious spatio-temporal parameters that define detection of the concentration evolution (radius, time, and signal), other important parameters for accurate data acquisition are temperature, rotor speed, and wavelength. Altogether, these parameters help define the experiment. Methodology to ensure the accuracy of several of these parameters has already been discussed in earlier sections. These are concisely recapitulated and referenced here for completeness, noting that we can only rely on the *accuracy* of the results obtained when *all* these data dimensions are calibrated and jointly corrected for.

6.1.1 Temporal Accuracy

As described in Section 4.2.4, there are significant scan time errors with current AUC instruments. This appears to be due to their design, and it is now clear that temporal errors have persisted since their introduction [296,370]. The errors in time affect equally the elapsed time t and $\int_0^t (\omega(t'))^2 dt'$ entries reported in the scan file headers, and are of a magnitude that depends on the scan settings and firmware. Except for firmware 5.06, which was distributed by the manufacturer between 2011 and 2013, errors of 0.1% to 2% have been observed.[2] In the case of firmware 5.06, distributed in conjunction with the ProteomeLab GUI 6, time errors of up to $\sim$10% have occurred, depending on the rotor speed [370]. Only data from the fluorescence detection system, which is manufactured by AVIV, have shown no indication of scan time errors.

Fortunately, it is now possible to reconstruct accurate scan time intervals using the time-stamps associated with the file table entries made by the Windows operating system upon creation of the new scan files. As described in Section 4.2.4, this opportunity arises because of the very short and essentially constant (i.e., with little scan-to-scan variation) delay in file creation after each scan has completed. This requires a computer dedicated solely to AUC data collection, free of unnecessary computational and disk writing loads during data collection.

The accuracy of the computer real-time clock used to provide file time-stamps can be verified by disabling updates from a network time server (e.g., disconnecting it from a network), and measuring the drift in time relative to a standard clock, for example, that of NIST or the U.S. Naval Observatory accessed through a radio signal or the internet (http://time.gov or http://tycho.usno.navy.mil/what.html, respectively) on a different device. The magnitude of drifts may also be assessed from accumulative discrepancies between the computer time and the real time, for example, in minutes per year.[3] Excessive drifts, exceeding 1 min per day (corre-

---

[2]A factor 10 error in t (but not $\int_0^t (\omega(t'))^2 dt'$) entries was observed for instruments running on one of the earliest firmware versions from the 1990s [297].

[3]Therefore, if clock adjustments are made on the data acquisition computer, records of the time adjustments should be kept. (It is important that adjustments, if necessary, are not made during data acquisition.)

sponding to ~0.07%), may sometimes be addressed by replacement of the battery of the computer real-time clock running the BIOS on the computer motherboard.[4] In the default configuration of SEDFIT, scan time intervals from the file timestamps are compared with those file header entries, and in case of a discrepancy exceeding a user-defined threshold (e.g., 0.01%) the file header entries are automatically corrected, the files resaved under different filenames and reloaded (Section 4.2.4). Alternatively, time corrections can be simply applied as a multiplicative factor to the measured sedimentation and diffusion coefficients: if the time is recorded erroneously on a time scale $t' = \tau \times t$ with a dilation factor τ, leading to apparent parameters s' and D', then the correct values of s and D can be obtained from:

$$s = \frac{s'}{\tau} \quad \text{and} \quad D = \frac{D'}{\tau} \tag{6.1}$$

Trivially $M = M'$ [296], and the molar mass values obtained from SV analysis are not affected by scan time errors because they impact both s and D in the same way, such that they cancel out in the Svedberg equation Eq. (1.8). The value of the corrected τ can be obtained from a SEDFIT utility function.

6.1.1.1 Scan Velocity

Except for the interference optical system, which instantly images the entire solution column, radial data points are acquired consecutively, but associated with a single scan time. Depending on the relative sedimentation and scanning velocities, this may give rise to systematic errors following data analysis. A detailed analysis of the quantitative impact can be found in Section 4.2.3, along with approaches to rigorously account for this problem during data modeling, or for applying corrections to the final result. The standard absorbance scan settings, with a radial resolution of 0.003 cm and a single acquisition per radial data point, lead to a scan velocity of ~2.5 cm/min at 50,000 rpm. If not accounted for in the data analysis, such a scan velocity results in an error in s of ~0.18% for a 4.3 S particle.

The value of the scan velocity required for calculating the corrections can be obtained by timing the duration of the audible flash lamp operation during a scan across a defined radial interval. In the case of the fluorescence system, the rate of change in the position of the radial motor reported in the scan progress display of the AOS graphical user interface can be timed.

6.1.2 Temperature Calibration

Issues relating to the precision of the built-in radiometer and methods for executing external temperature calibrations using iButtons® have been described in Section 5.3.1.

---

[4]Highly accurate real-time clocks are inexpensive and ubiquitous. For instance, those based on the Maxim Integrated temperature compensated crystal oscillator chip DS3232 have an accuracy of 2 ppm.

Based on a study of 79 instruments [297], the average deviation of the rotor temperature from the set-point of 20 °C is approximately −0.4 °C, and the standard deviation of instrument temperatures is ∼0.4 °C (Fig. 6.2), consistent with the manufacturer specifications; however temperature deviations of up to 1.5 °C [297] and even 3.0 °C [296] have been measured in individual instruments. The temperature-dependence of this calibration error has only been studied in a single instrument, where a uniform offset was observed [366].

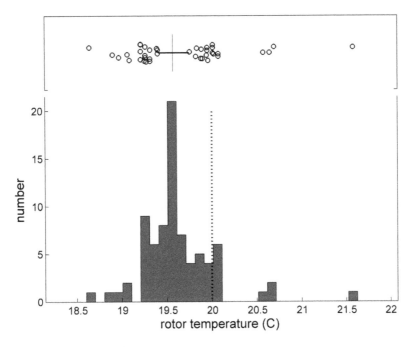

Figure 6.2 Equilibrium iButton® temperatures at 1,000 rpm obtained from 79 instruments in a multi-laboratory study [297] with an instrument set point of 20 °C (indicated as dotted vertical line). The box-and-whisker plot indicates the central 50% of the data as solid line, with the median displayed as a solid vertical line, and individual circles for data in the upper and lower 25% percentiles. The mean and standard deviation is 19.6 °C ± 0.4 °C.

The magnitude of this error has a huge impact on the solution viscosity, resulting in the propagation of a viscosity and ultimately $s_{20,w}$. Errors in $s_{20,w}$ would amount to ∼1.2% with a −0.5 °C temperature error at 20 °C, and increase to ∼2.4%, 4.9% or 7.5% with temperature errors of −1 °C, −2 °C, or −3 °C, respectively. These errors would go entirely unnoticed without independent calibration.[5]

As described, an accurate, external calibration for temperature can be achieved using the iButton®, which is an inexpensive miniature temperature logger manufactured by Maxim Integrated Products. It can be purchased from different suppliers pre-calibrated or custom-calibrated against a NIST-traceable reference thermome-

---

[5]Besides the hydrodynamic parameters, the temperature error is of concern also with regard to the accuracy of binding constants, which are defined at a given temperature.

ter [296, 366, 444]. There are two different approaches to their use. Using them as a real-time monitor of the actual rotor temperature during the experiment is optimal; this requires a modified rotor handle for placement of the iButton® in a manner that allows it to withstand centrifugation (Fig. 5.6 Panel A) [366].

Alternatively, they may be used on a resting rotor, to determine the real temperature at a given set point (Fig. 5.11). As described in [444], the rotor should be turned such that the radiometer is maximally covered by the metal of an 8-hole rotor An-50 Ti, and the monochromator should not be installed. Exhaustive temperature equilibration at an AUC console set point of 20 °C should follow, with the AUC in the 'RUNNING' mode at a target rotor speed of 0 rpm. Temperature equilibration should be carried out for at least two hours after nominal agreement between the set point and the temperature reading shown at the AUC console. The vacuum reading should show 0 micron.

Using either approach, temperature data collection has to be initiated and retrieved through an adaptor connected *via* USB interface to a computer where the iButton® control software is installed (which may be separate from the instrument computer). With the 11-bit resolution iButton® DS1922L a temperature resolution of 0.063 °C can be achieved, resulting in an accuracy of ∼0.1–0.2 °C [296, 366, 444], which translates to uncertainties of 0.23–0.47% in $s_{20,w}$. The remaining uncertainty in the temperature is the dominant source of systematic errors in absolute $s_{20,w}$-values after all external calibrations.

The appropriate temperature corrections for density, viscosity and partial-specific volume can be made with knowledge of the true temperature at a given console temperature. If a large temperature difference is observed, a compensatory and opposite offset in the set point for temperature regulation may be possible.

6.1.3   Radial Calibration

Based on a multi-laboratory study [297], calibration errors in the radial magnification of the absorbance system ranged from −6.5% to +3.1%, with a mean of −0.43% and a standard deviation of 1.36 % (Fig. 6.3). In the case of the interference optical system, radial magnification errors had a mean of −0.75% and a standard deviation of 0.82%, excluding outliers of up to 19%.[6] Because of the large possible errors it is critical to use an independent verification of the radial calibration, such as a precision mask.

In the same study, variations in the meniscus position of the identical solution column were measured. These variations, a consequence of errors in the absolute radial calibration (in addition to magnification errors), exhibited standard deviations of 0.015 cm for the absorbance optical system and 0.037 cm for the interference system. However, no correlation of the observed meniscus position with the measured

---

[6]These outliers occurred with the interference optical system, and exceed calibration errors that would arise from mis-identification of the reference marks during the user-dependent manual radial calibration protocol [207].

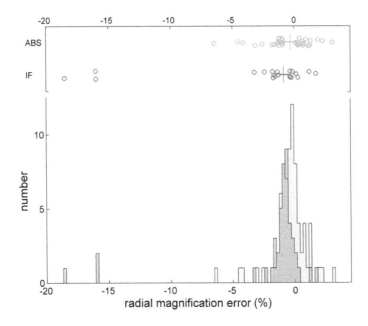

Figure 6.3 Radial magnification corrections obtained with the absorbance (green) and interference optical systems (magenta) from a multi-laboratory study [297]. The box-and-whisker plot above the histogram indicates the central 50% of the data as a solid horizontal line and depicts data in the smallest and highest 25% percentiles as individual circles. The median is displayed as vertical line. The mean and standard deviations are -0.43% $\pm1.36\%$ for the absorbance system, and $-0.75\% \pm 0.82\%$ for the interference system (once the three outliers are excluded).

s-values for the BSA monomer were observed, consistent with the expectation that errors in the absolute distance from the axis of rotation have a smaller impact than errors in the radial magnification (Section 4.1.1). The latter disproportionately influence the much smaller distances involved with boundary movement as compared to the distance from the center of rotation.

A detailed description of various methods for radial calibration and tests for scanner linearity can be found in Section 4.1.1. While initial efforts to measure radial magnification accuracy have relied on a precision steel mask with a regular pattern on the sample side, centered and sandwiched between window assemblies [296], a lithographically patterned window with calibrated dimensions is in development as a reference artifact in collaboration with NIST (unpublished)(Fig. 4.1 Panel B and C and Fig. 4.2 Panel B – D). Besides higher accuracy, the interleaved design of the dual sample and reference sector mask offers an opportunity for the compensation of variations in the angular alignment of the centerpiece and mask. The specially developed dedicated software package MARC can analyze the image of this mask with any of the optical systems (with the fluorescence system requiring the scan of a slowly-sedimenting fluorophore), and yield radial magnification correction factors (unpublished).

As in the case of temporal errors, small and constant radial magnification errors can be corrected with simple multiplicative factors applied to the apparent s''- ,

D''- and M''-values determined with a radial magnification error δ to arrive at corrected values [296]:

$$s = \frac{s''}{1+\delta}, \quad D = \frac{D''}{(1+\delta)^2}, \quad \text{and} \quad M = (1+\delta)M'' \tag{6.2}$$

When the radial magnification errors exceed a few percent it may be advantageous to radially transform the scan files directly prior to the data analysis, an optional task in the software MARC. Furthermore, in a minority of cases, imaging of the radial mask indicates nonlinear aberrations [297]. In such cases, a polynomial approximation in the radial magnification can be used for correction of the experimental radial scan files prior to sedimentation analysis.

To obtain the highest possible experimental accuracy, a cell assembly containing the radial calibration assembly should be run simultaneously in the same rotor side-by-side with the sample(s). In addition, for runs where both absorbance and interference data are acquired, it is good practice to compare the meniscus position of the solution column for consistency in order to identify the most obvious radial errors.

6.1.4 Spectral Calibration

The accuracy of the absorbance wavelength calibration is an important requisite for accurate photometric values as well as signal linearity (Section 4.3.3). Fortunately, the Hamamatsu xenon flashlamp that serves as a light source provides internal standards in the form of emission intensity peaks at known wavelengths (Fig. 4.32). The wavelength accuracy can be examined using the first UV peak at 229 nm and the 527 nm peak in the visible as markers (peaks labeled with stars in Fig. 4.32). To this end, an intensity scan as a function of wavelength should be acquired, preferably at the center (6.5 cm) of an empty rotor hole or empty cell (without windows). Wavelength calibrations are normally carried out during instrument service and rarely done by the user, except to troubleshoot instrument problems.

6.1.5 Photometric Calibration

The accurate photometric measurement of optical density does require verification and adjustment to match up against known standards. The photometric accuracy of the absorbance optical system is far lower than in most dual-beam laboratory bench-top spectrophotometers; after a certification service by the manufacturer (unfortunately photometric calibration is not part of standard instrument maintenance) current instrument specifications guarantee photometric accuracy to within ± 0.025 AU at an absorbance of 0.06 AU, and ± 0.045 AU at an absorbance of 1.0 AU [296]. The multi-laboratory benchmark study [297] showed instrument-to-instrument variations of ~6% in absorbance and ~4% in interference detection of the same sample at ~0.5 OD_{280}.

Glass filters commonly used for photometric reference materials cannot be used in AUC as they would probably not withstand the gravitational forces. However, holmium oxide, potassium dichromate and potassium nitrate solutions, or other photometric standard solutions are available from different manufacturers as NIST certified reference materials [465]. These may be measured at low rotor speed for self-calibration, or comparison against a calibrated bench-top spectrophotometer.

Fortunately, as long as the response is within the linear range of detection, the photometric accuracy will only influence the signal amplitudes and therefore will not directly affect the accuracy of the macromolecular parameters derived from the AUC analysis. It will, however, influence applications where the absorbance readings will be used to determine absolute sample concentrations, such as the determination of binding affinities or stoichiometries for interacting systems. AUC-based determinations of extinction coefficients (Section 4.3.2) will also be limited in their precision by the photometric accuracy, and based on the multi-laboratory benchmark study this is limited to a precision to ~5% [297].[7]

6.1.6   Rotor Speed

Independent measurement of the rotor speed requires access to the instrument electronic boards carrying Hall effect signals from magnets in the drive spindle. In a spot check of seven instruments at the National Institutes of Health, deviations from the nominal rotation frequency at 50,000 rpm were found to be ~0.01%, maximally 0.015% [296]. Considering the timing errors, it may perhaps seem surprising that rotor speeds are not subject to similar errors. This suggests the use of different timers for measuring the frequency of the rotor revolution and the elapsed time. At an error of 0.01%, uncertainties in the centrifugal field due to rotor speed errors will be negligible compared to other sources of error.

6.1.7   Mass and Density

Since a key quantity resulting from AUC experiments is the particle mass, or molar mass, it is worth recapitulating how the kg mass unit enters AUC measurements.

Provided the calibration in other dimensions is correct, the buoyant molar mass can be derived directly from the potential energy of the system in the centrifugal field at a given temperature. It is manifest in the shape of the macroscopic concentration profiles. These are analyzed either in thermodynamic equilibrium or with regard to their evolution with time.

However, a complication is that the measurement takes place in liquid, which requires consideration of buoyancy. In principle, for very large (macroscopic) particles with known and clearly-defined volume, the particle mass can be determined simply by adding the calculated mass of the displaced solvent to the measured

---

[7]This is distinct from the repeatability in a given instrument, which will usually be better, but not necessarily providing more accurate values.

buoyant molar mass.[8] This requires knowledge of the solvent density, for which usually a mass measurement using some form of a calibrated macroscopic scale will be required.

By contrast, for nanoscopic particles the molecular volume in solution cannot be directly imaged and simple geometric models of molecular volume are not suitably accurate. Instead, the buoyancy correction rests on the determination of their partial-specific volume. The process of experimental determination of the partial-specific volume, in turn, requires knowledge of the weight concentration, i.e., the total mass of dissolved particles, which is ultimately based on measurements with a macroscopic scale, as outlined in detail in Section 3.3.2. The theoretical compositional determination of partial-specific volumes from tabulated data is similarly based on previous measurements of weight concentrations with a macroscopic scale. Alternatively, if the partial-specific volume is determined by a density contrast method, at least a second solvent density measurement is required.

More intricate, but nonetheless crucial questions arise in the need to define clearly what the particle of interest is: Commonly in molecular biology, for example, it is silently assumed to be only the covalently bound polypeptide or polynucleotide chain. A more physically realistic definition would be the electroneutral particle comprised of the covalent polyelectrolyte chain including the invariably present counterion cloud. Its presence is inseparably linked to the fact that biomolecular events obligatorily take place in solution, and is often recognized to contribute in an essential way to macromolecular function. Counterions, however, will change dependent on the solvent used. Similar questions arise with regard to the accounting of water molecules surrounding the nanoscopic particle, which in its vicinity are disturbed in their structure and dynamics, and sometimes even weakly or tightly bound.

These issues are described at length in Chapter 2 and in Section 3.3.2: the key is to clearly define which of the alternative views of the particle is adopted, and to carry this definition consistently through both the interpretation of the AUC experiments, as well as the partial-specific volume and the weight concentration measurements (should weight concentration be necessary). Often, the most convenient definition is directly linked to the opportunities for concentration measurements. Either way, for nanoscopic particles in solution, it seems indispensable in AUC to rely ultimately on one or more density measurements and/or other measurements on a macroscopic scale.

---

[8]This approach was taken, for example, in the determination of the average hydrated mass of *Spiroplasma melliferum*, approximately 4 GDa, based on the known helical shape with a length of ~6 μm and volume of ~0.17 μm^3 of an average specimen as determined from electron microscopy [284].

6.2   STANDARD CONTROL EXPERIMENTS

In order to ensure the accuracy of the quantitative results from AUC experiments, a set of standard control experiments should be carried out regularly, especially after instrument repair or service. These experiments need to combine the external calibration of the temperature, time and radius with a sedimentation experiment of a stable reference sample. The reference sample can provide independent validation of the consistency of the calibration, and can be used independently to track many aspects of instrument performance.

A time-honored choice for a reference is BSA [296, 297, 462, 464]. Standard BSA samples dissolved into PBS[9] [297] consist of a mixture of monomer as the dominant species, oligomers, and trace degradation products, with a possible slowly time-dependent aggregation on the scale of many months [297]. Importantly, in conjunction with the $c(s)$ analysis [40] the monomer can be well separated from the other species. In practice, a sample solution of 400 μL BSA at $\sim$0.5 mg/ml in PBS can be kept in a sealed 12 mm centerpiece cell assembly (with PBS in the reference sector), stored in the refrigerator, and used for calibration control experiments at 50,000 rpm and 20 °C repeatedly with reproducible results. In between experiments, the sample can be fully resuspended by gentle inversion of the cell assembly. Integration of the monomer peak has been found to yield s-values that are consistent over long time periods [296, 297].[10]

To produce a full set of calibration and reference experiments at once, a rotor containing a BSA cell assembly along with a radial calibration cell assembly containing the precision mask (preferably also stored fully assembled), can be run with a counterbalance and empty cell (for balance) in a rotor containing the iButton® temperature logger in the rotor handle. After thorough temperature equilibration at 20 °C, the rotor should be accelerated from rest to 50,000 rpm, allowing for data acquisition of the ensuing concentration gradients to commence. Typical absorbance scan settings should be used with data collected in 'continuous mode' with a single acquisition at a radial increment of 0.003 cm, resulting in a nominal scan frequency of 1 per minute (Section 5.7.3). The data acquisition wavelength is 280 nm, and the radial range for scanning is chosen such as to encompass the entire solution column. The BSA monomer s-value should be determined with a standard $c(s)$ analysis in SEDFIT[11] [40], the pattern recorded from the radial precision mask should be analyzed in MARC (unpublished), and the temperature reading during the run be retrieved from the iButton® log.

---

[9]PBS used in [297] consists of 5.62 mM Na_2HPO_4, 1.06 mM KH_2PO_4, 154 mM NaCl, pH 7.40 (20 °C).

[10]The limit of stability has not been established, but is currently known to exceed two years.

[11]As usual, this analysis should be carried out loading scans representing the entire sedimentation process, allowing for scan time corrections, and with both the meniscus position and the average frictional ratio refined during nonlinear regression, while keeping the partial-specific volume, buffer density and viscosity at default settings (0.73 mL/g, 1.000 g/mL and 0.01002 P respectively).

Alternatively, in the absence of a rotor handle holder for the iButton® [366], the temperature calibration offset may be determined in a separate low-speed experiment carried out with the iButton® in a cell assembly [296], or with the resting rotor after placement of the iButton® on the counterbalance (without optical arm, following the protocol of [444] as summarized above in Section 6.1.2), both after exhaustive temperature equilibration of the evacuated rotor chamber at 20 °C.

While the calibration corrections may change adventitiously or gradually with time, one should always obtain the same s-value after application of corrections: In the multi-laboratory reference study, the s-value for BSA monomer in PBS was found to be 4.325 S ± 0.030 S after correction for all calibration corrections (but not for buffer viscosity and density) [297].

It is advantageous to perform these experiments with the same cell assemblies that have been stored without disassembly between control runs. In this way, in addition to providing an independent control for s-values and thereby the consistency of the calibration factors, consistency in the instrument performance with regard to the absolute radial position can be assessed by tracking the meniscus positions and the edges of the radial calibration cell from run to run. Furthermore, by collecting data of such repeated control experiments with the same cell assemblies over time, these experiments can help identify the emergence of aberrant scan behavior, such as sloping plateaus or irregularities in the boundary shapes that have been observed in individual instruments [297].

Macromolecular Partial-Specific Volumes

The data tables presented in the following appendices are a compilation of frequently used solvent and macromolecular properties required for data analysis in analytical ultracentrifugation. They are not a substitute for the primary sources referenced, but are intended as a convenient collection of common and relevant data, as well as a resource for more detailed references.

A.1 PROTEINS

The most widely used data sets for the determination of protein partial-specific volumes are those measured and computed by Cohn and Edsall [311], and the consensus values published by Perkins [111], which combine multiple experimental and computational residue volume values to best match a set of experimentally measured protein partial-specific volumes. Based on the Cohn-Edsall values, the average $\bar{v}$ of all known human proteins is 0.735 mL/g with a standard deviation of 0.01 mL/g [95]. Regarding compressibility, 2.5×10^{-4}/MPa represents an average value for globular proteins [214].

> A calculator function in SEDFIT will determine the compositional partial-specific volume, based on the Cohn-Edsall or Perkins consensus values, the latter temperature corrected based on composition.

As discussed in Chapter 2, preferential interactions in the presence of co-solutes may lead to effective partial-specific volumes ϕ' that are different from $\bar{v}_a$ calculated based on the amino acid composition. If preferential binding parameters are known, they may be used via Eq. (2.7) (or Eq. (3.13) for hydration only) to predict the effective partial-specific volume ϕ'. These contributions are indispensable for highly charged macromolecules, for situations in which a co-solute interacts strongly with

TABLE A.1 **Amino Acid Residue Partial-Specific Volumes and Temperature Coefficients**

amino acid residue	Cohn-Edsall[a] (mL/g)	Perkins[b] (mL/g)	$d\bar{v}/dT$ at 20°C [c] (10^{-3}mL/g × K^{-1})
Ile	0.90	0.884	0.67
Phe	0.77	0.776	0.79
Val	0.86	0.843	0.43
Leu	0.90	0.894	0.47
Trp	0.74	0.737	0.49
Met	0.75	0.758	0.77
Ala	0.74	0.744	0.57
Gly	0.64	0.632	0.55
Cys	0.63	0.615	0.54
Tyr	0.71	0.706	0.62
Pro	0.76	0.765	0.62
Thr	0.70	0.705	0.50
Ser	0.63	0.634	0.48
His	0.67	0.686	0.63
Glu	0.66	0.657	0.65
Asn	0.62	0.634	0.84
Gln	0.67	0.682	0.56
Asp	0.60	0.604	0.70
Lys	0.82	0.811	0.48
Arg	0.70	0.726	0.64

[a] From [311], with corrections for the Cys value from [111]; [b] Based on consensus volumes in [111]; [c] Calculated slope of temperature dependence at 20°C calculated from tabulated data and Eq. 7 of [325].

the particle of interest (such as detergents, or urea), and generally when the solution density is significantly different from that of the macromolecular hydration shell.

Preferential interactions are particularly important for co-solutes that stabilize or de-stabilize protein structure. Whereas Gekko and Timasheff found that proteins in glycerol-water mixtures are preferentially hydrated [119], the invariant particle model does not usually apply based on the dependence of the preferential binding parameter $(dw_3/dw_a)_{T,\mu_1,\mu_3}$ with glycerol concentration [120]. Depending on the protein, the preferential exclusion of glycerol exhibits a range of −0.02 to −0.04 g glycerol per g protein at 10% glycerol to −0.11 to −0.16 g glycerol per g protein at 40% glycerol [119], equivalent to a preferential hydration of 0.14–0.29 g water per g protein in 10% glycerol. Similarly, apparent partial-specific volumes of select proteins in urea [331], guanidine hydrochloride [332] and buffer salts [466] have been reported. See also Section 3.3.3.

A.2 NUCLEIC ACIDS

Apparent partial-specific volumes of nucleic acids depend strongly on the nature of the salt and buffer co-solutes due to their strong ion binding, coupled with the fact that their molar masses are usually determined for the counterion-free form.

Data for nucleotides in 100 mM sodium phosphate are listed in Table A.2.[1] A comprehensive data set of the salt concentration dependence of the apparent partial-specific volume of calf thymus DNA or linearized ColE1 plasmid in different buffer salts has been reported in [468], and is partially reproduced in Table A.3. A computational study on RNA volumes from crystal structures can be found in [469]. Values for a wider range of conditions and molecules are tabulated in [309].

TABLE A.2 **Apparent Partial-Specific Volumes of Nucleic Acids**

nucleotide	ϕ' (mL/g)
dA	0.56[a]
dC	0.58[a]
dT	0.61[a]
dG	0.54[a]
RNA (oligo duplexes, mixed base, A conformation)	0.508[b]
oligoDNA (oligo duplexes, mixed base, B conformation)	0.56[b]

[a]In 100 mM sodium phosphate pH 7.4, 20 °C from [336]; [b] 100 mM KCl, 20 mM sodium phosphate, pH 7.0 [337].

---

[1]Woodward and Lebowitz have published a revised equation relating the buoyant density of double stranded DNA to G+C content [467]. This relation, and others, has often been used to determine the partial-specific volume of DNA based simply on its G+C content. However, care needs to be taken, as these partial-specific volumes determined at high concentrations of CsCl may not correspond to the effective partial-specific volume in the solvent of interest. This is very clear from the data in Table A.3. Given that nucleic acids used for study are usually single species, an experimental determination of their effective partial-specific volume by sedimentation equilibrium is recommended, as discussed in Section 3.3.3.

TABLE A.3 **Partial-Specific Volumes and Preferential Binding Parameters of Linear DNA in Different Salt Solutions**

salt	concentration (mol/L)	$\bar{v}$ (mL/g)	ϕ' (mL/g)	ξ_3 (g/g)	ξ_1 (g/g)
NaCl	0	0.499			
	0.005		0.515	-0.0205	69.9
	0.01		0.515	-0.0213	36.6
	0.02		0.517	-0.0235	20
	0.05		0.522	-0.0285	9.71
	0.1		0.527	-0.034	5.79
	0.2	0.503	0.535	-0.0405	3.44
	0.5		0.544	-0.0425	1.437
	0.976	0.528			
	1		0.568	-0.0668	1.118
	1.5		0.571	-0.078	0.862
	2		0.584	-0.865	0.71
	3		0.596	-0.101	0.541
	3.16	0.539			
	4		0.596	-0.112	0.439
	4.19	0.543			
	4.99		0.596	-0.118	0.361
CsCl	0	0.44			
	0.2	0.446	0.503	-0.05	1.471
	0.5		0.517	-0.075	0.873
	0.97	0.46			
	0.99		0.53	-0.114	0.654
	1.396		0.548	-0.138	0.546
	1.86	0.471			
	2.292		0.563	-0.19	0.443
	2.325				
	2.445		0.566	-0.2	0.435
	3.44	0.471			
	3.666		0.587	-0.255	0.347
	4.996		0.584	-0.308	0.282
	7.46		0.566	-0.365	0.186
LiCl	0.2		0.563		
	2		0.596		
	3		0.604		
	4		0.608		
	5		0.615		
	6.55		0.624		
	7.38		0.623		
	9		0.621		
RbCl	0.2		0.514		

Data from [468].

A.3 CARBOHYDRATES

Carbohydrates have lower partial-specific volumes than proteins. Durchschlag has tabulated an extensive list of values, and values for carbohydrates that typically decorate glycosylated proteins are reproduced in Table A.4. Shire has reported calculated partial-specific volumes of different types of glycosylation [328]. Values for several sugars and proteoglycans have been measured by Perkins and colleagues, as reported in [470].

TABLE A.4 **Partial-Specific Volumes of Selected Carbohydrates**

carbohydrate	$\bar{v}_a'$ (mL/g)
mannose	0.607[c]
sialic acid	0.581[c]
n-acetyl neuraminic acid	0.584[a]
fucose	0.671[c]
fructose	0.614[a]
glucose	0.622[c]
n-acetyl glucose	0.684[c]
n-acetyl galactose	0.684[c]
galactose	0.622[c]
typical high mannose carbohydrate moiety	0.63[b]
hybrid	0.64[b]
complex	0.62–0.64[b]

[a] From [309]; [b] from [328]; [c] from [470].

A.4 DETERGENTS AND LIPIDS

Extensive tables of partial-specific volumes and other physical properties for detergents and lipids, including the critical micelle concentration and micelle size, have been published by Tanford, Reynolds and colleagues [178, 185, 471–473], and others [474–479]. Partial-specific volumes for detergents and lipids commonly used in AUC studies can be found in Table A.5.

It is possible to determine partial-specific volumes for detergents based on their atomic composition, and computational corrections for micellization have been reported by Durchschlag and Zipper [475]. However, a difficulty of partial-specific volumes of detergents and lipids is their dependence on phase, temperature and pressure, and the presence of dissolved ligands such as cholesterol. In addition, some detergents (or lipids) are electrolytes and their micelles (or bilayers) will exhibit preferential interactions with buffer salts. Therefore, if possible, in cases where published data cannot be found, the apparent partial-specific volume should be measured experimentally; this is best accomplished by density contrast SV or by establishing neutral buoyancy conditions in the buffer of interest (see Section 3.3).

TABLE A.5 **(Apparent) Partial-Specific Volumes of Detergents and Lipids**

detergent/lipid	$\bar{v}_a'$ (mL/g)
n-dodecyl--D-maltoside (DM)	0.815[f]
octyl- -D-glucoside (OG)	0.859[d]
octyl pentaethylene glycol ether (C8E5)	0.993[a]
dodecyl octaethylene glycol ether (C12E8)	0.95[f]
dodecyl nonaethylene glycol ether (C12E9)	0.952[g]
triton X-100 (reduced)	0.921[f]
lauryldimethyl amine-n-oxide (LDAO)	1.128[c]
3-laurylamido-n,n-dimethylpropyle amine oxide (LAPAO)	1.002[f]
n-dodecylphosphocholine (DPC)	0.944[h]
CHAPS	0.81[i]
3-(n,n-dimethylmyristyl-ammonio) propanesulfonate (C14SB)	0.96 [j]
n-dodecyl-n,n-dimethyl-3-ammonio-1-propanesulfonate (SB3-12)	0.957[k]
sodium lauryl sulfate (SDS)	0.870[b]
Tween 20	0.869[b]
amphipol A8-35	0.87[l]
(hemi)fluorinated (multi)glycosylated trishydroxymethyl acrylamidomethane	~0.6[m]
PC (egg yolk)	0.981[e]
DMPC	0.93[o]
PE	0. 965[e]
PS	0.93[e]
POPC	0.987[p]
SOPC	0.990[p]
SLPC	0.94[p]
SLnPC	0.922[p]
sphingomyelin	1.005[b]
MSP1D1 nanodisc (empty)	0.88 - 0.89[n]
POPC in nanodisc	0.981[n]
POPG in nanodisc	0.968[n]

[a]From [478]; [b] from [185]; [c] from [474]; [d] from [473]; [e] from tables in [471];[f] from [480]; [g] from [270]; [h] as the inverse of the matching density in [481]; [i] from [482]; [j] from [483]; [k] from [476]; [l] 100 mM NaCl, pH 7 [293]; [m] from [484]; [n] at 10 °C from [187]; [o] density contrast SV of 100 nm extruded vesicles at 18 °C in gel phase (P.S., unpublished); [p] at 23 °C from [485].

Solvent Properties

As described in Section 3.2, solvent properties may be measured experimentally, or determined from tabulated data. Data for most commonly used solvents are provided in the present section. The solvent density and viscosity are temperature dependent and thus need to be determined at the experimental temperature; these parameters also exhibit a dependence on pressure and this can be important, as pressure gradients reaching up to $\sim$40 MPa can be encountered during AUC experiments.

B.1 WATER

Both the density and viscosity of water are dependent on temperature, and this needs to be accounted for in the interpretation of AUC data. The density of H_2O as a function of temperature can be described with an accuracy of $\sim$0.001% by [486]:

$$\rho(t) = \rho_{max} \times \left\{ 1 - F(t - t_{max})^2 \left(1.74224 + \frac{482.502}{t - t_{max} + 77.861} \right) \times 10^{-6} \right\} \quad \text{(B.1)}$$

with t in this context denoting the temperature in °C, $\rho_{max} = 1.0$ mL/g, $t_{max} = 3.986$ °C, and $F = 1.0$.

It is critically important to recognize the significant temperature dependence of the viscosity of water, which changes by approximately a factor of two in the range of temperatures commonly used in the AUC, namely 4 to 30 °C. The following approximation, plotted in Fig. B.1, has an accuracy of 0.1% or better between 0 and 40 °C [487]:

$$log_{10}\left(\frac{\eta(t)}{\eta(20°C)}\right) = \frac{20 - t}{t + 96} \times \left[1.2364 - 1.37 \times 10^{-3} \times (20 - t) \right.$$
$$\left. + 5.7 \times 10^{-6} \times (20 - t)^2 \right] \quad \text{(B.2)}$$

The density and viscosity of water at select temperatures are listed in Table B.1.

SEDFIT and SEDPHAT require solvent viscosities in Poise units.

Water has well-known pressure dependent anomalies in bulk density, viscosity, and other physical properties when compared to other liquids. Fortuitously, these

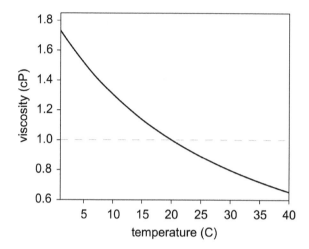

Figure B.1 Plot of the viscosity of ordinary water versus temperature, based on Eq. (B.2) [487], in units of cP, which is equivalent to mPa×s. The dashed line represents the standard viscosity.

TABLE B.1 **Density and Viscosity of Water as a Function of Temperature at Atmospheric Pressure**

temperature (°C)	density (g/mL)	viscosity (mPa×s[a])
4	0.99997	1.567
10	0.99970	1.306
15	0.99910	1.138
20	0.99820	1.002
25	0.99704	0.8901
30	0.99565	0.7974
36	0.99369	0.7052
40	0.99222	0.6530

Data from [488]. [a]1 mPa×s $= 10^{-3}$ Pa×s $= 10^{-2}$ Poise $= 1$ cP.

are generally favorable for AUC studies. The densities and viscosities of water at a range of relevant pressures are provided in Table B.2. As described in Section 2.3.1, water has a relatively low isothermal compressibility of $\kappa = 4.55 \times 10^{-4}$/MPa at 10 MPa (used as an average centrifugal pressure) and 20 °C [489]. At the highest rotor speeds, this leads to a density that is ~1% larger than atmospheric pressure density at the bottom of a typical 12 mm solution column. In SV experiments, however, the resulting increase in buoyancy is partially compensated for by the anomalous pressure dependence of the water viscosity: Below 33 °C, the viscosity decreases slightly with increasing pressure [217], resulting in a ~0.2–0.4% decrease in viscosity at room temperature in the range of common AUC pressures [218, 219] (Section 2.3.1).

TABLE B.2 Density and Viscosity of Water as a Function of Pressure at 20 °C

pressure (MPa)	density (g/L)	viscosity (mPa×s)
0.1	0. 99821	1.002
5	1.0004	0.9996
10	1.0027	0.9977
15	1.0049	0.9960
20	1.0071	0.9944
25	1.0093	0.9930
30	1.0115	0.9918

Data from [488].

B.2 HEAVY WATER: D$_2$O

Deuterated or heavy water with deuterium replacing hydrogen is often used for density contrast, typically to determine partial-specific volumes or to match the density of the solvent to that of bound detergent in protein-detergent complexes, allowing for a simple determination of the protein buoyant molar mass. The density of D$_2$O follows Eq. (B.1) with $\rho_{max} = 1.10602$, $t_{max} = 11.220$ C, and $F = 1.0555$ [486]. The viscosity of H$_2$O and D$_2$O are almost identical after a temperature shift of 6.5 °C and multiplication with a factor 1.0544 – the square root of their molar mass ratio [218]:

$$\eta_{D_2O}(t) = 1.0544 \times \eta_{H_2O}(t - 6.498) \tag{B.3}$$

(where η_{H_2O} can be evaluated using Eq. (B.2)). Values at 4 °C and 20 °C, often used in SE and SV experiments, respectively, are listed in Table B.3. In mixtures with ordinary water, the viscosity is nearly proportional (to within 0.1%) to the density of the mixture [490].

TABLE B.3 Density and Viscosity of Water and Heavy Water at Atmospheric Pressure

water	density (g/mL)		viscosity (mPa×s)	
	20 °C	4 °C	20 °C	4 °C
H$_2$O	0. 99821[a]	0.99997[a]	1.002[a]	1.567[a]
H$_2^{18}$O	1.11055[b]	1.11255[b]	1.056[c]	
D$_2$O	1.1054[a]	1.1056[a]	1.247[a]	2.071[a]
D$_2^{18}$O	1.2162[b]	1.2163[b]	1.304[d]	

[a] from [488]; [b] from [486]; [c] from [491]; [d] from [492].

B.3 HEAVY OXYGEN WATER: $H_2^{18}O$ AND $D_2^{18}O$

The significant and various effects that D_2O has on macromolecular properties can be largely avoided by instead using $H_2^{18}O$, which has a density similar to that of D_2O (due to an almost identical molar mass) and a viscosity similar to that of H_2O. The density of $H_2^{18}O$ is similar to that of ordinary water, after accounting for the higher mass and follows Eq. (B.1) with $\rho_{max} = 1.11255$, $t_{max} = 4.30$ °C, and $F = 1.0555$ [486]. The viscosity of $H_2^{18}O$ in the temperature range from 15 °C to 35 °C was found to be 1.19 to 1.15-fold lower than that of D_2O [491], and higher than that of H_2O by a factor approximately equaling the molar mass ratio [491] (Table B.3).

An even higher density can be achieved with deuterated heavy oxygen, $D_2^{18}O$. Unfortunately, this will have the same drawbacks regarding deuterium effects on macromolecules as D_2O. Analogous to $H_2^{18}O$, the density of $D_2^{18}O$ closely mirrors the density of D_2O after accounting for the higher mass, and is well-described by Eq. (B.1) with $\rho_{max} = 1.21691$, $t_{max} = 11.46$ °C, and $F = 1.0555$ [486]. The viscosity of $D_2^{18}O$ over the temperature range from 15 °C to 35 °C has been measured and found to be 1.28 to 1.23-fold higher than that of H_2O [492, 493].

In practical terms, the use of heavy water will require dilution of the sample into heavy water/buffer stocks, invariably leading to sample dilution. Density and viscosity data have been determined for phosphate buffered saline solutions prepared with different volume fractions of ordinary and $H_2^{18}O$ heavy oxygen water (Table B.4) [269].

TABLE B.4   **Density and Viscosity Values Measured for PBS Solutions Containing Different Fractions of $H_2^{18}O$**

$H_2^{18}O$ volume fraction[a]	0	0.5	0.9	0	0.5	0.9
temperature (°C)	density (g/mL)			viscosity (mPa×s)		
20	1.005584	1.059388	1.102733	1.0219	1.0456	1.0665
16	1.006406	1.060246	1.103606	1.2230	1.1493	1.1725
10	1.007304	1.061186	1.104581	1.3136	1.3436	1.3716
4	1.007758	1.061652	1.105048			

From [269]. Data were measured at atmospheric pressure. [a]$H_2^{18}O$ fraction by volume, using the 97% isotopically enriched heavy-oxygen water as a reference.

B.4 ORGANIC AND OTHER NON-AQUEOUS SOLVENTS

There is a long history of AUC experiments carried out in solvents other than water, and there are numerous sources for the density and viscosity of organic liquids. Solvent properties at a wide range of pressures and temperatures can be found in the interactive data base compiled by the U.S. National Institute of Standards and Technology (NIST) at http://webbook.nist.gov/chemistry/fluid/. Additionally, extensive data for organic solvents and their mixtures can be found in PubChem by the National Center of Biotechnology Information of the National Institutes of Health at https://pubchem.ncbi.nlm.nih.gov, as well as in commercial sources.

Table B.5 lists the densities and viscosities of some common solvents. Unfortunately, the corresponding properties for mixtures of solvents cannot be assumed to relate linearly to the mole, volume or mass fraction of the constituents.

TABLE B.5   **Density, Viscosity, and Compressibility of Non-Aqueous Solvents**

solvent	density (g/mL)	viscosity[b] (mPa×s)	compressibility (10^{-4}MPa^{-1})
acetone	0.7845[a]	0.327[d]	12.75[c]
acetonitrile	0.78745[a,h]	0.360[d]	
benzene	0.8789[c]	0.652[c]	9.44[c]
carbon tetrachloride	1.5940[a]	0.969[c]	10.46[c]
chloroform	1.489[c]	0.58[c]	9.98[c]
cyclohexane	0.7781[a]	1.02[e]	11.10[b,c]
dichloromethane	1.325[f]	0.425[f]	
dimethylformamide	0.9445[a]	0.802[a,b]	
dimethyl sulfoxide	1.0956[g]	1.99[g]	
1,4-dioxane	1.0329[a]	1.20[a,b]	
ethanol	0.791[c]	1.200[c]	11.19[c]
ethyl acetate	0.902[a]	0.455[c]	
ethylene glycol	1.11[a]	19.9[c]	
hexane	0.6606[a,b,i]	0.326[c]	16.06[b,c]
methanol	0.810[c]	0.597[c]	12.11[c]
toluene	0.8669[c]	0.590[c]	
ortho-xylene	0.8801[a]	0.810[c]	
para-xylene	0.86104[a]	0.620[c]	
meta-xylene	0.8684[a]	0.648[c]	8.46[c]

Data are usually at 20°C and atmospheric pressure, but the orginal source should be consulted. [a] From PubChem [494]; [b] at 25 °C; [c] from [495]; [d] interpolated between 15 °C and 25 °C from [495]; [e] at 17 °C from [495]; [f] from [496]; [g] from [497]; [h] at 15 °C; [i] at 25 °C.

Organic solvents commonly show a two- to three-fold higher compressibility than water (Table B.5), leading to greater changes in buoyancy driven by pressure gradients along the solution column. This is in contrast to DMSO/water mixtures, which over a range of mole fractions are actually less compressible than water [211].

As discussed in Section 2.3.1, some organic solvents exhibit a very strong pressure dependence of the viscosity [498]. This behavior is related to their high isothermal compressibility, low isobaric thermal expansion, and strong temperature dependence of the viscosity [499]. For instance, by increasing the pressure to 49 MPa at 30 °C, the viscosities of carbon tetrachloride and benzene increase by 21%, and 19%, respectively [215], whereas a 28% change in viscosity across a centrifugal solution column at 69,000 rpm was reported for acetone [216]. For the alkanes, hexane and octane, the viscosity increases by ∼60% over the same pressure range [500], and similar values have been measured for other organic solvents [500]. This is of significant consequence for SV experiments. Unfortunately the pressure dependence of viscosity cannot be easily measured with common benchtop viscometers. However, the effects may be partially compensated for by the use of shorter solution columns and lower rotor speeds than usual.

B.5 CO-SOLUTES

The solution density and viscosity can be substantially influenced by co-solutes. For example, aqueous solutions of 5% and 10% (by weight) glycerol at 20 °C have densities of 1.0097 g/mL and 1.0215 g/mL, respectively, and viscosities of 1.1273 cP and 1.2906 cP, respectively [495]. At the same temperature, 100 mM NaCl increases the density by 0.0041 g/mL, and the viscosity by a factor 1.009 [1]. The density and viscosity of aqueous solutions for a large selection of co-solutes over wide ranges of concentrations are listed in the CRC Handbook of Chemistry and Physics [495]. Generally these have sufficient data at closely spaced concentrations such that they can be linearly interpolated to the desired co-solute concentration.

At concentrations that are not too high, the combined impact of added co-solutes on solvent density and viscosity may be approximated as being linearly additive [1]. This approximation breaks down, however, at molar concentrations: For example, this leads to 1% errors in calculated density values for concentrated urea solutions containing 1 or 2 M NaCl [501].

Within the limits of dilute solutions, therefore, the independent contributions of different co-solutes may be added up. This can be accomplished by combining density increments $\Delta\rho = \rho(c) - \rho_0$ for each solute to yield the solution density $\rho = \rho_0 + \sum \Delta\rho$. Similarly, increments of relative viscosity $\Delta\eta_r = \eta(c,t)/\eta_0(t) - 1$ can be treated as approximately additive to yield a calculated solution viscosity $\eta = \eta_0 (1 + \sum \Delta\eta)$. As suggested by Svedberg [1], the increments of relative viscosity rather than absolute viscosity differences have the advantage of being not very temperature dependent, thus better capturing the temperature behavior of co-solute effects on viscosity.

Based largely on data from the CRC Handbook [495] and several user-supplied co-solute data sets, increments of density changes $\rho(c) - \rho_0$ and viscosity changes $\eta(c) - \eta_0$ as a function of molar co-solute concentration for a variety of common co-solutes in water have been fitted with a series expansion as a function of concentration [278] and implemented in the software SEDNTERP [502] for an interactive calculation of buffer densities and viscosities. This can be very convenient. However, it should be noted that viscosity data for heavy water are lacking, unfortunately resulting in incorrect viscosity values for heavy water solutions.[1]

---

[1]This may not be obvious to the uncritical user. As with all software, independent references and knowledge of the principles behind the computation of the data provided are indispensable.

Refractive Index Increments

As discussed in Section 4.3.2, refractive index increments depend on the wavelength of light, the solvent refractive index, and buffer composition where — in a manner analogous to the partial-specific volume (but with different quantitative impact) — preferential solvation might occur.

As an example for the effect of the solvent refractive index, since dn/dw arises from a refractive index contrast, the slightly lower refractive index of D_2O relative to H_2O leads to a slightly higher refractive index increment of ordinary macromolecules in solutions of D_2O. Similarly, the slightly higher refractive index of $H_2^{18}O$ relative to H_2O leads to a slightly lower macromolecular dn/dw in $H_2^{18}O$ solutions [503]. Solvent bulk refractive index effects may in practice usually be negligible for water isotopes, but will have large impact with organic solvents. Likewise, high concentrations of co-solutes that raise the solvent refractive index will decrease macromolecular refractive index increments.

Values reported in Table C.1 refer mostly to data acquired in dilute aqueous solutions with moderate ionic strength, at a wavelength between 540 and 690 nm, close to that of the interference optical system. References are provided for exact conditions. An extensive compilation of refractive index data can be found in [382, 383], which contains more specific information, especially on classes of compounds for which Table C.1 only provides an approximate range.

Protein refractive index increments (not accounting for any post-translational modifications) can be predicted on the basis of their amino acid composition, analogous to the partial-specific volume [94]. For large proteins this usually averages to a value of ~ 0.19 mL/g, but significant deviations from this value may occur for special classes of proteins and small peptides [95].

A calculator function in SEDFIT can be used to calculate protein refractive index increments based on their amino acid composition [95], using data from McMeekin [94].

TABLE C.1 **Refractive Index Increments**

		dn/dw **(mL/g)**
amino acids(e)[e]	Arg	0.216
	His	0.231
	Lys	0.187
	Asp	0.208
	Glu	0.191
	Ser	0.177
	Thr	0.179
	Asn	0.202
	Gln	0.195
	Cys	0.218
	Gly	0.183
	Pro	0.170
	Ala	0.173
	Ile	0.185
	Leu	0.178
	Met	0.214
	Phe	0.257
	Trp	0.297
	Tyr	0.254
	Val	0.178
detergents	n-dodecyl--D-maltoside (DM)	0.143[a]
	dodecyl octaethylene glycol ether (C12E8)	0.121[a]
	dodecyl nonaethylene glycol ether (C12E9)	0.109[b]
	Triton X-100 (reduced)	0.115[c]
	lauryldimethyl amine-n-oxide (LDAO)	0.148[b]
	amphipol A8-35	0.15[c]
	(hemi)fluorinated (multi)glycosylated trishydroxymethyl acrylamidomethane	0.07–0.094[d]
other classes of macromolecules	lipids	∼0.08–0.164[l,m,n]
	nucleic acids	∼0.20–0.22[j]
	carbohydrates	∼0.15[f,g,h,i,k,l]
	PEG	∼0.09-0.15[f,i]

[a]From [480]; [b] from [504]; [c] from [505] ; [d] from [484]; [e] at 590 nm for hypothetical polypeptide in 150 mM NaCl [95]; [f] from [383]; [g] from [382]; [h] from [495]; [i] from [506]; [j] from [507]; [k] from [508]; [l] from [182]; [m] from [509]; [n] from [510].

Solution Column

The relationship between loading volume and solution column height is useful for designing and setting up experiments, following a simulation of sedimentation profiles. The values shown in Table D.1 are for standard Beckman Coulter double sector centerpieces. The meniscus position does not account for rotor stretching, and actual experimental values may vary due to radial calibration errors.

TABLE D.1 **Relationship Between Solution Column Height, Meniscus Position, and Sample Volume.** Data are for standard 12 mm or 3 mm sector-shaped centerpieces with sides subtending an angle of 2.4° from the center of rotation.

column height (mm)	volume in 12 mm CP (μL)	volume in 3 mm CP (μL)	meniscus position[a] (cm)
3	106	27	6.90
3.5	123	31	6.85
4	140	35	6.80
4.5	157	39	6.75
5	174	44	6.70
5.5	190	48	6.65
6	207	52	6.60
6.5	223	56	6.55
7	240	60	6.50
7.5	256	64	6.45
8	272	68	6.40
8.5	288	72	6.35
9	304	76	6.30
9.5	319	80	6.25
10	335	84	6.20
10.5	350	88	6.15
11	366	92	6.10
11.5	381	95	6.05
12	396	99	6.00
12.5	411	103	5.95
13	426	107	5.90

[a] The meniscus position does not account for rotor stretching, which can lead to displacements of as much as 0.02–0.03 cm, depending on the rotor speed.

Bibliography

[1] T. Svedberg and K.O. Pedersen, *The Ultracentrifuge*. London: Oxford University Press, 1940.

[2] H.K. Schachman, *Ultracentrifugation in Biochemistry*. New York: Academic Press, 1959.

[3] J.M. Creeth and R.H. Pain, "The determination of molecular weights of biological macromolecules by ultracentrifuge methods." *Progr. Biophys. Chem. Sci.*, vol. 17, pp. 217–87, 1967.

[4] J.W. Williams, *Ultracentrifugation of Macromolecules*. New York: Academic Press, 1972.

[5] H. Fujita, *Foundations of Ultracentrifugal Analysis*. New York: John Wiley & Sons, 1975.

[6] J. Lebowitz, M.S. Lewis, and P. Schuck, "Modern analytical ultracentrifugation in protein science: A tutorial review." *Protein Sci.*, vol. 11, no. 9, pp. 2067–2079, 2002. doi: 10.1110/ps.0207702

[7] G.J. Howlett, A.P. Minton, and G. Rivas, "Analytical ultracentrifugation for the study of protein association and assembly." *Curr. Opin. Chem. Biol.*, vol. 10, no. 5, pp. 430–436, 2006. doi: 10.1016/j.cbpa.2006.08.017

[8] P. Schuck, "Analytical ultracentrifugation as a tool for studying protein interactions." *Biophys. Rev.*, vol. 5, no. 2, pp. 159–171, 2013. doi: 10.1007/s12551-013-0106-2

[9] T. Svedberg, "The ultracentrifuge. Nobel Lecture." 1926. Online at: http://www.nobelprize.org/nobel_prizes/chemistry/laureates/1926/svedberg-lecture.pdf

[10] B. Lindman and L.-O. Sundelöf, *A Tribute to the Memory of Theodor Svedberg*, A. Lindberg, Ed. Stockholm: Royal Swedish Academy of Engineering Sciences, 2010. ISBN 9789170828263

[11] J.W. Beams, A.J. Weed, and E.G. Pickels, "Scientific apparatus and laboratory methods. The ultracentrifuge." *Science*, vol. 78, no. 2024, pp. 338–340, 1933. doi: 10.1126/science.78.2024.338

[12] B. Elzen, "Two ultracentrifuges: A comparative study of the social construction of artefacts." *Soc. Stud. Sci.*, vol. 16, no. 4, pp. 621–662, 1986. doi: 10.1177/030631286016004004

[13] B. Elzen, *Scientists and Rotors. The Development of Biochemical Ultracentrifuges.* Enschede: Dissertation, University Twente, 1988.

[14] T. Svedberg and J.B. Nichols, "Determination of size and distribution of size of particle by centrifugal methods." *J. Am. Chem. Soc.*, vol. 45, pp. 2910–2917, 1923. doi: 10.1021/ja01665a016

[15] T. Svedberg and H. Rinde, "The ultra-centrifuge, a new instrument for the determination of size and distribution of size of particle in amicroscopic colloids." *J. Am. Chem. Soc.*, vol. 46, no. 1923, pp. 2677–2693, 1924.

[16] M.J. Perrin, *Brownian Movement and Molecular Reality.* London: Taylor & Francis, 1910.

[17] J.B. Perrin, "Discontinuous Structure of Matter. Nobel Lecture." 1926. Online at: http://www.nobelprize.org/nobel_prizes/physics/laureates/1926/perrin-lecture.html

[18] J.W. Beams and E.G. Pickels, "The production of high rotational speeds." *Rev. Sci. Instrum.*, vol. 6, no. 10, p. 299, 1935. doi: 10.1063/1.1751877

[19] S. Berres and R. Bürger, "On gravity and centrifugal settling of polydisperse suspensions forming compressible sediments." *Int. J. Solids Struct.*, vol. 40, pp. 4965–4987, 2003. doi: 10.1016/S0020-7683(03)00249-X

[20] H. Eisenberg, "Thermodynamics and the structure of biological macromolecules. Rozhinkes mit mandeln." *Eur. J. Biochem.*, vol. 187, no. 1, pp. 7–22, 1990. doi: 10.1111/j.1432-1033.1990.tb15272.x

[21] P.Y. Cheng and H.K. Schachman, "Studies on the validity of the Einstein viscosity law and Stokes law of sedimentation." *J. Polym. Sci.*, vol. 16, no. 81, pp. 19–30, 1955. doi: 10.1002/pol.1955.120168102

[22] I.N. Serdyuk, N.R. Zaccai, and J. Zaccai, *Methods In Molecular Biophysics: Structure, Dynamics, Function.* Cambridge, U.K.: Cambridge University Press, 2007. ISBN 9780521815246

[23] C.R. Cantor and P R Schimmel, *Biophysical Chemistry. II. Techniques for the study of biological structure and function.* New York: W.H. Freeman, 1980.

[24] H. Zhao, Y. Chen, L. Rezabkova, Z. Wu, G. Wistow, and P. Schuck, "Solution properties of γ-crystallins: Hydration of fish and mammal γ-crystallins." *Protein Sci.*, vol. 23, no. 1, pp. 88–99, 2014. doi: 10.1002/pro.2394

[25] Y. Chen, H. Zhao, P. Schuck, and G. Wistow, "Solution properties of γ-crystallins: Compact structure and low frictional ratio are conserved properties of diverse γ-crystallins." *Protein Sci.*, vol. 23, no. 1, pp. 76–87, 2014. doi: 10.1002/pro.2395

[26] T.M. Laue and W.F. Stafford, "Modern applications of analytical ultracentrifugation." *Annu. Rev. Biophys. Biomol. Struct.*, vol. 28, pp. 75–100, 1999. doi: 10.1146/annurev.biophys.28.1.75

[27] T. Svedberg, "Molecular weight analysis in centrifugal fields." *Science*, vol. 79, no. 2050, pp. 327–332, 1934. doi: 10.1126/science.79.2050.327

[28] P. Schuck, "Sedimentation Velocity Movies." 2011. Online at: https://sedfitsedphat.nibib.nih.gov/tools/SedimentationDiffusion%20Movies/Forms/AllItems.aspx

[29] R. Trautman and V.N. Schumaker, "Generalization of the radial dilution square law in ultracentrifugation." *J. Chem. Phys.*, vol. 22, no. 3, pp. 551–554, 1954. doi: 10.1063/1.1740105

[30] R.J. Goldberg, "Sedimentation in the ultracentrifuge." *J. Phys. Chem.*, vol. 57, no. 2, pp. 194–202, 1953. doi: 10.1021/j150503a014

[31] J.W. Williams, R.L. Baldwin, M. Saunders, and P.G. Squire, "Boundary spreading in sedimentation velocity experiments. I. The enzymatic degradation of serum globulins." *J. Am. Chem. Soc.*, vol. 74, no. 6, pp. 1542–1548, 1952. doi: 10.1021/ja01126a059

[32] O. Lamm, "Die Differentialgleichung der Ultrazentrifugierung." *Ark. Mat. Astr. Fys.*, vol. 21B(2), pp. 1–4, 1929.

[33] P.H. Brown and P. Schuck, "A new adaptive grid-size algorithm for the simulation of sedimentation velocity profiles in analytical ultracentrifugation." *Comput. Phys. Commun.*, vol. 178, no.2, pp. 105–120, 2008. doi: 10.1016/j.cpc.2007.08.012

[34] P. Schuck, "Sedimentation patterns of rapidly reversible protein interactions." *Biophys. J.*, vol. 98, no. 9, pp. 2005–2013, 2010. doi: 10.1016/j.bpj.2009.12.4336

[35] G.A. Gilbert and L.M. Gilbert, "Ultracentrifuge studies of interactions and equilibria: impact of interactive computer modelling." *Biochem. Soc. Trans.*, vol. 8, no. 5, pp. 520–522, 1980. doi: 10.1042/bst0080520

[36] D.J. Cox, "Computer simulation of sedimentation in the ultracentrifuge I. Diffusion." *Arch. Biochem. Biophys.*, vol. 112, no. 2, pp. 249–258, 1965. doi: 10.1016/0003-9861(65)90043-3

[37] W.B. Goad and J.R. Cann, "Theory of sedimentation of interacting systems." *Ann. N.Y. Acad. Sci.*, vol. 164, no. 1, pp. 172–182, 1969. doi: 10.1111/j.1749-6632.1969.tb14039.x

[38] J.M. Claverie, H. Dreux, and R. Cohen, "Sedimentation of generalized systems of interacting particles. I. Solution of systems of complete Lamm equations." *Biopolymers*, vol. 14, no. 8, pp. 1685–1700, 1975. doi: 10.1002/bip.1975.360140811

[39] M. Dishon, G.H. Weiss, and D.A. Yphantis, "Numerical solutions of the Lamm equation. I. Numerical procedure." *Biopolymers*, vol. 4, no. 4, pp. 449–455, 1966. doi: 10.1002/bip.1966.360040406

[40] P. Schuck, "Size-distribution analysis of macromolecules by sedimentation velocity ultracentrifugation and Lamm equation modeling." *Biophys. J.*, vol. 78, no. 3, pp. 1606–1619, 2000. doi: 10.1016/S0006-3495(00)76713-0

[41] M.N. Berberan-Santos, E.N. Bodunov, and L. Pogliani, "On the barometric formula." *Am. J. Phys.*, vol. 65, no. 5, pp. 404–412, 1997. doi: 10.1119/1.18555

[42] T. Svedberg and R. Fåhraeus, "A new method for the determination of the molecular weight of the proteins." *J. Am. Chem. Soc.*, vol. 48, no. 2, pp. 430–438, 1926. doi: 10.1021/ja01413a019

[43] H. Eisenberg, *Biological Macromolecules and Polyelectrolytes in Solution.* Oxford: Clarendon Press, 1976.

[44] J.S.L. Philpot, "Direct photography of ultracentrifuge sedimentation curves." *Nature*, vol. 141, no. 3563, pp. 283–284, 1938. doi: 10.1038/141283a0

[45] H. Svensson, "Theorie der Beobachtungsmethode der gekreuzten Spalte." *Kolloid-Zeitschrift*, vol. 90, no. 2, pp. 145–160, 1940.

[46] L.G. Longsworth, "A modification of the Schlieren methods for use in electrophoretic analysis." *J. Am. Chem. Soc.*, vol. 61, no. 2, pp. 529–530, 1939. doi: 10.1021/ja01871a511

[47] W. Mächtle, "The installation of an eight-cell schlieren optics multiplexer in a Beckman optima XLI/XL analytical ultracentrifuge used to measure steep refractive index gradients." *Progr. Colloid Polym. Sci.*, vol. 113, pp. 1–9, 1999. doi: 10.1007/3-540-48703-4_1

[48] H. Cölfen, P. Husbands, and S.E. Harding, "Alternative light sources for the Schlieren optical system of analytical ultracentrifuges." *Progr. Colloid Polym. Sci.*, vol. 99, pp. 193–198, 1995. doi: 10.1007/BFb0114089

[49] H. Cölfen and S.E. Harding, "A study on Schlieren patterns derived with the Beckman Optima XL-A UV-absorption optics." *Progr. Colloid Polym. Sci.*, vol. 99, pp. 167–168, 1995. doi: 10.1007/BFb0114087

[50] Z. Bozóky, L. Fülöp, and L. Köhidai, "A short-run new analytical ultracentrifugal micromethod for determining low-density lipoprotein sub-fractions using Schlieren refractometry." *Eur. Biophys. J.*, vol. 29, no. 8, pp. 621–627, 2001. doi: 10.1007/s002490000095

[51] L. Börger, M.D. Lechner, and M. Stadler, "Development of a new digital camera setup for the online recording of Schlieren optical pictures in a modified Beckman Optima XL analytical ultracentrifuge." *Progr. Colloid Polym. Sci.*, vol. 127, pp. 19–25, 2004. doi: 10.1007/b94246

[52] D. Kisters and W. Borchard, "New facilities to improve the properties of Schlieren optics." *Progr. Colloid Polym. Sci.*, vol. 113, pp. 10–13, 1999. doi: 10.1007/3-540-48703-4_2

[53] B. Ortlepp and D. Panke, "Analytical ultracentrifuges with multiplexer and video systems for measuring particle size and molecular mass distribution." *Progr. Colloid Polym. Sci.*, vol. 86, pp. 57–61, 1991. doi: 10.1007/BFb0115007

[54] J.W. Beams, "Some interferometer techniques for observing sedimentation." *Rev. Sci. Instrum.*, vol. 34, no. 2, pp. 139–142, 1963. doi: 10.1063/1.1718287

[55] H. Svensson, "An interferometric method for recording the refractive index derivative in concentration gradients." *Acta Chem. Scand.*, vol. 3, no. 8, pp. 1170–1177, 1949. doi: 10.3891/acta.chem.scand.03-1170

[56] L.G. Longsworth, "Interferometry in electrophoresis." *Anal. Chem.*, vol. 23, no. 2, pp. 346–348, 1951. doi: 10.1021/ac60050a032

[57] T. Kondo and M. Kawakami, "Differential interferometry in the analytical ultracentrifuge. I. Design, construction, and alignment of differential interferometer." *Anal. Biochem.*, vol. 117, no. 2, pp. 366–73, 1981. doi: 10.1016/0003-2697(81)90793-4

[58] C.H. Paul and D.A. Yphantis, "Pulsed laser interferometry (PLI) in the analytical ultracentrifuge: 1. System design." *Anal. Biochem.*, vol. 48, no. 2, pp. 588–604, 1972. doi: 10.1016/0003-2697(72)90114-5

[59] D.A. Yphantis, J.W. Lary, W.F. Stafford, S. Liu, P.H. Olsen, D.B. Hayes, T.P. Moody, T.M. Ridgeway, D.A. Lyons, and T.M. Laue, "On-line data acquisition for the Rayleigh interference optical system of the analytical ultracentrifuge." in *Modern Analytical Ultracentrifugation*, T.M. Schuster and T.M. Laue, Eds. Boston: Birkhäuser, 1994, pp. 209–226.

[60] T.M. Laue, L.A. Anderson, and P.D. Demaine, "An on-line interferometer for the XL-A ultracentrifuge." *Progr. Colloid Polym. Sci.*, vol. 94, pp. 74–81, 1994. doi: 10.1007/BFb0115604

[61] P. Lavrenko, V. Lavrenko, and V. Tsvetkov, "Shift interferometry in analytical ultracentrifugation of polymer solutions." *Prog. Colloid Polym. Sci.*, vol. 113, pp. 14–22, 1999. doi: 10.1007/3-540-48703-4_3

[62] K. Schilling and F. Krause, "Analysis of antibody aggregate content at extremely high concentrations using sedimentation velocity with a novel interference optics." *PLoS One*, vol. 10, no. 3, p. e0120820, 2015. doi: 10.1371/journal.pone.0120820

[63] S. Hanlon, K. Lamers, G. Lauterbach, R. Johnson, and H.K. Schachman, "Ultracentrifuge studies with absorption optics. I. An automatic photoelectric scanning absorption system." *Arch. Biochem. Biophys.*, vol. 99, no. 1, pp. 157–174, 1962. doi: 10.1016/0003-9861(62)90258-8

[64] H.K. Schachman, L. Gropper, S. Hanlon, and F. Putney, "Ultracentrifuge studies with absorption optics. II. Incorporation of a monochromator and its application to the study of proteins and interacting systems." *Arch. Biochem. Biophys.*, vol. 99, no. 1, pp. 175–190, 1962. doi: 10.1016/0003-9861(62)90259-X

[65] R. Giebeler, "The Optima XL-A: A new analytical ultracentrifuge with a novel precision absorption optical system." In *Analytical Ultracentrifugation in Biochemistry and Polymer Science*, S.E. Harding, A.J. Rowe, and J.C. Horton, Eds. Cambridge, U.K.: The Royal Society of Chemistry, 1992, pp. 16–25.

[66] J. Walter, K. Löhr, E. Karabudak, W. Reis, J. Mikhael, W. Peukert, W. Wohlleben, and H. Cölfen, "Multidimensional analysis of nanoparticles with highly disperse properties using multiwavelength analytical ultracentrifugation." *ACS Nano*, vol. 8, no. 9, pp. 8871–8886, 2014. doi: 10.1021/nn503205k

[67] W. Scholtan and H. Lange, "Bestimmung der Teilchengrößenverteilung von Latices mit der Ultrazentrifuge." *Kolloid-Z. u. Z. Polym.*, vol. 250, no. 8, pp. 782–796, 1972. doi: 10.1007/BF01498571

[68] H.G. Müller, "Automated determination of particle-size distributions of dispersions by analytical ultracentrifugation." *Colloid Polym. Sci.*, vol. 267, no. 12, pp. 1113–1116, 1989. doi: 10.1007/BF01496933

[69] R.H. Crepeau, R.H. Conrad, and S.J. Edelstein, "UV laser scanning and fluorescence monitoring of analytical ultracentrifugation with an on-line computer system." *Biophys. Chem.*, vol. 5, no. 1-2, pp. 27–39, 1976. doi: 10.1016/0301-4622(76)80024-5

[70] B. Schmidt and D. Riesner, "A fluorescence detection system for the analytical ultracentrifuge and its application to proteins, nucleic acids, viroids and viruses." in *Analytical Ultracentrifugation in Biochemistry and Polymer Science*, S.E. Harding, A.J. Rowe, and J.C. Horton, Eds. Cambridge: The Royal Society of Chemistry, 1992, pp. 176–207.

[71] T.M. Laue, A.L. Anderson, and B.W. Weber, "Prototype fluorimeter for the XLA/XLI analytical ultracentrifuge." in *Ultrasensitive Biochem. Diagnostics II. SPIE Proceedings.*, G.E. Cohn and S.A. Soper, Eds. Bellingham, WA: SPIE, 1997, vol. 2985, pp. 196–204.

[72] P.H. Lloyd, *Optical Methods in Ultracentrifugation, Electrophoresis, and Diffusion. With a Guide to the Interpretation of Records.* Oxford: Clarendon Press, 1974. ISBN 019854605X

[73] A. Tiselius, K.O. Pedersen, and T. Svedberg, "Analytical measurements of ultracentrifugal sedimentation." *Nature*, vol. 140, no. 3550, pp. 848–849, 1937. doi: 10.1038/140848a0

[74] M.A. Bothwell, G.J. Howlett, and H.K. Schachman, "A sedimentation equilibrium method for determining molecular weights of proteins with a tabletop high speed air turbine centrifuge." *J. Biol. Chem.*, vol. 253, no. 7, pp. 2073–2077, 1978.

[75] G.J. Howlett, P.J. Roche, and G. Schreiber, "Protein-protein interactions: Analysis of the interaction of concanavalin A with serum glycoproteins by sedimentation equilibrium using an air-driven ultracentrifuge." *Arch. Biochem. Biophys.*, vol. 224, no. 1, pp. 178–185, 1983. doi: 10.1016/0003-9861(83)90202-3

[76] A.K. Attri and A.P. Minton, "An automated method for determination of the molecular weight of macromolecules via sedimentation equilibrium in a preparative ultracentrifuge." *Anal. Biochem.*, vol. 133, no. 2, pp. 142–152, 1983. doi: 10.1016/0003-2697(84)90236-7

[77] S. Darawshe, G. Rivas, and A.P. Minton, "Rapid and accurate microfractionation of the contents of small centrifuge tubes: application in the measurement of molecular weight of proteins via sedimentation equilibrium." *Anal. Biochem.*, vol. 209, no. 1, pp. 130–135, 1993. doi: 10.1006/abio.1993.1092

[78] A. Furst, "The XL-I analytical ultracentrifuge with Rayleigh interference optics." *Eur. Biophys. J.*, vol. 25, no. 5-6, pp. 307–310, 1997. doi: 10.1007/s002490050043

[79] I.K. MacGregor, A.L. Anderson, and T.M. Laue, "Fluorescence detection for the XLI analytical ultracentrifuge." *Biophys. Chem.*, vol. 108, no. 1-3, pp. 165–185, 2004. doi: 10.1016/j.bpc.2003.10.018

[80] H. Zhao, A.J. Berger, P.H. Brown, J. Kumar, A. Balbo, C.A. May, E. Casillas, T.M. Laue, G.H. Patterson, M.L. Mayer, and P. Schuck, "Analysis of high-affinity assembly for AMPA receptor amino-terminal domains." *J. Gen. Physiol.*, vol. 139, no. 5, pp. 371–388, 2012. doi: 10.1085/jgp.201210770

[81] C.A. Brautigam, "Calculations and publication-quality illustrations for analytical ultracentrifugation data." *Methods Enzymol.*, vol. 562, pp. 109-133, 2015. doi: 10.1016/bs.mie.2015.05.001

[82] C.N. Pace, F. Vajdos, L. Fee, G. Grimsley, and T. Gray, "How to measure and predict the molar absorption coefficient of a protein." *Protein Sci.*, vol. 4, no. 11, pp. 2411–2423, 1995. doi: 10.1002/pro.5560041120

[83] M.S. Lewis, R.I. Shrager, and S.-J. Kim, "Analysis of protein-nucleic acid and protein-protein interactions using multi-wavelength scans from the XL-A analytical ultracentrifuge." in *Modern Analytical Ultracentrifugation*, T.M. Schuster and T.M. Laue, Eds. Boston: Birkhäuser, 1994, pp. 94–115.

[84] P. Schuck, "Simultaneous radial and wavelength analysis with the Optima XL-A analytical ultracentrifuge." *Progr. Colloid Polym. Sci.*, vol. 94, pp. 1–13, 1994. doi: 10.1007/BFb0115597

[85] A. Balbo, K.H. Minor, C.A. Velikovsky, R.A. Mariuzza, C.B. Peterson, and P. Schuck, "Studying multi-protein complexes by multi-signal sedimentation velocity analytical ultracentrifugation." *Proc. Nat. Acad. Sci. USA*, vol. 102, no. 1, pp. 81–86, 2005. doi: 10.1073/pnas.0408399102

[86] D. Noy, B.M. Discher, I.V. Rubtsov, R.M. Hochstrasser, and P.L. Dutton, "Design of amphiphilic protein maquettes: enhancing maquette functionality through binding of extremely hydrophobic cofactors to lipophilic domains." *Biochemistry*, vol. 44, no. 37, pp. 12 344–12 354, 2005. doi: 10.1021/bi050696e

[87] P. Schuck, "Sedimentation equilibrium analytical ultracentrifugation for multicomponent protein interactions." in *Protein Interactions: Biophysical Approaches for the Study of Complex Reversible Systems*, P. Schuck, Ed. New York: Springer, 2007, pp. 289–316.

[88] P. Schuck, "Sedimentation velocity in the study of reversible multiprotein complexes." in *Protein Interactions: Biophysical Approaches for the Study of Complex Reversible Systems*, P. Schuck, Ed. New York: Springer, 2007, pp. 469–518.

[89] H.M. Strauss, E. Karabudak, S. Bhattacharyya, A. Kretzschmar, W. Wohlleben, and H. Cölfen, "Performance of a fast fiber based UV/Vis multiwavelength detector for the analytical ultracentrifuge." *Colloid Polym. Sci.*, vol. 286, no. 2, pp. 121–128, 2008. doi: 10.1007/s00396-007-1815-5

[90] S.B. Padrick, R.K. Deka, J.L. Chuang, R.M. Wynn, D.T. Chuang, M.V. Norgard, M.K. Rosen, and C.A. Brautigam, "Determination of protein complex stoichiometry through multisignal sedimentation velocity experiments." *Anal. Biochem.*, vol. 407, no. 1, pp. 89–103, 2010. doi: 10.1016/j.ab.2010.07.017

[91] S.B. Padrick and C.A. Brautigam, "Evaluating the stoichiometry of macromolecular complexes using multisignal sedimentation velocity." *Methods*, vol. 54, no. 1, pp. 39–55, 2011. doi: 10.1016/j.ymeth.2011.01.002

[92] C.A. Brautigam, S.B. Padrick, and P. Schuck, "Multi-signal sedimentation velocity analysis with mass conservation for determining the stoichiometry of protein complexes." *PLoS One*, vol. 8, no. 5, p. e62694, 2013. doi: 10.1371/journal.pone.0062694

[93] E.G. Richards, D.C. Teller, and H.K. Schachman, "Ultracentrifuge studies with Rayleigh interference optics. II. Low-speed sedimentation equilibrium of homogeneous systems." *Biochemistry*, vol. 7, no. 3, pp. 1054–1076, 1968. doi: 10.1021/bi00843a026

[94] T.L. McMeekin, M.L. Groves, and N.J. Hipp, "Refractive indices of amino acids, proteins, and related substances." in *Amino Acids and Serum Proteins*, J. Stekol, Ed. Washington DC: American Chemical Society, 1964, pp. 54–66.

[95] H. Zhao, P.H. Brown, and P. Schuck, "On the distribution of protein refractive index increments." *Biophys. J.*, vol. 100, no. 9, pp. 2309–2317, 2011. doi: 10.1016/j.bpj.2011.03.004

[96] A. Böhm, S. Kielhorn-Bayer, and P Rossmanith, "Working with multidetection in the analytical ultracentrifuge: the benefits of the combination of a refractive index detector and an absorption detector for the analysis of colloidal systems." *Progr. Colloid Polym. Sci.*, vol. 113, pp. 121–128, 1999. doi: 10.1007/3-540-48703-4_17

[97] A.G. Salvay, M. Santamaria, M. le Maire, and C. Ebel, "Analytical ultracentrifugation sedimentation velocity for the characterization of detergent-solubilized membrane proteins Ca^{++}-ATPase and ExbB." *J Biol. Phys.*, vol. 33, no. 5-6, pp. 399–419, 2007. doi: 10.1007/s10867-008-9058-3

[98] R.R. Kroe and T.M. Laue, "NUTS and BOLTS: Applications of fluorescence-detected sedimentation." *Anal. Biochem.*, vol. 390, no. 1, pp. 1–13, 2009. doi: 10.1016/j.ab.2008.11.033

[99] H. Zhao, M.L. Mayer, and P. Schuck, "Analysis of protein interactions with picomolar binding affinity by fluorescence-detected sedimentation velocity." *Anal. Chem.*, vol. 18, no. 6, pp. 3181–3187, 2014. doi: 10.1021/ac500093m

[100] H. Zhao, E. Casillas, H. Shroff, G.H. Patterson, and P. Schuck, "Tools for the quantitative analysis of sedimentation boundaries detected by fluorescence optical analytical ultracentrifugation." *PLoS One*, vol. 8, no. 10, p. e77245, 2013. doi: 10.1371/journal.pone.0077245

[101] J.S. Kingsbury and T.M. Laue, "Fluorescence-detected sedimentation in dilute and highly concentrated solutions." *Methods Enzymol.*, vol. 492, pp. 283–304, 2011. doi: 10.1016/B978-0-12-381268-1.00021-5

[102] H. Zhao, S. Lomash, C. Glasser, M.L. Mayer, and P. Schuck, "Analysis of high affinity self-association by fluorescence optical sedimentation velocity analytical ultracentrifugation of labeled proteins: opportunities and limitations." *PLoS One*, vol. 8, no. 12, p. e83439, 2013. doi: 10.1371/journal.pone.0083439

[103] M.A. Olshina, L.M. Angley, Y.M. Ramdzan, J. Tang, M.F. Bailey, A.F. Hill, and D.M. Hatters, "Tracking mutant huntingtin aggregation kinetics in cells reveals three major populations that include an invariant oligomer pool." *J. Biol. Chem.*, vol. 285, no. 28, pp. 21807–21816, 2010. doi: 10.1074/jbc.M109.084434

[104] B. Demeule, S.J. Shire, and J. Liu, "A therapeutic antibody and its antigen form different complexes in serum than in phosphate-buffered saline: A study by analytical ultracentrifugation." *Anal. Biochem.*, vol. 388, no. 2, pp. 279–287, 2009. doi: 10.1016/j.ab.2009.03.012

[105] M.D.D. Miell, A.F Straight, and R.C. Allshire, "Reply to 'CENP-A octamers do not confer a reduction in nucleosome height by AFM'." *Nat. Struct. Mol. Biol.*, vol. 21, no. 1, pp. 5–8, 2014. doi: 10.1038/nsmb.2744

[106] A. Tardieu, P. Vachette, A. Gulik, and M. le Maire, "Biological macromolecules in solvent of variable density: Characterization of sedimentation equilbrium, densimetry, and x-ray forward scattering and an application to the 50S ribosomal subunit from Escherichia coli." *Biochemistry*, vol. 20, no. 15, pp. 4399–4406, 1981. doi: 10.1021/bi00518a026

[107] J.J. Virtanen, L. Makowski, T.R. Sosnick, and K.F. Freed, "Modeling the hydration layer around proteins: HyPred." *Biophys. J.*, vol. 99, no. 5, pp. 1611–1619, 2010. doi: 10.1016/j.bpj.2010.06.027

[108] B.M. Pettitt, V.A. Makarov, and B.K. Andrews, "Protein hydration density: theory, simulations and crystallography." *Curr. Opin. Struct. Biol.*, vol. 8, no. 2, pp. 218–221, 1998. doi: 10.1016/S0959-440X(98)80042-0

[109] D.I. Svergun, S. Richard, M.H.J. Koch, Z. Sayers, S. Kuprin, and G. Zaccai, "Protein hydration in solution: experimental observation by x-ray and neutron scattering." *Proc. Natl. Acad. Sci. USA*, vol. 95, no. 5, pp. 2267–2272, 1998.

[110] F. Merzel and J.C. Smith, "Is the first hydration shell of lysozyme of higher density than bulk water?" *Proc. Nat. Acad. Sci. USA*, vol. 99, no. 8, pp. 5378–5383, 2002. doi: 10.1073/pnas.082335099

[111] S.J. Perkins, "Protein volumes and hydration effects. The calculations of partial specific volumes, neutron scattering matchpoints and 280-nm absorption coefficients for proteins and glycoproteins from amino acid sequences." *Eur. J. Biochem.*, vol. 157, no. 1, pp. 169–180, 1986. doi: 10.1111/j.1432-1033.1986.tb09653.x

[112] B. Halle, "Protein hydration dynamics in solution: a critical survey." *Philos. Trans. R. Soc. Lond. B. Biol. Sci.*, vol. 359, no. 1448, pp. 1207–1224, 2004. doi: 10.1098/rstb.2004.1499

[113] J.C. Smith, F. Merzel, C.S. Verma, and S. Fischer, "Protein hydration water: Structure and thermodynamics." *J. Mol. Liq.*, vol. 101, no. 1-3, pp. 27–33, 2002. doi: 10.1016/S0167-7322(02)00100-9

[114] S.J. Perkins, "X-ray and neutron scattering analyses of hydration shells: A molecular interpretation based on sequence predictions and modelling fits." *Biophys. Chem.*, vol. 93, no. 2-3, pp. 129–139, 2001. doi: 10.1016/S0301-4622(01)00216-2

[115] J. Lebowitz, M.S. Lewis, and P. Schuck, "Back to the future: A rebuttal to Henryk Eisenberg." *Protein Sci.*, vol. 12, no. 11, pp. 2649–2650, 2003.

[116] H.K. Schachman and M.A. Lauffer, "The hydration, size and shape of tobacco mosaic virus." *J. Phys. Chem.*, vol. 71, no. 2, pp. 536–541, 1949. doi: 10.1021/ja01170a047

[117] F. Bonneté, C. Ebel, G. Zaccai, and H. Eisenberg, "Biophysical study of halophilic malate dehydrogenase in solution: revised subunit structure and solvent interactions of native and recombinant enzyme." *J. Chem. Soc. Faraday Trans.*, vol. 89, no. 15, p. 2659–2666, 1993. doi: 10.1039/ft9938902659

[118] S.N. Timasheff, "Protein-solvent preferential interactions, protein hydration, and the modulation of biochemical reactions by solvent components." *Proc. Nat. Acad. Sci. USA*, vol. 99, no. 15, pp. 9721–9726, 2002. doi: 10.1073/pnas.122225399

[119] K. Gekko and S.N. Timasheff, "Mechanism of protein stabilization by glycerol: Preferential hydration in glycerol-water mixtures." *Biochemistry*, vol. 20, no. 16, pp. 4667–4676, 1981. doi: 10.1021/bi00519a023

[120] C. Ebel, H. Eisenberg, and R. Ghirlando, "Probing protein-sugar interactions." *Biophys. J.*, vol. 78, no. 1, pp. 385–393, 2000. doi: 10.1016/S0006-3495(00)76601-X

[121] T. Arakawa, R. Bhat, and S.N. Timasheff, "Preferential interactions determine protein solubility in three-component solutions: the magnesium chloride system." *Biochemistry*, vol. 29, no. 7, pp. 1914–1923, 1990. doi: 10.1021/bi00459a036

[122] E.F. Casassa and H. Eisenberg, "Thermodynamic analysis of multicomponent solutions." *Adv. Protein Chem.*, vol. 19, pp. 287–395, 1964.

[123] L. Costenaro, G. Zaccai, and C. Ebel, "Link between protein-solvent and weak protein-protein interactions gives insight into halophilic adaptation." *Biochemistry*, vol. 41, no. 44, pp. 13 245–13 252, 2002. doi: 10.1021/bi025830z

[124] C. Ebel, "Solvent mediated protein-protein interactions." in *Protein Interactions: Biophysical Approaches for the Study of Complex Reversible Systems.*, P Schuck, Ed. New York: Springer, 2007, pp. 255–288.

[125] M.T. Record and C.F. Anderson, "Interpretation of preferential interaction coefficients of nonelectrolytes and of electrolyte ions in terms of a two-domain model." *Biophys. J.*, vol. 68, no. 3, pp. 786-7-94, 1995. doi: 10.1016/S0006-3495(95)80254-7

[126] H. Inoue and S.N. Timasheff, "Preferential and absolute interactions of solvent components with proteins in mixed solvent systems." *Biopolymers*, vol. 11, no. 4, pp. 737–743, 1972. doi: 10.1002/bip.1972.360110402

[127] E.F. Casassa and H. Eisenberg, "Partial specific volumes and refractive index increments in multicomponent systems." *J. Phys. Chem.*, vol. 65, no. 3, pp. 427–433, 1961. doi: 10.1021/j100821a010

[128] J.G. Kirkwood and F.P. Buff, "The statistical mechanical theory of solutions. I." *J. Chem. Phys.*, vol. 19, no. 6, p. 774–777, 1951. doi: 10.1063/1.1748352

[129] J.A. Schellman, "A simple model for solvation in mixed solvents. Applications to the stabilization and destabilization of macromolecular structures." *Biophys. Chem.*, vol. 37, no. 1-3, pp. 121–140, 1990. doi: 10.1016/0301-4622(90)88013-I

[130] S.N. Timasheff, "Water as ligand: Preferential binding and exclusion of denaturants in protein unfolding." *Biochemistry*, vol. 31, no. 41, pp. 9857–9864, 1992. doi: 10.1021/bi00156a001

[131] B. Halle and M. Davidovic, "Biomolecular hydration: from water dynamics to hydrodynamics." *Proc. Natl. Acad. Sci. USA*, vol. 100, no. 21, pp. 12 135–12 140, 2003. doi: 10.1073/pnas.2033320100

[132] S.R. Aragon, "Accurate hydrodynamic modeling with the boundary element method." in *Analytical Ultracentrifugation: Instrumentation, Software, and Applications*, S. Uchiyama, F. Arisaka, W.F. Stafford, and T.M. Laue, Eds. (in press).

[133] S.R. Aragon and D.K. Hahn, "Precise boundary element computation of protein transport properties: Diffusion tensors, specific volume, and hydration." *Biophys. J.*, vol. 91, no. 5, pp. 1591–1603, 2006. doi: 10.1529/biophysj.105.078188

[134] J García de la Torre, M.L. Huertas, and B. Carrasco, "Calculation of hydrodynamic properties of globular proteins from their atomic-level structure." *Biophys. J.*, vol. 78, no. 2, pp. 719–730, 2000. doi: 10.1016/S0006-3495(00)76630-6

[135] N. Rai, M. Nöllmann, B. Spotorno, G. Tassara, O. Byron, and M. Rocco, "SOMO (SOlution MOdeler) differences between X-Ray- and NMR-derived bead models suggest a role for side chain flexibility in protein hydrodynamics." *Structure*, vol. 13, no. 5, pp. 723–734, 2005. doi: 10.1016/j.str.2005.02.012

[136] P.G. Squire and M.E. Himmel, "Hydrodynamics and protein hydration." *Arch. Biochem. Biophys.*, vol. 196, no. 1, pp. 165–177, 1979. doi: 10.1016/0003-9861(79)90563-0

[137] H. Eisenberg, "Sedimentation in the ultracentrifuge and diffusion of macromolecules carrying electrical charges." *Biophys. Chem.*, vol. 5, no. 1 2, pp. 243–251, 1976. doi: 10.1016/0301-4622(76)80037-3

[138] Z. Alexandrowicz and E. Daniel, "Sedimentation and diffusion of polyelectrolytes. Part I. Theoretical description." *Biopolymers*, vol. 1, no. 5, pp. 447–471, 1963. doi: 10.1002/bip.360010505

[139] K.O. Pedersen, "On charge and specific ion effects on sedimentation in the ultracentrifuge." *J. Phys. Chem.*, vol. 62, no. 10, pp. 1282–1290, 1958. doi: 10.1021/j150568a028

[140] J.W. Williams, K.E. van Holde, R.L. Baldwin, and H. Fujita, "The theory of sedimentation analysis." *Chem. Rev.*, vol. 58, no. 4, pp. 715–744, 1958. doi: 10.1021/cr50022a005

[141] G. Cohen and H. Eisenberg, "Deoxyribonucleate solutions: sedimentation in a density gradient, partial specific volumes, density and refractive index increments, and preferential interactions." *Biopolymers*, vol. 6, no. 8, pp. 1077–1100, 1968. doi: 10.1002/bip.1968.360060805

[142] E.H. Braswell, "Polyelectrolyte charge corrected molecular weight and effective charge by sedimentation." *Biophys. J.*, vol. 51, no. 2, pp. 273–281, 1987. doi: 10.1016/S0006-3495(87)83333-7

[143] E.F. Casassa and H. Eisenberg, "On the definition of components in solutions containing charged macromolecular species." *J. Phys. Chem.*, vol. 64, no. 6, pp. 753–756, 1960. doi: 10.1021/j100835a011

[144] E. Daniel and Z. Alexandrowicz, "Sedimentation and diffusion of polyelectrolytes. Part II. Experimental studies with poly-L-lysine hydrohalides." *Biopolymers*, vol. 1, no. 5, pp. 473–495, 1963. doi: 10.1002/bip.360010506

[145] H. Ohshima, T.W. Healy, L.R. White, and R.W. O'Brien, "Sedimentation velocity and potential in a dilute suspension of charged spherical colloidal particles." *J. Chem. Soc. Faraday Trans 2*, vol. 80, no. 10, pp. 1299–1317, 1984. doi: 10.1039/F29848001299

[146] F. Keller, M. Feist, H. Nirschl, and W. Dörfler, "Investigation of the nonlinear effects during the sedimentation process of a charged colloidal particle by direct numerical simulation." *J. Colloid Interface Sci.*, vol. 344, no. 1, pp. 228–236, 2010. doi: 10.1016/j.jcis.2009.12.032

[147] W.E. Werner, J.R. Cann, and H.K. Schachman, "Boundary spreading in sedimentation velocity experiments on partially liganded aspartate transcarbamoylase." *J. Mol. Biol.*, vol. 206, no. 1, pp. 231–237, 1989. doi: 10.1016/0022-2836(89)90536-6

[148] W.E. Werner and H.K. Schachman, "Analysis of the ligand-promoted global conformational change in aspartate transcarbamoylase. Evidence for a two-state transition from boundary spreading in sedimentation velocity experiments." *J. Mol. Biol.*, vol. 206, no. 1, pp. 221–230, 1989. doi: 10.1016/0022-2836(89)90535-4

[149] J.C. Gerhart and H.K. Schachman, "Allosteric interactions in aspartate transcarbamylase. II. Evidence for different conformational states of the protein in the presence and absence of specific ligands." *Biochemistry*, vol. 7, no. 2, pp. 538–552, 1968. doi: 10.1021/bi00842a600

[150] C.A. Brautigam, C.A. Wakeman, and W.C. Winkler, "Methods for analysis of ligand-induced RNA conformational changes." *Methods Mol. Biol.*, vol. 540, pp. 77–95, 2009. doi: 10.1007/978-1-59745-558-9_7

[151] P. Schuck, Z. Taraporewala, P. McPhie, and J.T. Patton, "Rotavirus nonstructural protein NSP2 self-assembles into octamers that undergo ligand-induced conformational changes." *J. Biol. Chem.*, vol. 276, no. 13, pp. 9679–9687, 2001. doi: 10.1074/jbc.M009398200

[152] A. Tiselius, "The moving-boundary method of studying the electrophoresis of proteins." *Nov. Acta Regiae Soc. Sci. Ups. Ser. IV.*, vol. 7, no. 4, pp. 1–107, 1930.

[153] G.A. Gilbert, "Sedimentation and electrophoresis of interacting substances. I. Idealized boundary shape for a single substance aggregating reversibly." *Proc. R. Soc. Lond. A.*, vol. 250, no. 1262, pp. 377–388, 1959. doi: 10.1098/rspa.1959.0070

[154] P. Schuck, "On the analysis of protein self-association by sedimentation velocity analytical ultracentrifugation." *Anal. Biochem.*, vol. 320, no. 1, pp. 104–124, 2003. doi: 10.1016/S0003-2697(03)00289-6

[155] R.L. Baldwin, "Sedimentation coefficients of small molecules: Methods of measurement based on the refractive-index gradient curve; the sedimentation coefficient of polyglucose A." *Biochem. J.*, vol. 55, no. 4, pp. 644–648, 1953. doi: 10.1042/bj0550644

[156] P. Schuck, "Sedimentation analysis of noninteracting and self-associating solutes using numerical solutions to the Lamm equation." *Biophys. J.*, vol. 75, no. 3, pp. 1503–1512, 1998. doi: 10.1016/S0006-3495(98)74069-X

[157] C.A. Sontag, W.F. Stafford, and J.J. Correia, "A comparison of weight average and direct boundary fitting of sedimentation velocity data for indefinite polymerizing systems." *Biophys. Chem.*, vol. 108, no. 1–3, pp. 215–230, 2004. doi: 10.1016/j.bpc.2003.10.029

[158] J.L. Oncley, E. Ellenbogen, D. Gitlin, and F.R.N. Gurd, "Protein-protein interactions." *J. Phys. Chem.*, vol. 56, no. 1, pp. 85–92, 1952. doi: 10.1021/j150493a017

[159] R.F. Steiner, "Reversible association processes of globular proteins. V. The study of associating systems by the methods of macromolecular physics." *Arch. Biochem. Biophys.*, vol. 49, no. 2, pp. 400–416, 1954. doi: 10.1016/0003-9861(54)90209-X

[160] J. Vistica, J. Dam, A. Balbo, E. Yikilmaz, R.A. Mariuzza, T.A. Rouault, and P. Schuck, "Sedimentation equilibrium analysis of protein interactions with global implicit mass conservation constraints and systematic noise decomposition." *Anal. Biochem.*, vol. 326, no. 2, pp. 234–256, 2004. doi: 10.1016/j.ab.2003.12.014

[161] D. Canzio, M. Liao, N. Naber, E. Pate, A. Larson, S. Wu, D.B. Marina, J.F. Garcia, H.D. Madhani, R. Cooke, P. Schuck, Y. Cheng, and G.J. Narlikar, "A conformational switch in HP1 releases auto-inhibition to drive heterochromatin assembly." *Nature*, vol. 496, no. 7445, pp. 377–381, 2013. doi: 10.1038/nature12032

[162] N.A. May, Q. Wang, A. Balbo, S.L. Konrad, R. Buchli, W.H. Hildebrand, P. Schuck, and A.W. Hudson, "Human herpesvirus 7 U21 tetramerizes to associate with class I major histocompatibility complex molecules." *J. Virol.*, vol. 88, no. 6, pp. 3298 3308, 2014. doi: 10.1128/JVI.02639-13

[163] D.E. Roark, "Sedimentation equilibrium techniques: Multiple speed analyses and an overspeed procedure." *Biophys. Chem.*, vol. 5, no. 1-2, pp. 185–196, 1976. doi: 10.1016/0301-4622(76)80034-8

[164] L. Servillo, H.B. Brewer, and J.C. Osborne, "Evaluation of the mixed interaction between apolipoproteins A-II and C-I by equilibrium sedimentation." *Biophys. Chem.*, vol. 13, no. 1, pp. 29–38, 1981. doi: 10.1016/0301-4622(81)80022-1

[165] J.S. Philo, "Sedimentation equilibrium analysis of mixed associations using numerical constraints to impose mass or signal conservation." *Methods Enzymol.*, vol. 321, pp. 100–120, 2000. doi: 10.1016/S0076-6879(00)21189-0

[166] R. Ghirlando, "The analysis of macromolecular interactions by sedimentation equilibrium." *Methods*, vol. 54, no. 1, pp. 145–156, 2011. doi: 10.1016/j.ymeth.2010.12.005

[167] H. Zhao, C.A. Brautigam, R. Ghirlando, and P. Schuck, "Current methods in sedimentation velocity and sedimentation equilibrium analytical ultracentrifugation." *Curr. Protoc. Protein Sci.*, vol. 7, p. 20.12.1, 2013. doi: 10.1002/0471140864.ps2012s71

[168] G.A. Gilbert and R.C.L. Jenkins, "Boundary problems in the sedimentation and electrophoresis of complex systems in rapid reversible equilibrium." *Nature*, vol. 177, no. 4514, pp. 853–854, 1956. doi: 10.1038/177853a0

[169] H. Zhao, A. Balbo, P.H. Brown, and P. Schuck, "The boundary structure in the analysis of reversibly interacting systems by sedimentation velocity." *Methods*, vol. 54, no. 1, pp. 16–30, 2011. doi: 10.1016/j.ymeth.2011.01.010

[170] P. Schuck, "Diffusion of the reaction boundary of rapidly interacting macromolecules in sedimentation velocity." *Biophys. J.*, vol. 98, no. 11, pp. 2741–2751, 2010. doi: 10.1016/j.bpj.2010.03.004

[171] L. Zhi, J. Mans, M.J. Paskow, P.H. Brown, P. Schuck, S. Jonjić, K. Natarajan, and D.H. Margulies, "Direct interaction of the mouse cytomegalovirus m152/gp40 immunoevasin with RAE-1 isoforms." *Biochemistry*, vol. 49, no. 11, pp. 2443–2453, 2010. doi: 10.1021/bi902130j

[172] L.M. Gilbert and G.A. Gilbert, "Molecular transport of reversibly reacting systems: Asymptotic boundary profiles in sedimentation, electrophoresis, and chromatography." *Methods Enzymol.*, vol. 48, pp. 195–211, 1978. doi: 10.1016/S0076-6879(78)48011-5

[173] J. Dam and P. Schuck, "Sedimentation velocity analysis of protein-protein interactions: Sedimentation coefficient distributions c(s) and asymptotic boundary profiles from Gilbert-Jenkins theory." *Biophys. J.*, vol. 89, no. 1, pp. 651–666, 2005. doi: 10.1529/biophysj.105.059584

[174] J. Dam, C A Velikovsky, R.A. Mariuzza, C. Urbanke, and P. Schuck, "Sedimentation velocity analysis of heterogeneous protein-protein interactions: Lamm equation modeling and sedimentation coefficient distributions c(s)." *Biophys. J.*, vol. 89, no. 1, pp. 619–634, 2005. doi: 10.1529/biophysj.105.059568

[175] W.F. Stafford and P.J. Sherwood, "Analysis of heterologous interacting systems by sedimentation velocity: Curve fitting algorithms for estimation of sedimentation coefficients, equilibrium and kinetic constants." *Biophys. Chem.*, vol. 108, no. 1-3, pp. 231–243, 2004. doi: 10.1016/j.bpc.2003.10.028

[176] C.A. Brautigam, "Using Lamm-Equation modeling of sedimentation velocity data to determine the kinetic and thermodynamic properties of macromolecular interactions." *Methods*, vol. 54, no. 1, pp. 4–15, 2011. doi: 10.1016/j.ymeth.2010.12.029

[177] A. Irimia, C. Ebel, D. Madern, S.B. Richard, L.W. Cosenza, G. Zaccai, and F.M. Vellieux, "The oligomeric states of Haloarcula marismortui malate dehydrogenase are modulated by solvent components as shown by crystallographic and biochemical studies." *J. Mol. Biol.*, vol. 326, no. 3, pp. 859–873, 2003. doi: 10.1016/S0022-2836(02)01450-X

[178] C. Tanford and J.A. Reynolds, "Characterization of membrane proteins in detergent solutions." *Biochim. Biophys. Acta*, vol. 457, no. 2, pp. 133–170, 1976. doi: 10.1016/0304-4157(76)90009-5

[179] D. Schubert and P. Schuck, "Analytical ultracentrifugation as a tool for studying membrane proteins." *Progr. Colloid Polym. Sci.*, vol. 86, pp. 12–22, 1991. doi: 10.1007/BFb0115002

[180] G.J. Howlett, "Sedimentation analysis of membrane proteins." in *Analytical Ultracentrifugation in Biochemistry and Polymer Science*, S.E. Harding, A.J. Rowe, and J.C. Horton, Eds. Cambridge, U.K.: The Royal Society of Chemistry, 1992, pp. 470–483.

[181] K.G. Fleming, "Determination of membrane protein molecular weight using sedimentation equilibrium analytical ultracentrifugation." *Curr. Protoc. Protein Sci.*, vol. Chapter 7, pp. Unit 7.12.1–7.12.13, 2008.

[182] C. Ebel, "Sedimentation velocity to characterize surfactants and solubilized membrane proteins." *Methods*, vol. 54, no. 1, pp. 56–66, 2011. doi: 10.1016/j.ymeth.2010.11.003

[183] D.M. Mitrea, C.R. Grace, M. Buljan, M.-K. Yun, N.J. Pytel, J. Satumba, A. Nourse, C.-G. Park, M. Madan Babu, S.W. White, and R.W. Kriwacki, "Structural polymorphism in the N-terminal oligomerization domain of NPM1." *Proc. Natl. Acad. Sci. U.S.A.*, vol. 111, no. 12, pp. 4466–4471, 2014. doi: 10.1073/pnas.1321007111

[184] J.E. Herrera, J.J. Correia, A.E. Jones, and M.O.J. Olson, "Sedimentation analyses of the salt- and divalent metal ion-induced oligomerization of nucleolar protein B23." *Biochemistry*, vol. 35, no. 8, pp. 2668–2673, 1996. doi: 10.1021/bi9523320

[185] C. Tanford, Y. Nozaki, J.A. Reynolds, and S. Makino, "Molecular characterization of proteins in detergent solutions." *Biochemistry*, vol. 13, no. 11, pp. 2369–2376, 1974. doi: 10.1021/bi00708a021

[186] J.A. Reynolds and C. Tanford, "Determination of molecular weight of the protein moiety in protein-detergent complexes without direct knowledge of detergent binding." *Proc. Natl. Acad. Sci. USA*, vol. 73, no. 12, pp. 4467–4470, 1976.

[187] S. Inagaki, R. Ghirlando, and R. Grisshammer, "Biophysical characterization of membrane proteins in nanodiscs." *Methods*, vol. 59, no. 3, pp. 287–300, 2013. doi: 10.1016/j.ymeth.2012.11.006

[188] A. Le Roy, H. Nury, B. Wiseman, J. Sarwan, J.-M. Jault, and C. Ebel, "Sedimentation velocity analytical ultracentrifugation in hydrogenated and deuterated solvents for the characterization of membrane proteins." *Methods Mol. Biol.*, vol. 1033, pp. 219–251, 2013. doi: 10.1007/978-1-62703-487-6_15

[189] M. le Maire, B. Arnou, C. Olesen, D. Georgin, C. Ebel, and J.V. Møller, "Gel chromatography and analytical ultracentrifugation to determine the extent of detergent binding and aggregation, and Stokes radius of membrane proteins using sarcoplasmic reticulum Ca2+-ATPase as an example." *Nat. Protoc.*, vol. 3, no. 11, pp. 1782–1795, 2008. doi: 10.1038/nprot.2008.177

[190] G. Rivas and A.P. Minton, "Beyond the second virial coefficient: Sedimentation equilibrium in highly non-ideal solutions." *Methods*, vol. 54, no. 1, pp. 167–174, 2011. doi: 10.1016/j.ymeth.2010.11.004

[191] R.C. Chatelier and A.P. Minton, "Sedimentation equilibrium in macromolecular solutions of arbitrary concentration. I. Self-associating proteins." *Biopolymers*, vol. 26, no. 4, pp. 507–524, 1987. doi: 10.1002/bip.360260405

[192] S.B. Zimmerman and A.P. Minton, "Macromolecular crowding: biochemical, biophysical, and physiological consequences." *Ann. Rev. Biophys. Biomol. Struct.*, vol. 22, pp. 27–65, 1993. doi: 10.1146/annurev.bb.22.060193.000331

[193] A.P. Minton, "Molecular crowding: Analysis of effects of high concentrations of inert cosolutes on biochemical equilibria and rates in terms of volume exclusion." *Methods Enzymol.*, vol. 295, pp. 127–149, 1998. doi: 10.1016/S0076-6879(98)95038-8

[194] A.P. Minton, "The effective hard particle model provides a simple, robust, and broadly applicable description of nonideal behavior in concentrated solutions of bovine serum albumin and other nonassociating proteins." *J. Pharm. Sci.*, vol. 96, no. 12, pp. 3466–3469, 2007. doi: 10.1002/jps.20964

[195] G. Rivas, J.A. Fernández, and A.P. Minton, "Direct observation of the self-association of dilute proteins in the presence of inert macromolecules at high concentration via tracer sedimentation equilibrium: Theory, experiment, and biological significance." *Biochemistry*, vol. 38, no. 29, pp. 9379–9388, 1999. doi: 10.1021/bi990355z

[196] G.K. Batchelor, "Sedimentation in a dilute dispersion of spheres." *J. Fluid. Mech.*, vol. 52, no. 2, pp. 245–268, 1972. doi: 10.1017/S0022112072001399

[197] J.P. Johnston and A.G. Ogston, "A boundary anomaly found in the ultracentrifugal sedimentation of mixtures." *Trans. Faraday Soc.*, vol. 42, pp. 789–799, 1946. doi: 10.1039/TF9464200789

[198] G.K. Batchelor and R.W.J. Van Rensburg, "Structure formation in bidisperse sedimentation." *J. Fluid. Mech.*, vol. 166, pp. 379–407, 1986. doi: 10.1017/S0022112086000204

[199] R.H. Davis and A. Acrivos, "Sedimentation of noncolloidal particles at low Reynolds numbers." *Annu. Rev. Fluid Mech.*, vol. 17, no. 1, pp. 91–118, 1985. doi: 10.1146/annurev.fl.17.010185.000515

[200] V.N. Schumaker and B.H. Zimm, "Anomalies in sedimentation. III. A model for the inherent instability of solutions of very large particles in high centrifugal fields." *Biopolymers*, vol. 12, no. 4, pp. 877–894, 1973. doi: 10.1002/bip.1973.360120416

[201] A. Moncho-Jordá, A.A. Louis, and J.T. Padding, "Effects of interparticle attractions on colloidal sedimentation." *Phys. Rev. Lett.*, vol. 104, no. 6, pp. 068301, 2010. doi: 10.1103/PhysRevLett.104.068301

[202] G.K. Batchelor and C.-S. Wen, "Sedimentation in a dilute polydisperse system of interacting spheres. Part 2. Numerical results." *J. Fluid Mech.*, vol. 124, pp. 495–528, 1982. doi: 10.1017/S0022112082002602

[203] G. Pouyet and J. Dayantis, "Velocity sedimentation in the semidilute concentration range of polymers dissolved in good solvents." *Macromolecules*, vol. 12, no. 2, pp. 293–296, 1979. doi: 10.1021/ma60068a026

[204] C.A. MacRaild, D.M. Hatters, L.J. Lawrence, and G.J. Howlett, "Sedimentation velocity analysis of flexible macromolecules: self-association and tangling of amyloid fibrils." *Biophys. J.*, vol. 81, no. 4, pp. 2502–2509, 2003. doi: 10.1016/S0006-3495(03)75061-9

[205] M Gross and R. Jaenicke, "Proteins under pressure. The influence of high hydrostatic pressure on structure, function and assembly of proteins and protein complexes." *Eur. J. Biochem.*, vol. 221, no. 2, pp. 617–630, 1994. doi: 10.1111/j.1432-1033.1994.tb18774.x

[206] P.Y. Cheng and H.K. Schachman, "The effect of pressure on sedimentation, and compressibility measurements in the ultracentrifuge." *J. Am. Chem. Soc.*, vol. 77, no. 6, pp. 1498–1501, 1955. doi: 10.1021/ja01611a027

[207] H. Fujita, "Effects of hydrostatic pressure upon sedimentation in the ultracentrifuge." *J. Am. Chem. Soc.*, vol. 78, no. 15, pp. 3598–3604, 1956. doi: 10.1021/ja01596a012

[208] M. Dishon, G.H. Weiss, and D.A. Yphantis, "Numerical solutions of the Lamm equation. VI. Effects of hydrostatic pressure on velocity sedimentation of two-component systems." *J. Polym. Sci.*, vol. 8, no. 12, pp. 2163–2175, 1970. doi: 10.1002/pol.1970.160081212

[209] W.F. Harrington and G. Kegeles, "Pressure effects in ultracentrifugation of interacting systems." *Methods Enzymol.*, vol. 27, pp. 306–345, 1973. doi: 10.1016/S0076-6879(73)27016-7

[210] P. Schuck, "A model for sedimentation in inhomogeneous media. II. Compressibility of aqueous and organic solvents." *Biophys. Chem.*, vol. 108, no. 1–3, pp. 201–214, 2004. doi: 10.1016/j.bpc.2003.10.017

[211] K.H. Jung and J.B. Hyne, "The isothermal compressibility of dimethylsulfoxide-water mixtures at 35 C." *Can. J. Chem.*, vol. 48, no. 15, pp. 2423–2425, 1970. doi: 10.1139/v70-405

[212] B. Gavish, E. Gratton, and C.J. Hardy, "Adiabatic compressibility of globular proteins." *Proc. Natl. Acad. Sci. USA*, vol. 80, no. 3, pp. 750–754, 1983.

[213] K. Gekko and Y. Hasegawa, "Compressibility-structure relationship of globular proteins." *Biochemistry*, vol. 25, no. 21, pp. 6563–6571, 1986. doi: 10.1021/bi00369a034

[214] T.V. Chalikian, "Volumetric properties of proteins." *Annu. Rev. Biophys. Biomol. Struct.*, vol. 32, pp. 207–235, 2003. doi: 10.1146/annurev.biophys.32.110601.141709

[215] P.A. Witherspoon, "Studies on petroleum with the ultracentrifuge." *Illinois State Geol. Surv. Rep. Investig.*, vol. 206, 1958.

[216] H. Mosimann and R. Signer, "Über den Einfluss des hydrostatischen Druckes auf die Sedimentationsgeschwindigkeit in der Ultrazentrifuge." *Helv. Chim. Acta*, vol. 27, no. 1, pp. 1123–1127, 1944. doi: 10.1002/hlca.194402701142

[217] J. Wonham, "Effect of pressure on the viscosity of water." *Nature*, vol. 215, pp. 1053–1054, 1967. doi: 10.1038/2151053a0

[218] K.R. Harris and L.A. Woolf, "Temperature and volume dependence of the viscosity of water and heavy water at low temperatures." *J. Chem. Eng. Data*, vol. 49, no. 4, pp. 1064–1069, 2004. doi: 10.1021/je049918m

[219] R.A. Horne and D.S. Johnson, "The viscosity of water under pressure." *J. Phys. Chem.*, vol. 70, no. 7, pp. 2182–2190, 1966. doi: 10.1021/j100879a018

[220] G. Hummer, S. Garde, A.E. García, M.E. Paulaitis, and L.R. Pratt, "The pressure dependence of hydrophobic interactions is consistent with the observed pressure denaturation of proteins." *Proc. Natl. Acad. Sci. USA*, vol. 95, no. 4, pp. 1552–1555, 1998. doi: 10.1073/pnas.95.4.1552

[221] J.M. Marcum and G.G. Borisy, "Sedimentation velocity analyses of the effect of hydrostatic pressure on the 30 S microtubule protein oligomer." *J. Biol. Chem.*, vol. 253, no. 8, pp. 2852–2857, 1978.

[222] A. Gow, D.J. Winzor, and R. Smith, "Pressure-induced dissociation of aggregates of myelin proteolipid protein." *Biochem. Biophys. Acta*, vol. 828, no. 3, pp. 383–386, 1985. doi: 10.1016/0167-4838(85)90321-8

[223] V.N. Schumaker, A. Wlodawer, J.T. Courtney, and K.M. Decker, "Ultracentrifugation at moderate pressures." *Anal. Biochem.*, vol. 34, no. 2, pp. 359–365, 1970. doi: 10.1016/0003-2697(70)90120-X

[224] J.F. Marko and E.D. Siggia, "Stretching DNA." *Macromolecules*, vol. 28, no. 26, pp. 8759–8770, 1995. doi: 10.1021/ma00130a008

[225] A. Fukuzawa, M. Hiroshima, K. Maruyama, N. Yonezawa, M. Tokunaga, and S. Kimura, "Single-molecule measurement of elasticity of serine-, glutamate- and lysine-rich repeats of invertebrate connectin reveals that its elasticity is caused entropically by random coil structure." *J. Muscle Res. Cell Motil.*, vol. 23, no. 5–6, pp. 449–453, 2002. doi: 10.1023/A:1023406422275

[226] F. Schwesinger, R. Ros, T. Strunz, D. Anselmetti, H.-J. Güntherodt, A. Honegger, L. Jermutus, L. Tiefenauer, and A. Plueckthun, "Unbinding forces of single antibody-antigen complexes correlate with their thermal dissociation rates." *Proc. Natl. Acad. Sci. USA*, vol. 97, no. 18, pp. 9972–9977, 2000. doi: 10.1073/pnas.97.18.9972

[227] A.J. Jin, K. Prasad, P.D. Smith, E.M. Lafer, and R.J. Nossal, "Measuring the elasticity of clathrin-coated vesicles via atomic force microscopy." *Biophys. J.*, vol. 90, no. 9, pp. 3333–3344, 2006. doi: 10.1529/biophysj.105.068742

[228] M. Grandbois, M. Beyer, M. Rief, H. Clausen-Schaumann, and H.E. Gaub, "How strong is a covalent bond?" *Science*, vol. 283, no. 5408, pp. 1727–1730, 1999. doi: 10.1126/science.283.5408.1727

[229] B.H. Zimm, "Anomalies in sedimentation. IV. Decrease in sedimentation coefficients of chains at high fields." *Biophys. Chem.*, vol. 1, no. 4, pp. 279–291, 1974. doi: 10.1016/0301-4622(74)80014-1

[230] B.H. Zimm, V.N. Schumaker, and C.B. Zimm, "Anomalies in sedimentation. V. Chains at high fields, practical consequences." *Biophys. Chem.*, vol. 5, no. 1-2, pp. 265–270, 1976. doi: 10.1016/0301-4622(76)80039-7

[231] X. Schlagberger and R.R. Netz, "Anomalous polymer sedimentation far from equilibrium." *Phys. Rev. Lett.*, vol. 98, no. 12, p. 128301, 2007. doi: 10.1103/PhysRevLett.98.128301

[232] X. Schlagberger and R.R. Netz, "Anomalous sedimentation of self-avoiding flexible polymers." *Macromolecules*, vol. 41, no. 5, pp. 1861–1871, 2008. doi: 10.1021/ma070947m

[233] H. Brenner and D.W Condiff, "Transport mechanics in systems of orientable particles. III. Arbitrary particles." *J. Colloid Interface Sci.*, vol. 41, no. 2, pp. 228–274, 1972. doi: 10.1016/0021-9797(72)90111-7

[234] J.E. Hearst and J. Vinograd, "The effect of angular velocity on the sedimentation behavior of deoxyribonucleic acid and tobacco mosaic virus in the ultracentrifuge." *Arch. Biochem. Biophys.*, vol. 92, no. 2, pp. 206–215, 1961. doi: 10.1016/0003-9861(61)90338-1

[235] Z. Dogic, A.P. Philipse, S. Fraden, and J.K.G. Dhont, "Concentration-dependent sedimentation of colloidal rods." *J. Chem. Phys.*, vol. 113, no. 18, pp. 8368–8380, 2000. doi: 10.1063/1.1308107

[236] C.A. Silvera Batista, M. Zheng, C.Y. Khripin, X. Tu, and J.A. Fagan, "Rod hydrodynamics and length distributions of single-wall carbon nanotubes using analytical ultracentrifugation." *Langmuir*, vol. 30, no. 17, pp. 4895–4904, 2014. doi: 10.1021/la404892k

[237] H. Zhao, P.H. Brown, A. Balbo, M.C. Fernandez Alonso, N. Polishchuck, C. Chaudhry, M.L. Mayer, R. Ghirlando, and P. Schuck, "Accounting for solvent signal offsets in the analysis of interferometric sedimentation velocity data." *Macromol. Biosci.*, vol. 10, no. 7, pp. 736–745, 2010. doi: 10.1002/mabi.200900456

[238] P. Schuck, "A model for sedimentation in inhomogeneous media. I. Dynamic density gradients from sedimenting co-solutes." *Biophys. Chem.*, vol. 108, no. 1–3, pp. 187–200, 2004. doi: 10.1016/j.bpc.2003.10.016

[239] M. Meselson and F.W. Stahl, "The replication of DNA in Escherichia coli." *Proc. Nat. Acad. Sci. USA*, vol. 44, no. 7, pp. 671–682, 1958. doi: 10.1073/pnas.44.7.671

[240] J.P. Gabrielson, K.K. Arthur, B.S. Kendrick, T.W. Randolph, and M.R. Stoner, "Common excipients impair detection of protein aggregates during sedimentation velocity analytical ultracentrifugation." *J. Pharm. Sci.*, vol. 98, no. 1, pp. 50–62, 2009. doi: 10.1002/jps.21403

[241] D. Prosperi, C. Morasso, F. Mantegazza, M. Buscaglia, L. Hough, and T. Bellini, "Phantom nanoparticles as probes of biomolecular interactions." *Small*, vol. 2, no. 8-9, pp. 1060–1067, 2006. doi: 10.1002/smll.200600106

[242] H.J. Schönfeld, B. Pöschl, and F. Müller, "Quasi-elastic light scattering and analytical ultracentrifugation are indispensable tools for the purification and characterization of recombinant proteins." *Biochem. Soc. Trans.*, vol. 26, no. 4, pp. 753–758, 1998.

[243] J.R. Cann, "Effects of microheterogeneity on sedimentation patterns of interacting proteins and the sedimentation behavior of systems involving two ligands." *Methods Enzymol.*, vol. 130, no. 1964, pp. 19–35, 1986. doi: 10.1016/0076-6879(86)30005-3

[244] J. Dam and P. Schuck, "Calculating sedimentation coefficient distributions by direct modeling of sedimentation velocity concentration profiles." *Methods Enzymol.*, vol. 384, no. 301, pp. 185–212, 2004. doi: 10.1016/S0076-6879(04)84012-6

[245] O. Wiener, "Darstellung gekrümmter Lichtstrahlen und Verwerthung derselben zur Untersuchung von Diffusion und Wärmeleitung." *Ann. Phys.*, vol. 49, pp. 105–149, 1893.

[246] H. Svensson, "The second-order aberrations in the interferometric measurement of concentration gradients." *Opt. Acta Int. J. Opt.*, vol. 1, no. 1, pp. 25–32, 1954. doi: 10.1080/713818656

[247] R. Forsberg and H. Svensson, "The second-order aberrations in the interferometric measurement of concentration gradients. II. Experimental verification of theory." *Opt. Acta Int. J. Opt.*, vol. 1, no. 2, pp. 90–93, 1954. doi: 10.1080/713818666

[248] A.P. Minton and M.S. Lewis, "Self-association in highly concentrated solutions of myoglobin: a novel analysis of sedimentation equilibrium of highly nonideal solutions." *Biophys. Chem.*, vol. 14, no. 4, pp. 317–324, 1981. doi: 10.1016/0301-4622(81)85033-8

[249] J. Ma, M. Metrick, R. Ghirlando, H. Zhao, and P. Schuck, "Variable-field analytical ultracentrifugation: I. Time optimized sedimentation equilibrium." *Biophys. J.*, vol. 109, no. 4, pp. 827–837, 2015. doi: 10.1016/j.bpj.2015.07.015

[250] G.M. Pavlov, A.J. Rowe, and S.E. Harding, "Conformation zoning of large molecules using the analytical ultracentrifuge zones." *Trends Anal. Chem.*, vol. 16, no. 7, pp. 401–405, 1997. doi: 10.1016/S0165-9936(97)00038-1

[251] S.E. Harding, J.C. Horton, S. Jones, J.M. Thornton, and D.J. Winzor, "COVOL: an interactive program for evaluating second virial coefficients from the triaxial shape or dimensions of rigid macromolecules." *Biophys. J.*, vol. 76, no. 5, pp. 2432–2438, 1999. doi: 10.1016/S0006-3495(99)77398-4

[252] C.Y. Huang, "Determination of binding stoichiometry by the continuous variation method: the Job plot." *Methods Enzymol.*, vol. 87, pp. 509–525, 1982. doi: 10.1016/S0076-6879(82)87029-8

[253] D.W. Kupke and T.E. Dorrier, "Protein concentration measurements: The dry weight." *Methods Enzymol.*, vol. 48, pp. 155–162, 1978. doi: 10.1016/S0076-6879(78)480108-5

[254] C.M. Stoscheck, "Quantitation of protein." *Methods Enzymol.*, vol. 182, pp. 50–68, 1990. doi: 10.1016/0076-6879(90)82008-P

[255] J. Vinograd, R. Bruner, R. Kent, and J Weigle, "Band-centrifugation of macromolecules and viruses in self-generating density gradients." *Proc. Nat. Acad. Sci. USA*, vol. 49, no. 6, pp. 902–910, 1963.

[256] J. Lebowitz, M Teale, and P. Schuck, "Analytical band centrifugation of proteins and protein complexes." *Biochem. Soc. Trans.*, vol. 26, no. 4, pp. 745–749, 1998. doi: 10.1042/bst0260745

[257] D.L. Sackett and R.E. Lippoldt, "Thermodynamics of reversible monomer-dimer association of tubulin." *Biochemistry*, vol. 30, no. 14, pp. 3511–3517, 1991. doi: 10.1021/bi00228a023

[258] P. Schuck and D.B. Millar, "Rapid determination of molar mass in modified Archibald experiments using direct fitting of the Lamm equation." *Anal. Biochem.*, vol. 259, no. 1, pp. 48–53, 1998. doi: 10.1006/abio.1998.2638

[259] W.J. Archibald, "A demonstration of some new methods of determining molecular weights from the data of the ultracentrifuge." *J. Phys. Chem.*, vol. 51, no. 5, pp. 1204–1214, 1947. doi: 10.1021/j150455a014

[260] S.R. Liber, S. Borohovich, A.V. Butenko, A.B. Schofield, and E. Sloutskin, "Dense colloidal fluids form denser amorphous sediments." *Proc. Natl. Acad. Sci. U.S.A.*, vol. 110, no. 15, pp. 5769–5773, 2013. doi: 10.1073/pnas.1214945110

[261] C.E. Dann, C.A. Wakeman, C.L. Sieling, S.C. Baker, I. Irnov, and W.C. Winkler, "Structure and mechanism of a metal-sensing regulatory RNA." *Cell*, vol. 130, no. 5, pp. 878–892, 2007. doi: 10.1016/j.cell.2007.06.051

[262] R.F. Henderson, T.R. Henderson, and B.M. Woodfin, "Effects of D_2O on the association-dissociation equilibrium in subunit proteins." *J. Biol. Chem.*, vol. 245, no. 15, pp. 3733–3737, 1970.

[263] Y. Cho, L.B. Sagle, S. Iimura, Y. Zhang, J. Kherb, A. Chilkoti, J.M. Scholtz, and P.S. Cremer, "Hydrogen bonding of beta-turn structure is stabilized in D_2O." *J. Am. Chem. Soc.*, vol. 131, no. 42, pp. 15188–15193, 2009. doi: 10.1021/ja9040785

[264] M. Jasnin, M. Tehei, M. Moulin, M. Haertlein, and G. Zaccai, "Solvent isotope effect on macromolecular dynamics in E. coli." *Eur. Biophys. J.*, vol. 37, no. 5, pp. 613–617, 2008. doi: 10.1007/s00249-008-0281-4

[265] C. Eginton and D. Beckett, "A large solvent isotope effect on protein association thermodynamics." *Biochemistry*, vol. 52, no. 38, pp. 6595–6600, 2013. doi: 10.1021/bi400952m

[266] S.J. Edelstein and H.K. Schachman, "Measurement of partial specific volume by sedimentation equilibrium in H2O-D2O solutions." *Methods Enzymol.*, vol. 27, pp. 82–98, 1973. doi: 10.1016/S0076-6789(73)27006-4

[267] S. Filitti-Wurmser and M. Tempête-Gaillourdet, "Partial specific volume and molecular weight determinations from sedimentation equilibrium experiments in H2O and H2-18O. Study of an immunomacroglobulin type lambda, (IgM) Lambda." *Biochimie*, vol. 56, no. 9, pp. 1183–1189, 1974. doi: 10.1016/S0300-9084(74)80009-X

[268] N.P. Mullin, A. Yates, A.J. Rowe, B. Nijmeijer, D. Colby, P.N. Barlow, M.D. Walkinshaw, and I. Chambers, "The pluripotency rheostat Nanog functions as a dimer." *Biochem J.*, vol. 411, no. 2, pp. 227–231, 2008. doi: 10.1042/BJ20080134

[269] P.H. Brown, A. Balbo, H. Zhao, C. Ebel, and P. Schuck, "Density contrast sedimentation velocity for the determination of protein partial-specific volumes." *PLoS One*, vol. 6, no. 10, p. e26221, 2011. doi: 10.1371/journal.pone.0026221

[270] G. Mayer, B. Ludwig, H.W. Müller, J.A. van den Broek, R.H.E. Friesen, and D. Schubert, "Studying membrane proteins in detergent solution by analytical ultracentrifugation: Different methods for density matching." *Prog. Colloid Polym. Sci*, vol. 113, pp. 176–181, 1999. doi: 10.1007/3-540-48703-4_25

[271] A. Lustig, A. Engel, G. Tsiotis, E.M. Landau, and W. Baschong, "Molecular weight determination of membrane proteins by sedimentation equilibrium at the sucrose or nycodenz-adjusted density of hydrated detergent micelle." *Biochim. Biophys. Acta.*, vol. 1464, no. 2, pp. 199–206, 2000. doi: 10.1016/S0005-2736(99)00254-0

[272] K.S. Iyer and W.A. Klee, "Direct spectrophotometric measurement of the rate of reduction of disulfide bonds. The reactivity of the disulfide bonds of bovine α-lactalbumin." *J. Biol. Chem.*, vol. 248, no. 2, pp. 707–710, 1973.

[273] T.M. Laue, D.F. Senear, S. Eaton, and J.B. Ross, "5-hydroxytryptophan as a new intrinsic probe for investigating protein- DNA interactions by analytical ultracentrifugation. Study of the effect of DNA on self-assembly of the bacteriophage lambda cI repressor." *Biochemistry*, vol. 32, no. 10, pp. 2469–2472, 1993. doi: 10.1021/bi00061a003

[274] J.B.A. Ross, D.F. Senear, E. Waxman, B.B. Kombo, E. Rusinova, Y.T. Huang, W.R. Laws, and C.A. Hasselbacher, "Spectral enhancement of proteins: biological incorporation and fluorescence characterization of 5-hydroxytryptophan in bacteriophage lambda cI repressor." *Proc. Natl. Acad. Sci. U.S.A.*, vol. 89, no. 24, pp. 12 023–12 027, 1992.

[275] R.Y. Tsien, "The green fluorescent protein." *Annu. Rev. Biochem.*, vol. 67, pp. 509–544, 1998. doi: 10.1146/annurev.biochem.67.1.509

[276] J.M. González, M. Jiménez, M. Vélez, J. Mingorance, J.M. Andreu, M. Vicente, and G. Rivas, "Essential cell division protein FtsZ assembles into one monomer-thick ribbons under conditions resembling the crowded intracellular environment." *J. Biol. Chem..*, vol. 278, no. 39, pp. 37 664–37 671, 2003. doi: 10.1074/jbc.M305230200

[277] B. Monterroso, C. Alfonso, S. Zorrilla, and G. Rivas, "Combined analytical ultracentrifugation, light scattering and fluorescence spectroscopy studies on the functional associations of the bacterial division FtsZ protein." *Methods*, vol. 59, no. 3, pp. 349–362, 2013. doi: 10.1016/j.ymeth.2012.12.014

[278] T.M. Laue, B.D. Shah, T.M. Ridgeway, and S.L. Pelletier, "Computer-aided interpretation of analytical sedimentation data for proteins." in *Analytical Ultracentrifugation in Biochemistry and Polymer Science*, S.E. Harding, A.J. Rowe, and J.C. Horton, Eds. Cambridge: The Royal Society of Chemistry, 1992, pp. 90–125.

[279] O. Kratky, H. Leopold, and H. Stabinger, "The determination of the partial specific volume of proteins by the mechanical oscillator technique." *Methods Enzymol.*, vol. 27, pp. 98–110, 1973. doi: 10.1016/S0076-6879(73)27007-6

[280] S.E. Harding and P. Johnson, "The concentration-dependence of macromolecular parameters." *Biochem. J.*, vol. 231, no. 3, pp. 543–547, 1985. doi: 10.1042/bj2310543

[281] H. Durchschlag, "Determination of the partial specific volume of conjugated proteins." *Colloid Polym Sci*, vol. 267, no. 12, pp. 1139–1150, 1989. doi: 10.1007/BF01496937

[282] M.A. Perugini, P. Schuck, and G.J. Howlett, "Self-association of human apolipoprotein E3 and E4 in the presence and absence of phopholipid." *J. Biol. Chem.*, vol. 275, no. 47, pp. 36 758–36 765, 2000. doi: 10.1074/jbc.M005565200

[283] E.A. Englund, D. Wang, H. Fujigaki, H. Sakai, C.M. Micklitsch, R. Ghirlando, G. Martin-Manso, M.L. Pendrak, D.D. Roberts, S.R. Durell, and D.H. Appella, "Programmable multivalent display of receptor ligands using peptide nucleic acid nanoscaffolds." *Nat. Commun.*, vol. 3, p. 614, 2012. doi: 10.1038/ncomms1629

[284] S. Trachtenberg, P. Schuck, T.M. Phillips, S.B. Andrews, and R.D. Leapman, "A structural framework for a near-minimal form of life: Mass and compositional analysis of the helical mollicute *Spiroplasma melliferum BC3*." *PLoS One*, vol. 9, no. 2, p. e87921, 2014. doi: 10.1371/journal.pone.0087921

[285] H.K. Schachman and S.J. Edelstein, "Ultracentrifuge studies with absorption optics. IV. Molecular weight determinations at the microgram level." *Biochemistry*, vol. 5, no. 8, pp. 2681–2705, 1966. doi: 10.1021/bi00872a029

[286] S.J. Edelstein and H.K. Schachman, "The simultaneous determination of partial specific volumes and molecular weights with microgram quantities." *J. Biol. Chem.*, vol. 242, no. 2, pp. 306–311, 1967.

[287] D. Schubert, C. Tziatzios, J.A. van den Broeck, P. Schuck, L. Germeroth, and H. Michel, "Determination of the molar mass of pigment-containing complexes of intrinsic membrane proteins: Problems, solutions and application to the light-harvesting complex B800/820 of Rhodospirillum molischianum." *Progr. Colloid Polym. Sci.*, vol. 94, pp. 14–19, 1994. doi: 10.1007/BFb0115598

[288] R.J. Center, P. Schuck, R.D. Leapman, L.O. Arthur, P.L. Earl, B. Moss, and J. Lebowitz, "Oligomeric structure of virion-associated and soluble forms of the simian immunodeficiency virus envelope protein in the prefusion activated conformation." *Proc. Nat. Acad. Sci. USA*, vol. 98, no. 26, pp. 14877–14882, 2001. doi: 10.1073/pnas261573898

[289] D. Noy, J.R. Calhoun, and J.D. Lear, "Direct analysis of protein sedimentation equilibrium in detergent solutions without density matching." *Anal. Biochem.*, vol. 320, no. 2, pp. 185–192, 2003. doi: 10.1016/S0003-2697(03)00347-6

[290] D.G. Sharp and J.W. Beams, "Size and density of polystyrene particles measured by ultracentrifugation." *J. Biol. Chem.*, vol. 185, no. 1, pp. 247–253, 1950.

[291] W.G. Martin, C.A. Winkler, and W. H. Cook, "Partial specific volume measurements by differential sedimentation." *Can. J. Chem.*, vol. 37, no. 10, pp. 1662–1670, 1959. doi: 10.1139/v59-241

[292] W.L. Gagen, "The significance of the "partial specific volume" obtained from sedimentation data." *Biochemistry*, vol. 5, no. 8, pp. 2553–2557, 1966. doi: 10.1021/bi00872a010

[293] Y. Gohon, G.M. Pavlov, P. Timmins, C. Tribet, J-L Popot, and C. Ebel, "Partial specific volume and solvent interactions of amphipol A8-35." *Anal. Biochem.*, vol. 334, no. 2, pp. 318–334, 2004. doi: 10.1016/j.ab.2004.07.033

[294] H.B. Bull and K. Breese, "Concentrations and partial volumes of proteins." *Arch. Biochem. Biophys.*, vol. 197, no. 1, pp. 199–204, 1979. doi: 10.1016/0003-9861(79)90237-6

[295] N. Errington and A.J. Rowe, "Probing conformation and conformational change in proteins is optimally undertaken in relative mode." *Eur. Biophys. J.*, vol. 32, no. 5, pp. 511–517, 2003. doi: 10.1007/s00249-003-0315-x

[296] R. Ghirlando, A. Balbo, G. Piszczek, P.H. Brown, M.S. Lewis, C.A. Brautigam, P. Schuck, and H. Zhao, "Improving the thermal, radial, and temporal accuracy of the analytical ultracentrifuge through external references." *Anal. Biochem.*, vol. 440, no. 1, pp. 81–95, 2013. doi: 10.1016/j.ab.2013.05.011

[297] H. Zhao, R. Ghirlando, C. Alfonso, F. Arisaka, I. Attali, D.L. Bain, M.M. Bakhtina, D.F. Becker, G.J. Bedwell, A. Bekdemir, T.M.D. Besong, C. Birck, C.A. Brautigam, W. Brennerman, O. Byron, A. Bzowska, J.B. Chaires, C.T. Chaton, H. Cölfen, K.D. Connaghan, K.A. Crowley, U. Curth, T. Daviter, W.L. Dean, A.I. Diez, C. Ebel, D.M. Eckert, L.E. Eisele, E. Eisenstein, P. England, C. Escalante, J.A. Fagan, R. Fairman, R.M. Finn, W. Fischle, J. García de la Torre, J. Gor, H. Gustafsson, D. Hall, S.E. Harding, J.G. Hernandez Cifre, A.B. Herr, E.E. Howell, R.S. Isaac, S.-C. Jao, D. Jose, S.-J. Kim, B. Kokona, J.A. Kornblatt, D. Kosek, E. Krayukhina, D. Krzizike, E.A. Kusznir, H. Kwon, A. Larson, T.M. Laue, A. Le Roy, A.P. Leech, H. Lilie, K. Luger, J.R. Luque-Ortega, J. Ma, C.A. May, E.L. Maynard, A. Modrak-Wojcik, Y.-F. Mok, N. Mücke, L. Nagel-Steger, G.J. Narlikar, M. Noda, A. Nourse, T. Obsil, C.K. Park, J.-K. Park, P.D. Pawalek, E.E. Perdue, S.J. Perkins, M.A. Perugini, C.L. Peterson, M.G. Peverelli, G. Piszczek, G. Prag, P.E. Prevelige, B.D.E. Raynal, L. Rezabkova, K. Richter, A.E. Ringel, R. Rosenberg, A.J. Rowe, A.C. Rufer, D.J. Scott, J.G. Seravalli, A.S. Solovyova, R. Song, D. Staunton, C. Stoddard, K. Stott, H.M. Strauss, W.W. Streicher, J.P. Sumida, S.G. Swygert, R.H. Szczepanowski, I. Tessmer, R.T. Toth IV, A. Tripathy, S. Uchiyama, S.F.W. Uebel, S. Unzai, A. Vitlin Gruber, P.H. von Hippel, C. Wandrey, S.-H. Wang, S.E. Weitzel, B. Wielgus-Kutrowska, C. Wolberger, M. Wolff, E. Wright, Y.-S. Wu, J.M. Wubben, P. Schuck, "A multilaboratory comparison of calibration accuracy and the performance of external references in analytical ultracentrifugation." *PLoS One*, vol. 10, no. 5, p. e0126420, 2015. doi: 10.1371/journal.pone.0126420

[298] J.A. Fagan, M. Zheng, V. Rastogi, J.R. Simpson, C.Y. Khripin, C.A. Silvera Batista, and A.R. Hight Walker, "Analyzing surfactant structures on length and chirality resolved (6,5) single-wall carbon nanotubes by analytical ultracentrifugation." *ACS Nano*, vol. 7, no. 4, pp. 3373–3387, 2013. doi: 10.1021/nn4002165

[299] M.S. Arnold, J. Suntivich, S.I. Stupp, and M.C. Hersam, "Hydrodynamic characterization of surfactant encapsulated carbon nanotubes using an analytical ultracentrifuge." *ACS Nano*, vol. 2, no. 11, pp. 2291–2300, 2008. doi: 10.1021\nn800512t

[300] A. Lustig, A. Engel, and M. Zulauf, "Density determination by analytical ultracentrifugation in a rapid dynamical gradient: application to lipid and detergent aggregates containing proteins." *Biochem. Biophys. Acta*, vol. 1115, no. 2, pp. 89–95, 1991. doi: 10.1016/0304-4165(91)90016-A

[301] M. Meselson, F.W. Stahl, and J. Vinograd, "Equilibrium sedimentation of macromolecules in density gradients." *Proc. Nat. Acad. Sci. USA*, vol. 43, no. 7, pp. 581–588, 1957.

[302] C. Tziatzios, H. Durchschlag, J.J. Gonzales, E. Albertini, P. Prados, J. de Mendoza, C. Eschbaumer, U.S. Schubert, P. Schuck, and D. Schubert, "Characterization of supramolecular assemblies by analytical ultracentrifugation: Potential, problems, and application to Co coordination arrays and calixarenes." *Abstr. Pap. Am. Chem. Soc.*, vol. 219, no. Part 2, pp. U456–U457, 2000.

[303] M. Raşa, B.G.G. Lohmeijer, H. Hofmeier, H.M.L. Thijs, D. Schubert, U.S. Schubert, and C. Tziatzios, "Characterization of metallo-supramolecular block copolymers by analytical ultracentrifugation." *Macromol. Chem. Phys.*, vol. 207, no. 22, pp. 2029–2041, 2006. doi: 10.1002/macp.200600235

[304] J.P. Elder, "Density measurements by the mechanical oscillator." *Methods Enzymol.*, vol. 61, pp. 12–25, 1979. doi: 10.1016/0076-6879(79)61004-2

[305] H. Eisenberg, "Analytical ultracentrifugation in a Gibbsian perspective." *Biophys. Chem.*, vol. 88, no. 1–3, pp. 1–9, 2000. doi: 10.1016/S0301-4622(00)00205-2

[306] J.C. Lee, K. Gekko, and S.N. Timasheff, "Measurements of preferential solvent interactions by densimetric techniques." *Methods Enzymol.*, vol. 61, no. 1, pp. 26–49, 1979.

[307] S. Pundak and H. Eisenberg, "Structure and activity of malate dehydrogenase from the extreme halophilic bacteria of the dead sea 1. Conformation and interaction with water and salt between 5 M and 1 M NaCl concentration." *Eur. J. Biochem.*, pp. 463–470, 1981. doi: 10.1111/j.1432-1033.1981.tb05542.x

[308] P.H. Stothart, "Determination of partial specific volume and absolute concentration by densimetry." *Biochem. J.*, vol. 219, no. 3, pp. 1049–1052, 1984. doi: 10.1042/bj2191049

[309] H. Durchschlag, "Specific volumes of biological macromolecules and some other molecules of biological interest." in *Thermodynamic Data for Biochemistry and Biotechnology*, H.-J. Hinz, Ed. Berlin: Springer, 1986, pp. 45–128.

[310] A.A. Zamyatnin, "Amino acid, peptide, and protein volume in solution." *Annu. Rev. Biophys. Bioeng*, vol. 13, pp. 145–165, 1984. doi: 10.1146/annurev.bb.13.060184.001045

[311] E.J. Cohn and J.T. Edsall, "Density and apparent specific volume of proteins." in *Proteins, Amin. Acids Pept.*, E.J. Cohn and J.T. Edsall, Eds. Princeton, NJ: Van Nostrand-Reinhold, 1943, pp. 370–381.

[312] T.L. McMeekin and K. Marshall, "Specific volumes of proteins and the relationship to their amino acid contents." *Science*, vol. 116, no. 3006, pp. 142–143, 1952. doi: 10.1126/science.116.3006.142

[313] T. Arakawa and S.N. Timasheff, "Calculation of the partial specific volume of proteins in concentrated salt and amino acid solutions." *Methods Enzymol.*, vol. 117, pp. 60–65, 1985. doi: 10.1016/S0076-6879(85)17007-2

[314] T. Arakawa, "Calculation of the partial specific volumes of proteins in concentrated salt, sugar, and amino acid solutions." *J. Biochem.*, vol. 100, no. 6, pp. 1471–1475, 1986.

[315] R. DeVane, C. Ridley, R.W. Larsen, B. Space, P.B. Moore, and S.I. Chan, "A molecular dynamics method for calculating molecular volume changes appropriate for biomolecular simulation." *Biophys. J.*, vol. 85, no. 5, pp. 2801–2807, 2003. doi: 10.1016/S0006-3495(03)74703-1

[316] S. Sarupria, T. Ghosh, A.E. García, and S. Garde, "Studying pressure denaturation of a protein by molecular dynamics simulations." *Proteins*, vol. 78, no. 7, pp. 1641–51, 2010. doi: 10.1002/prot.22680

[317] J. Spooner, H. Wiebe, N. Boon, E. Deglint, E. Edwards, B. Yanciw, B. Patton, L. Thiele, P. Dance, and N. Weinberg, "Molecular dynamics calculation of molecular volumes and volumes of activation." *Phys. Chem. Chem. Phys.*, vol. 14, no. 7, pp. 2264–2277, 2012. doi: 10.1039/c2cp22949h

[318] J. Traube, "Ueber Atom- und Molekularräume." *Ann. Phys.*, vol. 310, no. 7, pp. 548–564, 1901. doi: 10.1002/andp.19013100705

[319] H. Høiland, "Partial molar volumes of biochemical model compounds in aqueous solutions." in *Thermodynamic Data for Biochemistry and Biotechnology*, H.-J. Hinz, Ed. Berlin: Springer, 1986, pp. 17–44.

[320] H. Durchschlag and P. Zipper, "Calculation of partial specific volumes and other volumetric properties of small molecules and polymers." *J. Appl. Crystallogr.*, vol. 30, no. 5-2, pp. 803–807, 1997. doi: 10.1107/S0021889897003348

[321] H. Durchschlag and P. Zipper, "Calculation of the partial specific volume of organic compounds and polymers." *Progr. Colloid Polym. Sci.*, vol. 94, pp. 20–39, 1994. doi: 10.1007/BFb0115599

[322] E.J. Cohn, T.L. McMeekin, J.T. Edsall, and M.H. Blanchard, "Studies in the physical chemistry of amino acids, peptides and related substances. I. The apparent molal volume and the electrostriction of the solvent." *J. Am. Chem. Soc.*, vol. 56, no. 4, pp. 784–794, 1934. doi: 10.1021/ja01319a006

[323] Y. Harpaz, M. Gerstein, and C. Chothia, "Volume changes on protein folding." *Structure*, vol. 2, no. 7, pp. 641–649, 1994. doi: 10.1016/S0969-2126(00)00065-4

[324] G.I. Makhatadze, V.N. Medvedkin, and P.L. Privalov, "Partial molar volumes of polypeptides and their constituent groups in aqueous solution over a broad temperature range." *Biopolymers*, vol. 30, no. 11-12, pp. 1001–1010, 1990. doi: 10.1002/bip.360301102

[325] G.R. Hedwig and H.J. Hinz, "Group additivity schemes for the calculation of the partial molar heat capacities and volumes of unfolded proteins in aqueous solution." *Biophys. Chem.*, vol. 100, no. 1-3, pp. 239–260, 2003. doi: 10.1016/S0301-4622(02)00284-3

[326] R. Fairman, W. Fenderson, M.E. Hail, Y. Wu, and S.-Y. Shaw, "Molecular weights of CTLA-4 and CD80 by sedimentation equilibrium ultracentrifugation." *Anal. Biochem.*, vol. 270, no. 2, pp. 286–295, 1999. doi: 10.1006/abio.1999.4095

[327] R. Ghirlando, M.B. Keown, G.A. Mackay, M.S. Lewis, J.C. Unkeless, and H.J. Gould, "Stoichiometry and thermodynamics of the interaction between the Fc fragment of human IgG1 and its low-affinity receptor Fc gamma RIII." *Biochemistry*, vol. 34, no. 41, pp. 13320–13327, 1995. doi: 10.1021/bi00041a007

[328] S.J. Shire, "Determination of molecular weight of glycoproteins by analytical ultracentrifugation." *Beckman Tech. Information, Publ. DS-837.*, pp. 1–4, 1992. Online at: https://www.beckmancoulter.com/wsrportal/bibliography?docname=DS_837.pdf

[329] M.S. Lewis and R.P. Junghans, "Ultracentrifugal analysis of molecular mass of gly-coproteins of unknown or ill-defined carbohydrate composition." *Methods Enzymol.*, vol. 321, pp. 136–149, 2000. doi: 10.1016/S0076-6879(00)21191-9

[330] I.D. Kuntz, "Hydration of macromolecules. III. Hydration of polypeptides." *J. Am. Chem. Soc.*, vol. 93, no. 2, pp. 514–516, 1971. doi: 10.1021/ja00731a036

[331] V. Prakash and S.N. Timasheff, "The calculation of partial specific volumes of proteins in 8 M urea solution." *Anal. Biochem.*, vol. 117, no. 2, pp. 330–335, 1981. doi: 10.1016/0003-2697(81)90788-0

[332] J.C. Lee and S.N. Timasheff, "The calculation of partial specific volumes of proteins in guanidine hydrochloride." *Arch. Biochem. Biophys.*, vol. 165, no. 1, pp. 268–273, 1974. doi: 10.1016/0003-9861(74)90164-7

[333] O. Clay, C.J. Douady, N. Carels, S. Hughes, G. Bucciarelli, and G. Bernardi, "Using analytical ultracentrifugation to study compositional variation in vertebrate genomes." *Eur. Biophys. J.*, vol. 32, no. 5, pp. 418–426, 2003. doi: 10.1007/s00249-003-0294-y

[334] G. Bucciarelli, G. Bernardi, and G. Bernardi, "An ultracentrifugation analysis of two hundred fish genomes." *Gene*, vol. 295, no. 2, pp. 153–162, 2002. doi: 10.1016/S0378-1119(02)00733-3

[335] L.M. Hellman, D.W. Rodgers, and M.G. Fried, "Phenomenological partial-specific volumes for G-quadruplex DNAs." *Eur. Biophys. J.*, vol. 39, no. 3, pp. 389–396, 2010. doi: 10.1007/s00249-009-0411-7

[336] D.M. Hatters, L. Wilson, B.W. Atcliffe, T.D. Mulhern, N. Guzzo-Pernell, and G.J. Howlett, "Sedimentation analysis of novel DNA structures formed by homo-oligonucleotides." *Biophys. J.*, vol. 81, no. 1, pp. 371–381, 2001. doi: 10.1016/S0006-3495(01)75706-2

[337] G.F. Bonifacio, T. Brown, G.L. Conn, and A.N. Lane, "Comparison of the electrophoretic and hydrodynamic properties of DNA and RNA oligonucleotide duplexes." *Biophys. J.*, vol. 73, no. 3, pp. 1532–1538, 1997. doi: 10.1016/S0006-3495(97)78185-2

[338] B. Husain, I. Mukerji, and J.L. Cole, "Analysis of high-affinity binding of protein kinase R to double-stranded RNA." *Biochemistry*, vol. 51, no. 44, pp. 8764–8770, 2012. doi: 10.1021/bi301226h

[339] P.H. Brown, A. Balbo, and P. Schuck, "On the analysis of sedimentation velocity in the study of protein complexes." *Eur. Biophys. J.*, vol. 38, no. 8, pp. 1079–1099, 2009. doi: 10.1007/s00249-009-0514-1

[340] H. Zhao, J. Ma, M. Ingaramo, E. Andrade, J. MacDonald, G. Ramsay, G. Piszczek, G.H. Patterson, and P. Schuck, "Accounting for photophysical processes and specific signal intensity changes in fluorescence-detected sedimentation velocity." *Anal. Chem.*, vol. 86, no. 18, pp. 9286–9292, 2014. doi: 10.1021/ac502478a

[341] W. Mächtle, "High-resolution, submicron particle size distribution analysis using gravitational-sweep sedimentation." *Biophys. J.*, vol. 76, no. 2, pp. 1080–1091, 1999. doi: 10.1016/S0006-3495(99)77273-5

[342] R. Trautman, "Optical fine-structure of a meniscus in analytical ultracentrifugation in relation to moelcular-weight determinations using the Archibald principle." *Biochim. Biophys. Acta*, vol. 28, pp. 417–431, 1958. doi: 10.1016/0006-3002(58)90490-6

[343] D.A. Yphantis, "Equilibrium ultracentrifugation of dilute solutions." *Biochemistry*, vol. 3, no. 3, pp. 297–317, 1964. doi: 10.1021/bi00891a003

[344] A. Ginsburg, P. Appel, and H.K. Schachman, "Molecular-weight determinations during the approach to sedimentation equilibrium." *Arch. Biochem. Biophys.*, vol. 65, no. 2, pp. 545–566, 1956. doi: 10.1016/0003-9861(56)90213-2

[345] K.E. van Holde and R.L. Baldwin, "Rapid attainment of sedimentation equilibrium." *J. Phys. Chem.*, vol. 62, no. 6, pp. 734–743, 1958. doi: 10.1021/j150564a025

[346] D.A. Yphantis, "Rapid determination of molecular weights of peptides and proteins." *Ann. N. Y. Acad. Sci.*, vol. 88, no. 3, pp. 586–601, 1960. doi: 10.1111/j.1749-6632.1960.tb20055.x

[347] D.C. Teller, T.A. Horbett, E.G. Richards, and H.K. Schachman, "Ultracentrifuge studies with Rayleigh interference optics. III. Computational methods applied to high-speed sedimentation equilibrium experiments." *Ann. N. Y. Acad. Sci.*, vol. 164, no. 1, pp. 66–100, 1969. doi: 10.1111/j.1749-6632.1969.tb14033.x

[348] J.S. Philo, "An improved function for fitting sedimentation velocity data for low-molecular-weight solutes." *Biophys. J.*, vol. 72, no. 1, pp. 435–444, 1996. doi: 10.1016/S0006-3495(97)78684-3

[349] S.R. Erlander and G.E. Babcock, "Interface and meniscus widths and positions for low-speed ultracentrifuge runs." *Biochem. Biophys. Acta*, vol. 50, no. 2, pp. 205–212, 1961. doi: 10.1016/0006-3002(61)90318-3

[350] L. Gropper, "Optical alignment procedure for the analytical ultracentrifuge." *Anal. Biochem.*, vol. 428, no. 4, pp. 401–428, 1964. doi: 10.1016/0003-2697(64)90152-6

[351] M.F. Bailey, L.M. Angley, and M.A. Perugini, "Methods for sample labeling and meniscus determination in the fluorescence-detected analytical ultracentrifuge." *Anal. Biochem.*, vol. 390, no. 2, pp. 218–220, 2009. doi: 10.1016/j.ab.2009.03.045

[352] W. Cao and B. Demeler, "Modeling analytical ultracentrifugation experiments with an adaptive space-time finite element solution of the Lamm equation." *Biophys. J.*, vol. 89, no. 3, pp. 1589–1602, 2005. doi: 10.1529/biophysj.105.061135

[353] P. Schuck, "On computational approaches for size-and-shape distributions from sedimentation velocity analytical ultracentrifugation." *Eur. Biophys. J.*, vol. 39, no. 8, pp. 1261–1275, 2010. doi: 10.1007/s00249-009-0545-7

[354] H. Zhao and P. Schuck, "Global multi-method analysis of affinities and cooperativity in complex systems of macromolecular interactions." *Anal. Chem.*, vol. 84, no. 21, pp. 9513–9519, 2012. doi: 10.1021/ac302357w

[355] E.K. Dimitriadis and M.S. Lewis, "Optimal data analysis using transmitted light intensities in analytical ultracentrifuge." *Methods Enzymol.*, vol. 321, pp. 121–136, 2000. doi: 10.1016/S0076-6879(00)21190-7

[356] S.R. Kar, J.S. Kingsbury, M.S. Lewis, T.M. Laue, and P. Schuck, "Analysis of transport experiment using pseudo-absorbance data." *Anal. Biochem.*, vol. 285, no. 1, pp. 135–142, 2000. doi: 10.1006/abio.2000.4748

[357] M.J. Rosovitz, P. Schuck, M. Varughese, A.P. Chopra, V. Mehra, Y. Singh, L.M. McGinnis, and S.H. Leppla, "Alanine scanning mutations in anthrax toxin protective antigen domain 4 reveal residues important for binding to the cellular receptor and to a neutralizing monoclonal antibody." *J. Biol. Chem.*, vol. 278, pp. 30 936–30 944, 2003. doi: 10.1074/jbc.M301154200

[358] P. Schuck and B. Demeler, "Direct sedimentation analysis of interference optical data in analytical ultracentrifugation." *Biophys. J.*, vol. 76, no. 4, pp. 2288–2296, 1999. doi: 10.1016/S0006-3495(99)77384-4

[359] A.T. Ansevin, D.E. Roark, and D.A. Yphantis, "Improved ultracentrifuge cells for high-speed sedimentation equilibrium studies with interference optics." *Anal. Biochem.*, vol. 34, pp. 237–261, 1970. doi: 10.1016/0003-2697(70)90103-X

[360] R. Cohen, J. Cluzel, H. Cohen, P. Male, M. Moignier, and C. Soulie, "MaD, An automated precise analytical ultracentrifuge scanner system." *Biophys. Chem.*, vol. 5, no. 1-2, pp. 77–96, 1976. doi: 10.1016/0301-4622(76)80027-0

[361] P. Schuck, "Some statistical properties of differencing schemes for baseline correction of sedimentation velocity data." *Anal. Biochem.*, vol. 401, no. 2, pp. 280–287, 2010. doi: 10.1016/j.ab.2010.02.037

[362] P. Schuck and P Rossmanith, "Determination of the sedimentation coefficient distribution by least-squares boundary modeling." *Biopolymers*, vol. 54, no. 5, pp. 328–341, 2000. doi:10.1002/1097-0282(20001015)54:5<328::AID-BIP40>3.0.CO;2-P

[363] A. Ruhe and P.Å. Wedin, "Algorithms for separable nonlinear least squares problems." *SIAM Rev.*, vol. 22, no. 3, pp. 318–337, 1980. doi: 10.1137/1022057

[364] W.F. Stafford, "Protein-protein and ligand-protein interactions studied by analytical ultracentrifugation." in *Protein Structure, Stability, and Interactions*, ser. Methods in Molecular Biology, J.W. Shriver, Ed. Totowa, NJ: Humana Press, 2009, vol. 490, pp. 83–113. ISBN 978-1-58829-954-3, doi: 10.1007/978-1-59745-367-7_4

[365] T.M.D. Besong, S.E. Harding, and D.J. Winzor, "The effective time of centrifugation for the analysis of boundary spreading in sedimentation velocity experiments." *Anal. Biochem.*, vol. 421, no. 2, pp. 755–758, 2012. doi: 10.1016/j.ab.2011.11.035

[366] H. Zhao, A. Balbo, H. Metger, R. Clary, R. Ghirlando, and P. Schuck, "Improved measurement of the rotor temperature in analytical ultracentrifugation." *Anal. Biochem.*, vol. 451, pp. 69–75, 2014. doi: 10.1016/j.ab.2014.02.006

[367] R.J. Nossal and G.H. Weiss, "Sedimentation in a time-varying ultracentrifuge." *Anal. Biochem.*, vol. 38, no. 1, pp. 115–120, 1970. doi: 10.1016/0003-2697(70)90161-2

[368] W.F. Stafford, "Boundary analysis in sedimentation transport experiments: a procedure for obtaining sedimentation coefficient distributions using the time derivative of the concentration profile." *Anal. Biochem.*, vol. 203, no. 2, pp. 295–301, 1992. doi: 10.1016/0003-2697(92)90316-Y

[369] J.S. Philo, "A method for directly fitting the time derivative of sedimentation velocity data and an alternative algorithm for calculating sedimentation coefficient distribution functions." *Anal. Biochem.*, vol. 279, no. 2, pp. 151–163, 2000. doi: 10.1006/abio.2000.4480

[370] H. Zhao, R. Ghirlando, G. Piszczek, U. Curth, C.A. Brautigam, and P. Schuck, "Recorded scan times can limit the accuracy of sedimentation coefficients in analytical ultracentrifugation." *Anal. Biochem.*, vol. 437, no. 1, pp. 104–108, 2013. doi: 10.1016/j.ab.2013.02.011

[371] M. Andresen, A.C. Stiel, J. Fölling, D. Wenzel, A. Schönle, A. Egner, C. Eggeling, S.W. Hell, and S. Jakobs, "Photoswitchable fluorescent proteins enable monochromatic multilabel imaging and dual color fluorescence nanoscopy." *Nat. Biotechnol.*, vol. 26, no. 9, pp. 1035–40, 2008. doi: 10.1038/nbt.1493

[372] M.R. Dorwart, R. Wray, C.A. Brautigam, Y. Jiang, and P. Blount, "S. aureus MscL is a pentamer in vivo but of variable stoichiometries in vitro: Implications for detergent-solubilized membrane proteins." *PLoS Biol.*, vol. 8, no. 12, p. e1000555, 2010. doi: 10.1371/journal.pbio.1000555

[373] H. Eisenberg, R. Josephs, E. Reisler, and J.A. Schellman, "Scattering correction to the absorbance, wavelength dependence of the refractive index increment, and molecular weight of the bovine liver glutamate dehydrogenase oligomer and subunits." *Biopolymers*, vol. 16, no. 12, pp. 2773–2783, 1977. doi: 10.1002/bip.1977.360161214

[374] C.R. Cantor, M.M. Warshaw, and H. Shapiro, "Oligonucleotide interactions. III Circular dichroism studies of the conformation of deoxyoligonucleotides." *Biopolymers*, vol. 9, no. 9, pp. 1059–1077, 1970. doi: 10.1002/bjp.1970.360090909

[375] G.D. Fasman, Ed., *Handbook of Biochemistry and Molecular Biology, Volume 1: Nucleic Acids.* Cleveland: CRC Press, 1975.

[376] M.J. Cavaluzzi and P.N. Borer, "Revised UV extinction coefficients for nucleoside-5'-monophosphates and unpaired DNA and RNA." *Nucleic Acids Res.*, vol. 32, no. 1, p. e13, 2004. doi: 10.1093/nar/gnh015

[377] A.V. Tataurov, Y. You, and R. Owczarzy, "Predicting ultraviolet spectrum of single stranded and double stranded deoxyribonucleic acids." *Biophys. Chem.*, vol. 133, no. 1-3, pp. 66–70, 2008. doi: 10.1016/j.bpc.2007.12.004

[378] W.E. Cohn, "Methods of isolation and characterization of mono- and polynucleotides by ion exchange chromatography." *Methods Enzymol.*, vol. 3, pp. 724–743, 1957. doi: 10.1016/S0076-6879(57)03450-3

[379] G. Mie, "Beiträge zur Optik trüber Medien, speziell kolloidaler Metallösungen." *Ann. Phys.*, vol. 330, no. 3, pp. 377–445, 1908. doi: 10.1002/andp.19083300302

[380] W. Heller and W.J. Pangonis, "Theoretical investigations on the light scattering of colloidal spheres. I. The specific turbidity." *J. Chem. Phys.*, vol. 26, no. 3, pp. 498–506, 1957. doi: 10.1063/1.1743332

[381] W. Heller, "Theoretical investigations on the light scattering of colloidal spheres. II. Accurate interpolations of theoretical turbidity-data." *J. Chem. Phys.*, vol. 26, no. 4, pp. 920–922, 1957. doi: 10.1063/1.1743435

[382] A. Theisen, C. Johann, M.P. Deacon, and S.E. Harding, *Refractive increment data-book for polymer and biomolecular scientists.* Nottingham UK: Nottingham University Press, 2000.

[383] M.B. Huglin, "Specific refractive index increments." in *Light Scattering from Polymer Solutions*, M.B. Huglin, Ed. London: Academic Press, 1972, pp. 165–332.

[384] T.L. McMeekin, M. Wilensky, and M.L. Groves, "Refractive indices of proteins in relation to amino acid composition and specific volume." *Biochem. Biophys. Res. Commun.*, vol. 7, no. 2, pp. 151–156, 1962. doi: 10.1016/0006-291X(62)90165-1

[385] H. Zhao, P.H. Brown, M.T. Magone, and P. Schuck, "The molecular refractive function of lens γ-crystallins." *J. Mol. Biol.*, vol. 411, no. 3, pp. 680–699, 2011. doi: 10.1016/j.jmb.2011.06.007

[386] G.E. Perlman and L.G. Longsworth, "The specific refractive increment of some purified proteins." *J. Am. Chem. Soc.*, vol. 70, no. 8, pp. 2719–2724, 1948. doi: 10.1021/ja01188a027

[387] W. Heller, "Remarks on refractive index mixture rules." *J. Phys. Chem.*, vol. 69, no. 4, pp. 1123–1129, 1965. doi: 10.1021/j100888a006

[388] D.F. Lyons, J.W. Lary, B. Husain, J.J. Correia, and J.L. Cole, "Are fluorescence-detected sedimentation velocity data reliable?" *Anal. Biochem.*, vol. 437, no. 2, pp. 133–137, 2013. doi: 10.1016/j.ab.2013.02.019

[389] P.H. Brown, A. Balbo, and P. Schuck, "A Bayesian approach for quantifying trace amounts of antibody aggregates by sedimentation velocity analytical ultracentrifugation." *AAPS J.*, vol. 10, no. 3, pp. 481–493, 2008. doi: 10.1208/s12248-008-9058-z

[390] M. Straume and M.L. Johnson, "Analysis of residuals: Criteria for determining goodness-of-fit." *Methods Enzymol.*, vol. 210, pp. 87–105, 1992. doi: 10.1016/0076-6879(92)10007-Z

[391] J. Ma, H. Zhao, and P. Schuck, "A histogram approach to the quality of fit in sedimentation velocity analyses." *Anal. Biochem.*, vol. 483, pp. 1–3, 2015. doi: 10.1016/j.ab.2015.04.029

[392] E. Karabudak, W. Wohlleben, and H. Cölfen, "Investigation of β-carotene-gelatin composite particles with a multiwavelength UV/vis detector for the analytical ultracentrifuge." *Eur. Biophys. J.*, vol. 39, no. 3, pp. 397–403, 2010. doi: 10.1007/s00249-009-0412-6

[393] J.C.D. Houtman, H. Yamaguchi, M. Barda-Saad, A. Braiman, B. Bowden, E. Appella, P. Schuck, and L.E. Samelson, "Oligomerization of signaling complexes by the multipoint binding of GRB2 to both LAT and SOS1." *Nat. Struct. Mol. Biol.*, vol. 13, no. 9, pp. 798–805, 2006. doi: 10.1038/nsmb1133

[394] M. Barda-Saad, N. Shirasu, M.H. Pauker, N. Hassan, O. Perl, A. Balbo, H. Yamaguchi, J.C.D. Houtman, E. Appella, P. Schuck, and L.E. Samelson, "Cooperative interactions at the SLP-76 complex are critical for actin polymerization." *EMBO J.*, vol. 29, no. 14, pp. 2315–2328, 2010. doi: 10.1038/emboj.2010.133

[395] N.P. Coussens, R. Hayashi, P.H. Brown, L. Balagopalan, A. Balbo, I. Akpan, J.C.D. Houtman, V.A. Barr, P. Schuck, E. Appella, and L.E. Samelson, "Multipoint binding of the SLP-76 SH2 domain to ADAP is critical for oligomerization of SLP-76 signaling complexes in stimulated T cells." *Mol. Cell. Biol.*, vol. 33, no. 21, pp. 4140–4151, 2013. doi: 10.1128/MCB.00410-13

[396] M. Melikishvili, D.W. Rodgers, and M.G. Fried, "6-Carboxyfluorescein and structurally similar molecules inhibit DNA binding and repair by O(6)-alkylguanine DNA alkyltransferase." *DNA Repair (Amst).*, vol. 10, no. 12, pp. 1193–1202, 2011. doi: 10.1016/j.dnarep.2011.09.007

[397] G. Boestad, K.O. Pedersen, and T. Svedberg, "Design and operation of the oil-turbine ultracentrifuge." *Rev. Sci. Instrum.*, vol. 9, no. 11, pp. 346–353, 1938. doi: 10.1063/1.1752364

[398] R.T. Hersh and H.K. Schachman, "Ultracentrifuge studies with a synthetic boundary cell. III. Sedimentation of a slow component in the presence of a faster species." *J. Phys. Chem.*, vol. 62, no. 2, pp. 170–178, 1958. doi: 10.1021/j150560a008

[399] R.G. Martin and B.N. Ames, "A method for determining the sedimentation behavior of enzymes: application to protein mixtures." *J. Biol. Chem.*, vol. 236, no. 5, pp. 1372–1379, 1961.

[400] R.T. Hersh and H.K. Schachman, "Ultracentrifuge studies with a synthetic boundary cell. II. Differential sedimentation." *J. Am. Chem. Soc.*, vol. 77, no. 20, pp. 5228–5234, 1955. doi: 10.1021/ja01625a007

[401] J.P. Gabrielson, K.K. Arthur, M.R. Stoner, B.C. Winn, B.S. Kendrick, V. Razinkov, J. Svitel, Y. Jiang, P.J. Voelker, C.A. Fernandes, and R. Ridgeway, "Precision of protein aggregation measurements by sedimentation velocity analytical ultracentrifugation in biopharmaceutical applications." *Anal. Biochem.*, vol. 396, no. 2, pp. 231–241, 2009. doi: 10.1016/j.ab.2009.09.036

[402] G. Kegeles and J.R. Cann, "Kinetically controlled mass transport of associating-dissociating macromolecules." *Methods Enzymol.*, vol. 48, pp. 248–270, 1978. doi: 10.1016/S0076-6879(78)48014-0

[403] R. Josephs and W.F. Harrington, "On the stability of myosin filaments." *Biochemistry*, vol. 7, no. 8, pp. 2834–2847, 1968. doi: 10.1021/bi00848a020

[404] M.L. Johnson, D.A. Yphantis, and G.H. Weiss, "Instability in pressure-dependent sedimentation of monomer-polymer systems." *Biopolymers*, vol. 12, no. 11, pp. 2477–2490, 1973. doi: 10.1002/bip.1973.360121104

[405] R.P. Frigon and S.N. Timasheff, "Magnesium-induced self-association of calf brain tubulin. II. Thermodynamics." *Biochemistry*, vol. 14, no. 21, pp. 4567–4573, 1975. doi: 10.1021/bi00692a002

[406] H.E. Huppert, R.C. Kerr, J.R. Lister, and J.S. Turner, "Convection and particle entrainment driven by differential sedimentation." *J. Fluid Mech.*, vol. 226, pp. 349–369, 1991. doi: 10.1017/S0022112091002410

[407] J. Vinograd and R. Bruner, "Band centrifugation of macromolecules in self-generating density gradients. III. Conditions for convection-free band sedimentation." *Biopolymers*, vol. 4, no. 2, pp. 157–170, 1966. doi: 10.1002/bip.1966.360040203

[408] F. Rietz and R. Stannarius, "On the brink of jamming: Granular convection in densely filled containers." *Phys. Rev. Lett.*, vol. 100, no. 7, p. 078002, 2008. doi: 10.1103/PhysRevLett.100.078002

[409] J. Walter, T. Thajudeen, S. Süß, D. Segets, and W. Peukert, "New possibilities of accurate particle characterisation by applying direct boundary models to analytical centrifugation." *Nanoscale*, vol. 7, no. 15, pp. 6574–6587, 2015. doi: 10.1039/C5NR00995B

[410] G.H. Weiss and D.A. Yphantis, "Rectangular approximation for concentration-dependent sedimentation in the ultracentrifuge." *J. Chem. Phys.*, vol. 42, no. 6, pp. 2117–2123, 1965. doi: 10.1063/1.1696254

[411] G.A. Gilbert and R.C.L. Jenkins, "Sedimentation and electrophoresis of interacting substances. II. Asymptotic boundary shape for two substances interacting reversibly." *Proc. R. Soc. London Ser. A*, vol. 253, no. 1274, pp. 420–437, 1959. doi: 10.1098/rspa.1959.0204

[412] K.K. Arthur, J.P. Gabrielson, B.S. Kendrick, and M.R. Stoner, "Detection of protein aggregates by sedimentation velocity analytical ultracentrifugation (SV-AUC): Sources of variability and their relative importance." *J Pharm Sci*, vol. 98, no. 10, pp. 3522–3539, 2009. doi: 10.1002/jps.21654

[413] A. Balbo, H. Zhao, P.H. Brown, and P. Schuck, "Assembly, loading, and alignment of an analytical ultracentrifuge sample cell." *J. Vis. Exp.*, no. 33, p. e1530, 2009. doi: 10.3791/1530

[414] A.H. Pekar and M. Sukumar, "Quantitation of aggregates in therapeutic proteins using sedimentation velocity analytical ultracentrifugation: practical considerations that affect precision and accuracy." *Anal. Biochem.*, vol. 367, no. 2, pp. 225–237, 2007. doi: 10.1016/j.ab.2007.04.035

[415] E.G. Pickels, W.F. Harrington, and H.K. Schachman, "An ultracentrifuge cell for producing boundaries synthetically by a layering technique." *Pro. Natl. Acad. Sci. USA*, vol. 38, no. 11, pp. 943–948, 1952.

[416] R. Cohen and M. Mire, "Analytical-band centrifugation of an active enzyme-substrate complex. 2. Determination of active units of various enzymes." *Eur. J. Biochem.*, vol. 23, no. 2, pp. 276–281, 1971. doi: 10.1111/j.1432-1033.1971.tb01619.x

[417] D.J. Cox, "Sedimentation of an initially skewed boundary." *Science*, vol. 152, no. 3720, pp. 359–361, 1966. doi: 10.1126/science.152.3720.359

[418] P. Schuck, C.E. MacPhee, and G.J. Howlett, "Determination of sedimentation coefficients for small peptides." *Biophys. J.*, vol. 74, no. 1, pp. 466–474, 1998. doi: 10.1016/S0006-3495(98)77804-X

[419] D.C. Teller, "Characterization of proteins by sedimentation equilibrium in the analytical ultracentrifuge." *Methods Enzymol.*, vol. 27, pp. 346–441, 1973. doi: 10.1016/S0076-6879(73)27017-9

[420] L. Börger, H. Cölfen, and M. Antonietti, "Synthetic boundary crystallization ultracentrifugation: a new method for the observation of nucleation and growth of inorganic colloids and the determination of stabilizer efficiencies." *Colloids Surfaces A Physicochem. Eng. Asp.*, vol. 163, no. 1, pp. 29–38, 2000. doi: 10.1016/S0927-7757(99)00427-6

[421] C. Wandrey, G. Grigorescu, and D. Hunkeler, "Study of polyelectrolyte complex formation applying the synthetic boundary technique of analytical ultracentrifugation." *Progr. Colloid Polym. Sci.*, vol. 119, pp. 84–91, 2002. doi: 10.1007/3-540-44672-9_13

[422] C. Wandrey, U. Hasegawa, A.J. van der Vlies, C. O'Neil, N. Angelova, and J.A. Hubbell, "Analytical ultracentrifugation to support the development of biomaterials and biomedical devices." *Methods*, vol. 54, no. 1, pp. 92–100, 2011. doi: 10.1016/j.ymeth.2010.12.003

[423] D.L. Kemper and J. Everse, "Active enzyme centrifugation." *Methods Enzymol.*, vol. 27, pp. 67–82, 1973. doi: 10.1016/S0076-6879(73)27005-2

[424] J. Vinograd and R. Bruner, "Band centrifugation of macromolecules in self-generating density gradients. II. Sedimentation and diffusion of macromolecules in bands." *Biopolymers*, vol. 4, no. 2, pp. 131–156, 1966. doi: 10.1002/bip.1966.360040202

[425] S.R. Kar, J. Lebowitz, S. Blume, K.B. Taylor, and L.M. Hall, "SmtB-DNA and protein-protein interactions in the formation of the cyanobacterial metallothionein repression romplex: Zn2+ does not dissociate the protein-DNA complex in vitro." *Biochemistry*, vol. 40, no. 44, pp. 13 378–13 389, 2001. doi: 10.1021/bi011289f

[426] R. Cohen and M. Mire, "Analytical-band centrifugation of an active enzyme substrate complex 1. Principle and practice of the centrifugation." *Eur. J. Biochem*, vol. 23, no. 2, pp. 267–275, 1971. doi: 10.111/j.1432-1033.1971.tb01618.x

[427] W.T. Wolodko, C.M. Kay, and W.A. Bridger, "Active enzyme sedimentation, sedimentation velocity, and sedimentation equilibrium studies of succinyl-CoA synthetases of porcine heart and Escherichia coli." *Biochemistry*, vol. 25, no. 19, pp. 5420–5425, 1986. doi: 10.1021/bi00367a012

[428] R. Cohen and J.M. Claverie, "Sedimentation of generalized systems of interacting particles. II. Active enzyme centrifugation–theory and extensions of its validity range." *Biopolymers*, vol. 14, no. 8, pp. 1701–1716, 1975. doi: 10.1002/bip.1975.360140812

[429] *Instructions For Use An-50 Ti and An-60 Ti Analytical Rotor, Cells, and Counterbalance For Use in Beckman Coulter Analytical XL-A and XL-I Instruments.* Brea, CA: Beckman Coulter, Inc., 2011. Online at: https://www.beckmancoulter.com/wsrportal/techdocs?docname=LXLA-TB-003

[430] *Chemical Resistances for Beckman Coulter Centrifugation Products.* Palo Alto, CA: Beckman Coulter, Inc., 2001. Online at: https://www.beckmancoulter.com/wsrportal/bibliography?docname=chemres.pdf

[431] H.K. Schachman, "Those wonderful early years with the Model E ultracentrifuge and David Yphantis." *Biophys. Chem.*, vol. 108, no. 1-3, pp. 9–16, 2004. doi: 10.1016/j.bpc.2003.10.005

[432] R.C. Williams, "A laser light source for the analytical ultracentrifuge." *Anal. Biochem.*, vol. 48, no. 1, pp. 164–171, 1972. doi: 10.1016/0003-2697(72)90180-7

[433] G.J. Howlett and H.K. Schachman, "Allosteric regulation of aspartate transcarbamoylase. Changes in the sedimentation coefficient promoted by the bisubstrate analogue N-(phosphonacetyl)-L-aspartate." *Biochemistry*, vol. 16, no. 23, pp. 5077–5083, 1977. doi: 10.1021/bi00642a021

[434] M.S. Springer and H.K. Schachman, "A difference sedimentation equilibrium technique for measuring small changes in molecular weight. II. Experimental." *Biochemistry*, vol. 13, no. 18, pp. 3726–3733, 1974. doi: 10.1021/bi00715a017

[435] T.A. Horbett and D.C. Teller, "An experimental study of baseline reproducibility and its effect on high-speed sedimentation equilibrium data." *Anal. Biochem.*, vol. 45, no. 1, pp. 86–99, 1972. doi: 10.1016/0003-2697(72)90009-7

[436] G. Kegeles and F.J. Gutter, "The determination of sedimentation constants from Fresnel diffraction patterns." *J. Am. Chem. Soc.*, vol. 73, no. 8, pp. 3770–3777, 1951. doi: 10.1021/ja01152a061

[437] P.A. Baghurst and P.E. Stanley, "Determination of stretch of a titanium analytical ultracentrifuge rotor subjected to various centrifugal fields." *Anal. Biochem.*, vol. 33, no. 1, pp. 168–173, 1970. doi: 10.1016/0003-2697(70)90450-1

[438] D.F. Waugh and D.A. Yphantis, "Rotor temperature measurement and control in the ultracentrifuge." *Rev. Sci. Instrum.*, vol. 23, no. 11, pp. 609–614, 1952. doi: 10.1063/1.1746108

[439] A. Biancheria and G. Kegeles, "Thermodynamic measurements of ultracentrifuge rotor temperature." *J. Am. Chem. Soc.*, vol. 76, no. 14, pp. 3737–3741, 1954. doi: 10.1021/ja01643a047

[440] L. Gropper and W. Boyd, "Temperature measurement and control of analytical rotors in the ultracentrifuge." *Anal. Biochem.*, vol. 11, no. 2, pp. 238–245, 1965. doi: 10.1016/0003-2697(65)90011-4

[441] W.B. Stine, "Analysis of monoclonal antibodies by sedimentation velocity analytical ultracentrifugation." *Methods Mol. Biol.*, vol. 988, pp. 227–240, 2013. doi: 10.1007/978-1-62703-327-5_15

[442] S. Ang and A.J. Rowe, "Evaluation of the information content of sedimentation equilibrium data in self-interacting systems." *Macromol. Biosci.*, vol. 10, no. 7, pp. 798–807, 2010. doi: 10.1002/mabi.201000065

[443] G.J. Howlett, M.N. Blackburn, J.G. Compton, and H.K. Schachman, "Allosteric regulation of aspartate transcarbamoylase. Analysis of the structural and functional behavior in terms of a two-state model." *Biochemistry*, vol. 16, no. 23, pp. 5091–5099, 1977. doi: 10.1021/bi00642a023

[444] R. Ghirlando, H. Zhao, A. Balbo, G. Piszczek, U. Curth, C.A. Brautigam, and P. Schuck, "Measurement of the temperature of the resting rotor in analytical ultracentrifugation." *Anal. Biochem.*, vol. 458, pp. 37–39, 2014. doi: 10.1016/j.ab.2014.04.029

[445] R. Cecil and A.G. Ogston, "The accuracy of the Svedberg oil-turbine ultracentrifuge." *Biochem. J.*, vol. 43, no. 4, pp. 592–598, 1948. doi: 10.1042/bj0430592

[446] T.K. Robinson and J.W. Beams, "Radio telemetering from magnetically suspended rotors." *Rev. Sci. Instrum.*, vol. 34, no. 1, pp. 63–64, 1963. doi: 10.1063/1.1718126

[447] J.H. Bauer and E.G. Pickels, "An improved air-driven type of ultracentrifuge for molecular sedimentation." *J. Exp. Med.*, vol. 65, no. 4, pp. 565–586, 1937. doi: 10.1084/jem.65.4.565

[448] S. Liu and W.F. Stafford, "An optical thermometer for direct measurement of cell temperature in the Beckman instruments XL-A analytical ultracentrifuge." *Anal. Biochem.*, vol. 224, no. 1, pp. 199–202, 1995. doi: 10.1006/abio.1995.1030

[449] J.B. Johnson, K. Becker, and G. Edwards, "Pressure corrections for CoCl2 as a thermometer in an analytic ultracentrifuge." *Anal. Biochem.*, vol. 227, no. 2, pp. 385–387, 1995. doi: 10.1006/abio.1995.1295

[450] L.A. Holladay, "Simultaneous rapid estimation of sedimentation coefficient and molecular weight." *Biophys. Chem.*, vol. 11, no. 2, pp. 303–308, 1980. doi: 10.1016/0301-4622(80)80033-0

[451] P.F. Mijnlieff, P. van Es, and W.J.M. Jaspers, "Temperature gradients in ultracentrifuge cells due to adiabatic volume changes." *Recueil*, vol. 88, no. 2, pp. 220–224, 1969. doi: 10.1002/recl.19690880213

[452] W. Mächtle and L Börger, *Analytical Ultracentrifugation of Polymers and Nanoparticles.* Berlin: Springer, 2006.

[453] B. Lelj-Garolla and A.G. Mauk, "Self-association of a small heat shock protein." *J. Mol. Biol.*, vol. 345, no. 3, pp. 631–642, 2005. doi: 10.1016/j.jmb.2004.10.056

[454] B. Lelj-Garolla and A.G. Mauk, "Self-association and chaperone activity of Hsp27 are thermally activated." *J. Biol. Chem.*, vol. 281, no. 12, pp. 8169–8174, 2006. doi: 10.1074/jbc.M512553200

[455] S.-J. Kim, T. Tsukiyama, M.S. Lewis, and C. Wu, "Interaction of the DNA-binding domain of Drosophila heat shock factor with its cognate DNA site: A thermodynamic analysis using analytical ultracentrifugation." *Protein Sci.*, vol. 3, no. 7, pp. 1040–1051, 1994. doi: 10.1002/pro.5560030706

[456] J.A. Kornblatt and P. Schuck, "Influence of temperature on the conformation of canine plasminogen: An analytical ultracentrifugation and dynamic light scattering study." *Biochemistry*, vol. 44, no. 39, pp. 13 122–13 131, 2005. doi: 10.1021/bi050895y

[457] Z.F. Taraporewala, P. Schuck, R.F. Ramig, L. Silvestri, and J.T. Patton, "Analysis of a temperature-sensitive mutant rotavirus indicates that NSP2 octamers are the functional form of the protein." *J. Virol.*, vol. 76, no. 14, pp. 7082–7093, 2002. doi: 10.1128/JVI.76.14.7082-7093.2002

[458] N.A. Sieracki, H.J. Hwang, M.K. Lee, D.K. Garner, and Y. Lu, "A temperature independent pH (TIP) buffer for biomedical biophysical applications at low temperatures." *Chem. Commun. (Camb).*, no. 7, pp. 823–825, 2008. doi: 10.1039/b714446f

[459] *ProteomeLab XL-A/XL-I Instructions For Use.* Beckman Coulter, Inc., 2014.

[460] P.E. Hexner, L.E. Radford, and J.W. Beams, "Achievement of sedimentation equlibrium." *Proc. Natl. Acad. Sci. U.S.A.*, vol. 47, no. 11, pp. 1848–1852, 1961.

[461] I.H. Billick, M. Dishon, M. Schulz, G.H. Weiss, and D.A. Yphantis, "The effects of rotor deceleration on equilibrium sedimentation experiments." *Proc. Nat. Acad. Sci. USA*, vol. 56, no. 2, pp. 399–404, 1966. doi: 10.1073/pnas.56.2.399

[462] J.F. Taylor, "The determination of sedimentation constant with the Spinco ultracentrifuge." *Arch. Biochem. Biophys.*, vol. 36, no. 2, pp. 357–364, 1952. doi: 10.1016/0003-9861(52)90421-9

[463] G.L. Miller and R.H. Golder, "Sedimentation studies with the Spinco ultracentrifuge." *Arch. Biochem. Biophys.*, vol. 36, no. 2, pp. 249–258, 1952. doi: 10.1016/0003-9861(52)90409-8

[464] S. Shulman, "The determination of sedimentation constants with the oil-turbine and spinco ultracentrifuges." *Arch. Biochem. Biophys.*, vol. 44, no. 1, pp. 230–240, 1953. doi: 10.1016/0003-9861(53)90028-9

[465] J.C. Travis, M.V. Smith, S.D. Rasberry, and G.W. Kramer, "NIST Special Publication 260-140. Technical Specifications for Certification of Spectrophotometric NTRMs." Tech. Rep., 2000. Online at: http://www.nist.gov/srm/upload/SP260-140.PDF

[466] T. Arakawa and S.N. Timasheff, "Preferential interactions of proteins with salts in concentrated solutions." *Biochemistry*, vol. 21, no. 25, pp. 6545–6552, 1982. doi: 10.1021/bi00268a034

[467] R.S. Woodward and J. Lebowitz, "A revised equation relating DNA buoyant density to guanine plus cytosine content." *J. Biochem. Biophys. Methods*, vol. 2, no. 5, pp. 307 309, 1980. doi: 10.1016/0165-022X(80)90055-X

[468] H. Eisenberg, "Solution properties of DNA: sedimentation, scattering of light, X-rays and neutrons, and viscometry." in *Spectroscopic and Kinetic Data. Physical Data I*, W. Saenger, Ed. Berlin: Springer, 1990, pp. 257–276.

[469] N.R. Voss and M. Gerstein, "Calculation of standard atomic volumes for RNA and comparison with proteins: RNA is packed more tightly." *J. Mol. Biol.*, vol. 346, no. 2, pp. 477–492, 2005. doi: 10.1016/j.jmb.2004.11.072

[470] S.J. Perkins, A. Miller, T.E. Hardingham, and H. Muir, "Physical properties of the hyaluronate binding region of proteoglycan from pig laryngeal cartilage. Densitometric and small-angle neutron scattering studies of carbohydrates and carbohydrate-protein macromolecules." *J. Mol. Biol.*, vol. 150, no. 1, pp. 69–95, 1981. doi: 10.1016/0022-2836(81)90325-9

[471] J.C. Steele, C. Tanford, and J.A. Reynolds, "Determination of partial specific volumes for lipid-associated proteins." *Methods Enzymol.*, vol. 48, pp. 11–23, 1978. doi: 10.1016/S0076-6879(78)48004-8

[472] A. Helenius, D.R. McCaslin, E. Fries, and C. Tanford, "Properties of detergents." *Methods Enzymol.*, vol. 56, pp. 734–749, 1979. doi: 10.1016/0076-6879(79)56066-2

[473] J.A. Reynolds and D.R. McCaslin, "Determination of protein molecular weight in complexes with detergent without knowledge of binding." *Methods Enzymol.*, vol. 117, pp. 41–53, 1985. doi: 10.1016/S0076-6879(85)17005-9

[474] J.V. Møller and M. le Maire, "Detergent binding as a measure of hydrophobic surface area of integral membrane proteins." *J. Biol. Chem.*, vol. 268, no. 25, pp. 18 659–18 572, 1993.

[475] H. Durchschlag and P. Zipper, "Calculation of partial specific volumes of detergents and lipids." *Jorn. Com. Esp. Deterg.*, vol. 26, pp. 275–292, 1995.

[476] K.-J. Tiefenbach, H. Durchschlag, and R. Jaenicke, "Spectroscopic and hydrodynamic investigations of nonionic and zwitterionic detergents." *Prog. Colloid Polym. Sci.*, vol. 113, pp. 135–141, 1999. doi: 10.1007/3-540-48703-4_19

[477] H. Durchschlag, K.-J. Tiefenbach, R. Weber, B. Kuchenmüller, and R. Jaenicke, "Comparative investigations of the molecular properties of detergents and protein-detergent complexes." *Colloid Polym. Sci.*, vol. 278, no. 4, pp. 312–320, 2000. doi: 10.1007/s003960050519

[478] M. le Maire, P. Champeil, and J.V. Møller, "Interaction of membrane proteins and lipids with solubilizing detergents." *Biochim. Biophys. Acta*, vol. 1508, no. 1-2, pp. 86–111, 2000. doi: 10.1016/S0304-4157(00)00010-1

[479] L.E. Fisher, D.M. Engelman, and J.N. Sturgis, "Effect of detergents on the association of the glycophorin a transmembrane helix." *Biophys. J.*, vol. 85, no. 5, pp. 3097–3105, 2003. doi: 10.1016/S0006-3495(03)74728-6

[480] A.G. Salvay and C. Ebel, "Analytical ultracentrifuge for the characterization of detergent in solution." *Prog. Colloid Polym. Sci.*, vol. 131, pp. 74–82, 2006. doi: 10.1007/2882_006

[481] G.G. Kochendoerfer, D. Salom, J.D. Lear, R. Wilk-Orescan, S.B.H. Kent, and W.F. DeGrado, "Total chemical synthesis of the integral membrane protein influenza A virus M2: Role of its C-terminal domain in tetramer assembly." *Biochemistry*, vol. 38, no. 37, pp. 11 905–11 913, 1999. doi: 10.1021/bi990720m

[482] L.M. Hjelmeland, D.W. Nebert, and J.C. Osborne, "Sulfobetaine derivatives of bile acids: nondenaturing surfactants for membrane biochemistry." *Anal. Biochem.*, vol. 130, no. 1, pp. 72–82, 1983. doi: 10.1016/0003-2697(83)90651-6

[483] E.C. Smith, S.E. Smith, J.R. Carter, S.R. Webb, K.M. Gibson, L.M. Hellman, M.G. Fried, and R.E. Dutch, "Trimeric transmembrane domain interactions in paramyxovirus fusion proteins: Roles in protein folding, stability, and function." *J. Biol. Chem.*, vol. 288, no. 50, pp. 35 726–35 735, 2013. doi: 10.1074/jbc.M113.514554

[484] C. Breyton, F. Gabel, M. Abla, Y. Pierre, F. Lebaupain, G. Durand, J.L. Popot, C. Ebel, and B. Pucci, "Micellar and biochemical properties of (hemi)fluorinated surfactants are controlled by the size of the polar head." *Biophys. J.*, vol. 97, no. 4, pp. 1077–1086, 2009. doi: 10.1016/j.bpj.2009.05.053

[485] T. Hianik, M. Haburcák, K. Lohner, E. Prenner, F. Paltauf, and A. Hermetter, "Compressibility and density of lipid bilayers composed of polyunsaturated phospholipids and cholesterol." *Colloids Surfaces A Physicochem. Eng. Asp.*, vol. 139, no. 2, pp. 189–197, 1998. doi: 10.1016/S0927-7757(98)00280-5

[486] F. Steckel and S. Szapiro, "Physical properties of heavy oxygen water. Part 1.Density and thermal expansion." *Trans. Faraday Soc.*, vol. 59, pp. 331–343, 1963. doi: 10.1039/tf9635900331

[487] J. Kestin, M. Sokolov, and W.A. Wakeham, "Viscosity of liquid water in the range –8°C to 150°C." *J. Phys. Chem. Ref. Data*, vol. 7, no. 3, pp. 941–948, 1978. doi: 10.1063/1.555581

[488] NIST, "Thermophysical Properties of Fluid Systems." Online at: http://webbook.nist.gov/chemistry/fluid/

[489] R.A. Fine and F.J. Millero, "Compressibility of water as a function of temperature and pressure." *J. Chem. Phys.*, vol. 59, no. 10, pp. 5529–5536, 1973. doi: 10.1063/1.1679903

[490] G. Jones and H.J. Fornwalt, "The viscosity of deuterium oxide and its mixtures with water at 25 °C." *J. Chem. Phys.*, vol. 4, pp. 30–33, 1936. doi: 10.1063/1.1749743

[491] A.I. Kudish, D. Wolf, and F. Steckel, "Physical properties of heavy-oxygen water. Absolute viscosity of $H_2^{18}O$ between 15 and 35°C." *J. Chem. Soc. Faraday Trans. 1*, vol. 68, pp. 2041–2046, 1972. doi: 10.1039/f19726802041

[492] D. Wolf and A.I. Kudish, "Absolute viscosity of $D_2^{18}O$ between 15 and 35°." *J. Phys. Chem.*, vol. 79, no. 14, pp. 1481–1482, 1975. doi: 10.1021/j100581a028

[493] A.I. Kudish and D. Wolf, "Absolute viscosity of $D_2^{18}O$ between 15 and 35°." *J. Phys. Chem.*, vol. 79, no. 3, pp. 272–275, 1975. doi: 10.1021/j100570a016

[494] NCBI, "PubChem." Online at: https://pubchem.ncbi.nlm.nih.gov

[495] R.C. Weast, *Handbook of Chemistry and Physics*. Cleveland, OH: CRC Press, 1972.

[496] Dow Chemical Company, "Material Data Safety Sheet." Online at: http://www.dow.com/hyperlast/msds.htm

[497] R.G. LeBel and D.A.I. Goring, "Density, viscosity, refractive index, and hygroscopicity of mixtures of water and dimethyl sulfoxide." *J. Chem. Eng. Data*, vol. 7, no. 1, pp. 100–101, 1962. doi: 10.1021/je60012a032

[498] P.W. Bridgman, "The viscosity of liquids under pressure." *Proc. Natl. Acad. Sci. U.S.A.*, vol. 11, no. 10, pp. 603–606, 1925.

[499] J.W.P. Schmelzer, E.D. Zanotto, and V.M. Fokin, "Pressure dependence of viscosity." *J. Chem. Phys.*, vol. 122, no. 7, p. 074511, 2005. doi: 10.1063/1.1851510

[500] D.W. Brazier and G.R. Freeman, "The effects of pressure on the density, dielectric constant, and viscosity of several hydrocarbons and other organic liquids." *Can. J. Chem.*, vol. 47, no. 6, pp. 893–899, 1969.

[501] K. Kawahara and C. Tanford, "Viscosity and density of aqueous solutions of urea and guanidine hydrochloride." *J. Biol. Chem.*, vol. 241, pp. 3228–3232, 1966.

[502] T. Hurton, A. Wright, G. Deubler, and B. Bashir, "SEDNTERP." Online at: http://sednterp.unh.edu/

[503] G.N. Lewis and D.B. Luten, "The refractive index of H_2O^{18} and the complete isotopic analysis of water." *J. Am. Chem. Soc.*, vol. 55, no. 12, pp. 5061–5062, 1933. doi: 10.1021/ja01339a513

[504] P. Strop and A.T. Brunger, "Refractive index-based determination of detergent concentration and its application to the study of membrane proteins." *Protein Sci.*, vol. 14, no. 8, pp. 2207–2211, 2005. doi: 10.1110/ps.051543805

[505] H. Nury, F. Manon, B. Arnou, M. le Maire, E. Pebay-Peyroula, and C. Ebel, "Mitochondrial bovine ADP/ATP carrier in detergent is predominantly monomeric but also forms multimeric species." *Biochemistry*, vol. 47, no. 47, pp. 12319–12331, 2008. doi: 10.1021/bi801053m

[506] T. Tumolo, L. Angnes, and M.S. Baptista, "Determination of the refractive index increment (dn/dc) of molecule and macromolecule solutions by surface plasmon resonance." *Anal. Biochem.*, vol. 333, no. 2, pp. 273–279, 2004. doi: 10.1016/j.ab.2004.06.010

[507] T.M. Davis and W.D. Wilson, "Determination of the refractive index increments of small molecules for correction of surface plasmon resonance data." *Anal. Biochem.*, vol. 284, no. 2, pp. 348–353, 2000. doi: 10.1006/abio.2000.4726

[508] B.S. Kendrick, B.A. Kerwin, B.A. Chang, and J.S. Philo, "Online size-exclusion high-performance liquid chromatography light scattering and differential refractometry methods to determine degree of polymer conjugation to proteins and protein-protein or protein-ligand association states." *Anal. Biochem.*, vol. 299, no. 2, pp. 136–146, 2001. doi: 10.1006/abio.2001.5411

[509] Y. Hayashi, H. Matsui, and T. Takagi, "Membrane protein molecular weight determined by low-angle laser light-scattering photometry coupled with high-performance gel chromatography." *Methods Enzymol.*, vol. 172, pp. 514–528, 1989. doi: 10.1016/S0076-6879(89)72031-0

[510] T. Stora, Z. Dienes, H. Vogel, and C. Duschl, "Histidine-tagged amphiphiles for the reversible formation of lipid bilayer aggregates on chelator-functionalized gold surfaces." *Langmuir*, vol. 16, no. 12, pp. 5471–5478, 2000. doi: 10.1021/la991711h

Index

absorbance optics, **3**, **17**, 84, 186, 189, 190, 218, 224, *see also* monochromator

  buffer, 72, 73, 159

  calibration, 100–102, 104, 193, 232–234

  concentrations, 63–65, **98**

  data characteristics, 19, 60, 64, 69, 107, **107**, 112, 115, 117, 163, 175, 194, 200, 204, 221, 225, 230, 237

  design, 17, **18**, 65, 119, 146, 152, 170, 182, 191

  extinction coefficient, 67, 147, **147**

  intensity, *see* intensity data

  multi-signal, 42, 94, 146, 169, **176**, 178–180, 192

  multi-wavelength, 42, 72, 169, **171**, 173, **175**, **176**, 234

  wavelength, *see* wavelength

active enzyme centrifugation, 188, 190

activity coefficient, 47, 49

adiabatic rotor stretching, 127, 195, 196, 210, 221

aggregates, 69, 70, 126, 190, 225, 237

  density increment, 28

  re-suspension, 70, 88, 205

  scattering, 67, 149

  trace, 61, 62, 84, 168, 184, 201

Archibald, William J., 69, 106

Archimedes principle, 2, 25

asymptotic boundary, *see* Gilbert–Jenkins theory

back-diffusion, 11, 12, 14, 106, 112, 115, 226

band centrifugation, 68, 70, 182, 184, 187–189

Beams, Jesse Wakefield, 17, 212

Beer–Lambert law, 17, 147

binding constant, 16, 22, 40, 44, 46, 55, 62, 68, 70, 75, 147, 149, 152, 154, 159, 177, 211, 227, 231

blocking protein, 22, 60, 62, 73, 74, 146, 152

Boltzmann distribution, 15, 16, 40, 41, 47, 48, 82, 83, 123, 183, 184, 194, 215, 216, 222, 224

bovine serum albumin

  blocking agent, 73, 146

  calibration, 70, 75, 228, 237, 238

  example data, 4, 14, 84, 95, 105, 118, 128, 136, 138, 157, 158, 172, 189, 205

  sedimentation force, 56

Brownian diffusion, 55, 56

buffer mismatch, 63, 107, 108, 157–159, 221

buoyant molar mass, 42, 216, 217

  contributions, 25, 31, 34, 46, 75, 87, 95, 247

  measurement, 25, 26, 35, 42, 47, 69, 75–80, 82, 89, 91, 94, 96, 98, 102, 127, 214, 215, 218, 247

  nonideality, 49, 51, 54, *see also* nonideality, thermodynamic

  theory, 16, 25, **27**, 29, 41, 42, 47, 78, 90, 235, 236

c(s) distribution, 13, 14, 52, 60, 62, 63, 78, 83, 90, 94, 95, 98, 128, 136, 149, 151, 153, 154, 160–162, 165–167, 169, 170, 178, 189, 226, 237

carbohydrate, 94, 145, 243

cell assembly, 3, 71, 103, 123, 129, 191,

199, 200, 202–205, 208, 226, 237, 238
ageing, 122, 123, 202
external loading, 199, 203
torquing, 197–201, 203, 205
centerpiece, 3, 68, 106, 182, **182**, 183, 184, 196–198, 200–202, 205, *see also* sector shape
3-mm, 17, 165, 191
6-channel, 184, 185, 191, 199
alignment, 183, 184, 198, 199, 233
fluorescence, 111
material, 71, 190, 197
synthetic boundary, 55, 184–189
charge effect, **34–36**, 57, 72
conformation linked, 36
ligand linked, 36
primary, **34**
secondary, **35**
co-solute
density and viscosity, 32, 33, 57, 58, 74, 79, 80, 87, 181, 250
linkage, 46
preferential solvation, 27–29, 32–34, 46, 80, 89, 95, 96, 239, 251
sedimentation, **57**, 58, 87, 165, 181, 187
signal, 21, 63, 72, 145, 158
complete sedimenting particle
formalism, 24, 26, **29–31**, 32, 33, 77, 81, 86, 89, 96, *see also* partial-specific volume of the sedimenting particle
definition, **30**
complex lifetime, **37**, 38–40, **40**, 42, 43, 68, 223
compressibility, 53–54, 249, *see also* pressure
adiabatic heating, 210
density gradient, 87, 181
macromolecular, 87, 239
organic solvents, 71, 249
water, 71, 246

convection, 112, 131, 181, **181**, 182–184, 206, 208, 211
apparent meniscus, 211
counterbalance, 100–102, 104, 146, 147, 190, 192, 193, 209, 237, 238
counterion, 35, *see also* polyelectrolyte
CsCl, 57, 58, 96, 150, 157, 165, 241, 242
cutoff filter, 171–173

$D_2^{18}O$, 71, 80, 188, 247, 248
D_2O, 32, 54, 71, 80, 81, 85, 86, 90, 188, 247, 248
densimetry, 28, 33, 74, 76, 81, 84, **87**, 88–91, 243
density contrast, 25, 26, 31, 33, 46, 71, 72, 76, 78–87, 90, 96, 207, 236, 243, 244, 247
density gradient, 53, 54, 87, 181, 188
density increment, 27–29, 34, 76, 87, 88, 90, 91, 150, *see also* partial-specific volume
density matching, 30, 31, 71, 72, 78, 80, 86, 87, 244
detergent, 31, 32, 45, 46, 62, 78, 80, 81, 86–88, 95, 145, 146, 198, 240, 243, 244, 247
dialysis, 25, 28, 34, 68, 72, 87–91, 157
diffusion, 7, 13, 50, 62
apparent, 37, 45, 51, 61, 127, 129, 135, 154, 156, 214, 215
rotational, 55
translational, 7–15, 36, 39, 40, 42, 50, 51, 55, 57, 68, 70, 78, 83, 85, 127, 128, 132, 156, 158, 183, 185–187, 189, 214, 215, 218, 220, 223, 230
DNA, 34, 36, 56, 57, 97, 148, 150, 241, 242
Donnan equilibrium, 34
DTT, 73, 139, 141, 145, 159
dynamic light scattering, 8, 61, 78

Edelstein–Schachman method, 78, 80, 82, 83

effective particle theory, 44, **44**, 45, 67, 151

electroneutral, 34, 236

extinction coefficient, *see also* absorbance optics, extinction coefficient, **17**, 19, 22, 62, 67, 72, 89–91, 93, 94, 98, 99, 146–152, 173, 174, 177, 178, 235

Faxén approximation, 130, 155

Fick's law, 15

flotation, 2, **4**, 7, 11, 54, 58, 87, 151

flow field, 24, 51, 57

fluorescence detection, 60, 189
    buffer, 72, 145, 152
    concentrations, **22**, 62–64, 145, 151, 182, 206
    data characteristics, 22, 99, 103, 109–111, 113, 115, 122, 126, 128, 133, 134, 136, 139, 141, 142, 144, 151–154, 163–166, 177, 188, 200, 229, 233
    design, 17, 22, **22**, 64, 114, 144, 147, 169, 192, 193, 198, 230
    excitation and emission, 22, 169
    inner filter effect, 64, 98, 126, 151, 152, **153**
    photobleaching, 141–144
    photophysical processes, 22, 139, 141, 142, **142**, 143

fluorophore, 22, 60, 64, 98, 109, 139, 142, 145, 151, 152, 206

focal plane, 64, 106, 110, 111, 113, 126, 128, 151–154, 191, 192

frictional coefficient, **4–6**, 31, 32, **32**, 34, 46, 50, 56, 66, 74, 76, 78, 83, 130, 223

frictional force, 4, 5, 55, 56

frictional ratio, *see* frictional coefficient

Gilbert–Jenkins theory, 45, 183

glycerol, 27, 71, 74, 81, 86, 95, 96, 145, 157, 158, 240, 250

glycoprotein, 90, 94, 167, 243

H-D exchange, 80, 86, 90

$H_2^{18}O$, 71, 80, 82, 85, 86, 188, 247, 248, 251

HEPES, 73, 145

hydrodynamic interactions, 49–51, 53, 56, 57, 60, 62

hydrodynamic resolution, 13, 83, 131, 169, 178, 214, 215, *see also* mass resolution

hypo- and hyper-chromicity, 148, 149, 151, 176, 177

iButton, 95, 195, 196, **207**, 208–211, 230–232, **232**, 237, 238

intensity data, 17, 103, 104, 107, 108, 115, **119**, 120, 145, 163

interconversion kinetics, *see* complex lifetime

interference optics, 13, 21, 60, 145, 146, 201
    acquisition, 21, 131, 191–193, 200, 201, 203, 218, 220, 226
    buffer, 34, 73, 145
    buffer match, *see also* buffer mismatch, 21, 25, 58, 63, 69, 72, 107, 156–158, 202
    calibration, 100–104, 136, 232, 234
    concentrations, 63, 90, 98, 147
    data characteristics, 21, 64, 107–109, 114–118, 120–122, 125, 131, 134, 136, 138, 139, 145, 152, 156, 157, 159–162, 164, 165, 168, 188, 189, 202
    design, **19**, 20, 64, 118, 126, 138, 169, 182, 191–193, 200, 202–204
    fringe increment, 19, 60, 147, 148, **149**, 151, 251
    multi-signal, *see also* absorbance, multi-signal, 94, 133, 145, 146, 169, 177, 180, 192

invariant particle, 27, **29**, 30, 33, 81, 95, 240

Johnston–Ogston effect, 50, 52

Kirkwood–Buff integral, 29
Kratky oscillator, 74, **88**, *see also*
    densimetry

Lamm equation, 9, **11**, 13, 41, 60, 127,
    129, 130, 134, 135, 143, 151,
    155, 158, 165, 185–188, 190,
    211, 220, 222, 223, 225
leaks, 191, 196, 198, 199, **200–202**, 212,
    220, 221

macroion, *see also* polyelectrolyte,
    34–36
mass action law, 16, 40–43, 45, 151
mass conservation, 106, 118, 142, 179,
    185, 195, 200, 219
mass resolution, 41, 70, 178, *see also*
    hydrodynamic resolution
meniscus, 3, 7, 9, 19, 21, 68, 69, 100,
    102, 105–113, 127, 128, 131,
    137, 157–159, 187, 188, 194,
    195, 202, 205, 211, 221, 225,
    232, 234, 238, 253
meniscus depletion, 218
micro-heterogeneity, 61
Model E, 2, 182, 184, 192, 208, 228
monochromator, 169, 171, 172, 174,
    175, 178, 209, 232
monodispersity, 9, 45, 50, 61
multi-speed sedimentation equilibrium,
    16, 70, 118, 122, 124, 125, 192,
    194, 195, 203, 204, 225

neutral buoyancy, 2, 26, 30, 31, 78, 80,
    86–92, 244, *see also* density
    matching
Newton rings, 197
noise, **159–169**, *see also* residuals
    radial-invariant (RI), **137–141**,
    141, 146, 159–162
    rotor-speed-invariant, 194, 204
    signal-to-noise, 14, 17, 20, 22, 62,
    63, 65, 81, 83, 84, 98, 99, 112,
    132, 133, 135, 144, 145, 148,
    170, 178, 221

statistical, 17, 21, 62, 63, 102, 115,
    145, 159, 160, **162–169**, 170,
    187, 215, 222, 224
time-invariant (TI), 114, **117–126**,
    128, 146, 159–162, 188, 192, 194
    in SE, 192, 194, 195, 202–204
nonideality, 62
    excluded volume, 62, 66
    hydrodynamic, 46, **49**, 51, 65, 74,
    112, 113, *see also*
    hydrodynamic interactions and
    Johnston–Ogston effect
    onset, 46, 50, 65, 66
    thermodynamic, 28, 46, 47, **48**, 49,
    66, 113, *see also* buoyant molar
    mass, nonideality
nonlinear detection, 63, 152–156
nucleotide, 38, 72, 73, 96, 145, 241

Optima XL-A/I, 3, 18, 20, 127, 162,
    174, 175, 182, 191, 207, 208,
    212
osmotic pressure, 27, 28
overspeed disk, 193
overspeeding, 69, 206, **222–225**

partial boundary modeling, 124
partial-specific volume, 4–6, 16, **24–27**,
    34, 71, **75–97**, **239–244**, 247,
    see also density increment and
    density contrast
    effective or apparent, 27, **29–31**,
    31, 33, 68, 76, 77, 80, 89
    from composition, 91–97, 150
    from frictional ratio, 78
    in self-associating systems, 77, 91
    of the sedimenting particle, 30, **30**,
    32, 33, 77, 78, 81, 86, 89, *see*
    also complete sedimenting
    particle formalism
    temperature dependence, 211
    with deuteration, 71, 80, 86
play dough formalism, **5**, 23, 24, 26, 27,
    32, 33, 50

polychromicity, 152, 173, 174, 177
polydispersity, 11, 50, 61, 82, 85, 147, 178, 218
polyelectrolyte, *see also* macroion, 24, 28, 34–36, 72, 81, 88, **89**, 96, **150**, 236
preferential interaction, *see* co-solute, preferential solvation
pressure, 24, 51, **53–55**, 135, 182, 185, 201, 208, 210, *see also* compressibility
    convection, 182
    in rotor chamber, *see* vacuum
    on cell components, 192, 196, 202
    pressure-dependence of viscosity, 71, *see also* compressibility, organic solvents
primary charge effect, *see* charge effect, primary
pseudo-absorbance data, 120, 146, 147, 182, *see also* intensity data
    definition, **119**
    limitations, 147
    RI noise, 139
    TI noise, 119

radial calibration, **100–126**, 193, 232–234, 237, 238
    errors, 228, 233
    mask, 102–104, 232–234, 237
radial dilution, 10, **11**, 39, 40, 142, 143, 200
radial magnification error, 101–103, 232, 233
radial resolution, 69, 99, 108, 110–112, 114–117, 145, 146, 174, 218, 230
radial-invariant noise (RI noise), *see* noise, radial-invariant
radiometer, 196, 207, **207**, 208, 210–212, 230, 232
Raman scattering, 60, 109, 146
reaction boundary, **43**, 45, *see also* effective particle theory, Gilbert–Jenkins theory

reaction kinetics, *see* complex lifetime
reducing agents, 73, 139, 141, 159, *see also* DTT
refractive index increment, 19, 62, 93, 148, 149, **149**, 152, **251**, *see also* interference optics, fringe increment
    and solvation, 150
    of proteins, 149
    solvent contrast, 151
    wavelength dependence, 150
residuals, 222
    bitmap, 118, 161, **164–165**, 175
    histogram, 162, 168, **168**
    overlay, 118, 163
    runs test, 167
RNA, 70, 97, 148, 241
rotor acceleration, 70, 127–129, 131, 187, 189, 196, 201, 210, 211, 219, 220
rotor handle modification, 195, 208, 210, 211, 232, 237, 238
rotor speed, 70, 86, *see also* rotor stretching
    accuracy, 235
    back-diffusion, 14
    baseline, 123
    correction for rotor stretch, 194
    limits, 192
    loading concentration, 98
    meniscus shape, 106
    multi-speed, 16, 70, 113, 117, 122, 123, 194
    pelleting, 61
    pressure, 53, 55, 71, 246, *see also* pressure
    resolution, 61
    sedimentation force, 2, 56
      deformation, 56
    selection for SE, 15, 69, 106, 182, **215–219**
    selection for SV, **213–215**
    solution column in SE, 71

time-varying, 69, 127, 129–131, 202, 210, 220, 224, 225
rotor stretching, 100, 106, **193–200**
   adiabatic cooling, *see* adiabatic rotor stretching
   calibration, 100, 193
   meniscus and bottom, 106, 194
   TI noise, 118, 122–124, 204
   unstretch, 195
running mode, 208, 212, 220, 232

s-value, *see also* sedimentation coefficient
   accuracy and precision, 75, 83, 101–103, 134, 136, 156, 184, 191, 207, 211, 227, 228, 237, 238
   boundary, *see* reaction boundary, undisturbed boundary
   complex, 43
   correction, 6, 32, 74, 78, 83, 93, 206, *see also* standard condition
   correlation for small particles, 187
   definition, **4**
   size-dependence, 6
scan settings, 203, 220, 229, 230, 237
scan time
   delay, 130, 136
   duration, 134, *see also* scan velocity
   spacing, 131, 222
   time stamp, 133
     error, 136, 229, 230, 237
scan velocity, 99, 133–135, 230
scattering techniques, 2
Schachman, Howard, 5, 17, 26, 37, 54, 78, 87, 174, 184, 218
sector shape, 2, 3, 9, 12, 57, 143, 182, **183–185**, 191, *see also* radial dilution
sedimentation coefficient, 4, 7, *see also* flotation, *see also* s-value
   distribution, 14, 178, *see also* c(s) distribution
   hydrodynamic shape, 31, 38
   mass and friction, 5
   of macroion, 35
   oligomeric state, 41
   rotor-speed dependence, 56
   signal-weighted average, 39, 40, 45, 62, 86
   tangling, 51
   time-averaged, 38
   upper limit, 132
sedimenting particle, *see* complete sedimenting particle
self-association, 37, **38–41**, 43, 66
   binding isotherms, 42
   density contrast, 82, 85
   partial-specific volume, 28, 77, *see also* partial-specific volume, in self-associating systems
   pressure effect, 55, 182
   signal increment, 153
   solvent-mediated, 45
short-column equilibrium, 69, 218
signal linearity, *see* nonlinear detection
size distribution, *see* c(s) distribution, *see* polydispersity
size-exclusion chromatography
   buffer chemical equilibrium, 25, 61, 68, 71, 72, 88, 157, *see also* buffer match, *see also* dialysis
   difference from SV, 43
   hydrodynamic radius, 78
   molar mass, 94
   purification, 60, 61
solution plateau, **10**
   dilution, 9, 10, 12, 38, 86, *see also* radial dilution
   Johnston–Ogston effect, 50
   signal drift, 141–143
   signal-weighted s-value, 40
   sloping, 54, 58, 126, 133, 135, 238
solvent plateau, **10**
speedsteps, **130**, 131, 195
square dilution law, *see* radial dilution
standard condition, **6**, 33, 74, *see also* s-value, correction

Stokes law, 5
Stokes radius, 8, 31, 36
Stokes–Einstein relationship, 7
sucrose, 58, 74, 79, 81, 86, 145, 158, 181
Svedberg equation, **7**, 15, 57, 78, 214, 230
Svedberg, Theodor (The), 1, 35, 183, 250
synthetic boundary, 182, *see also* centerpiece, synthetic boundary
 artificial bottom, 106
  FC-43, 106
 creating interface, 187
 increasing pressure, 55
 initial distribution, 186, 187
 small molecules, 185

temperature, **206–212**
 adiabatic cooling, *see* adiabatic rotor stretching
 calibration, 75, 206, 228, **230–232**, *see also* iButton, *see also* rotor handle modification
 control
  aluminum can, 207
  iButton, *see* iButton
  radiometer, *see* radiometer
  vacuum, 2, 207, 208, *see also* vacuum
 convection, *see* convection
 correction, *see* s-value, correction
 equilibration, 69, *see also* running mode
  3,000 rpm, 129, 201, 220
  at rest, 210
  chemical equilibration, 68
  running mode, *see* running mode
  sedimentation equilibrium, 223
  time, 208, 211, 220
 partial-specific volume, *see* partial-specific volume, temperature dependence
 sedimentation equilibrium, 70, 206
  time-optimized, 206

sedimentation velocity, 70, 75, 206
standard condition, *see* standard condition
temporal resolution, 131–133, 145, 146, 177
time-invariant noise (TI noise), *see* noise, time-invariant
Traube's rules, **92**, 94, 243
TRIS, 73, 145, 212
turbidity
 apparent absorbance, 67, 145, 149, *see also* aggregates, scattering
 detector, 17

undisturbed boundary, 43–45, *see also* effective particle theory

vacuum, **212**
 and temperature control, *see* temperature control, vacuum
 chamber, 2
  pre-equilibration, 206
 leaks, *see* leaks
 limits, 207, 208, 212, 219, 220, 232
 oil diffusion pump, 123, 201, 211, 212
  engaging, *see* running mode
 roughing pump, 212
 sample aspiration, 199, 202, 203, 205
 turbomolecular pump, 212
viscometry, **74–75**
viscosity, 5, 245, *see also* co-solute, density and viscosity, *see also* s-value correction, *see also* standard condition
 pressure dependence, 54
 sample preparation, 68, 71, 79

water
 anomaly, 54, 245
 structure, 25
water blank, 123, 203, 204, 226
wavelength, **169–179**
 bandwidth, 99, 169, **171**, 174

calibration, 67, 148, **234**

excitation and emission, *see* fluorescence, excitation and emission

monochromator, *see* monochromator

range, 17

refractive index, 19, 150, 251

    laser, 150

reproducibility, 175, *see also* absorbance, data characteristics

scans, **175**

selection, 63, 65, 72, 132, 163, *see also* absorbance, multi-wavelength

Xenon lamp intensity, **170**

Wiener skewing, 64, 105, 145

windows, 2, 122, **191–192**

    assembly, **198**

    breakage, 200

    calibration mask, 101, 103, 104

    cleaning, 197, 205

    gasket, 106, 192, 197

    holders, 196

    imperfections, 117, 122, 123, 161, 163, 192, 194

    liners, 161, 192, 197, 202

    quartz, 163, 191

    sapphire, 163, 191

    wedged, 192

    weight, 192

Xenon flashlamp, 17, 169, 170, 174, 175, 177, 234

Yphantis, David, 191, 202, 203, 218, 222

Milton Keynes UK
Ingram Content Group UK Ltd.
UKHW050453071024
449327UK00015B/353